Emmy Noether's Wonderful Theorem

Emmy Noether's Wonderful Theorem

Revised and Updated Edition

Dwight E. Neuenschwander

Southern Nazarene University

Johns Hopkins University Press
Baltimore

© 2011, 2017 Johns Hopkins University Press
All rights reserved. Published 2017
Printed in the United States of America on acid-free paper
9 8 7 6 5 4 3 2 1

Johns Hopkins University Press
2715 North Charles Street
Baltimore, Maryland 21218-4363
www.press.jhu.edu

Library of Congress Control Number: 2016953397

A catalog record for this book is available from the British Library.

ISBN-13: 978-1-4214-2267-1
ISBN-10: 1-4214-2267-0

Frontispiece: Emmy Noether (1882–1935). Photo courtesy Bryn Mawr College Library.

Special discounts are available for bulk purchases of this book. For more information, please contact Special Sales at 410-516-6936 or specialsales@press.jhu.edu.

Johns Hopkins University Press uses environmentally friendly book materials, including recycled text paper that is composed of at least 30 percent post-consumer waste, whenever possible.

To my wife,

Rhonda

with love and appreciation

"Yesterday I received from Miss Noether a very interesting paper on the generation of invariants. I am impressed by the fact that these things can be understood from so general a point of view. It would have done the Old Guard of Göttingen no harm to be sent back to school under Miss Noether. She really seems to know her trade!"

—Albert Einstein, letter to David Hilbert, May 24, 1918* in Yvette Kosmann-Schwarzbach, *The Noether Theorems: Invariance and Conservation Laws in the Twentieth Century*, translated by Betram E. Schwarzbach, 2010, 71–72

Pure mathematics is, in its way, the poetry of logical ideas. One seeks the most general ideas of operation which will bring together in simple, logical and unified form the largest possible circle of formal relationships. In this effort toward logical beauty spiritual formulas are discovered necessary for the deeper penetration into the laws of nature.

—Albert Einstein, epitaph for Emmy Noether, *The New York Times*, May 5, 1935

---

*[Calaprice (1996), Byers (1999)]

Contents

II   WHEN FUNCTIONALS ARE INVARIANT

III   THE INVARIANCE OF FIELDS

IV TRANS-NOETHER INVARIANCE

Preface

This is a pedagogical book. It aims to introduce Noether's theorem to students who do not understand it already.

When as an undergraduate student I first met the elegant connections between symmetries and conservation laws through Lagrangian mechanics, I changed my major to physics and never looked back. In due time I learned how those connections in mechanics are special cases of a deeply profound theorem published by Emmy Noether in 1918. Noether's theorem as a work of mathematical physics is also a work of poetic beauty.

The creative act, wrote Jacob Bronowski, occurs twice. The first occurs in the mind of the original creator or discoverer. The second occurs in the mind of the appreciator, who re-creates the discovery afresh and sees its significance:

> The poem or the discovery exists in two moments of vision: the moment of appreciation as much as that of creation; for the appreciator must see the movement, wake to the echo which was started in the creation of the work...In the moment of appreciation we live again the moment when the creator saw and held the hidden likeness....When a theory is at once fresh and convincing, we do not merely nod over someone else's work. We re-enact the creative act, and we ourselves make the discovery again. [Bronowski (1956) 19]

As noted above, this book is written especially for physics students to whom Noether's theorem and its related topics are new. The reader I have in mind is a junior or senior undergraduate physics major, or a beginning physics graduate student. Well do I remember being one of those students myself. Those memories include the frustration of trying to read a manuscript loaded with jargon that assumed a fluency I was still struggling to master. At that point in one's career, details that are incidental trifles to experts can become major sticking points for novices (see the following list of questions for examples). If some passages herein seem pedantic or repetitious to experts, I offer to them my apologies. But the offending passages are here because they address questions that were sticking points in my mind when I was a novice, and/or were questions raised by my students. Once fluency is approached, it's easy to forget the struggles. I have tried to remember them.

The Noether theorem—or, as some would say, the two Noether theorems—studied here [Noether (1918); Noether & Tavel (1971)] was for Emmy Noether a special case of her expertise in the abstract algebra of Lie groups and the study of invariants. In 1915 Albert Einstein unveiled his new general theory of relativity, and the mathematicians at Göttingen, notably David Hilbert and Felix Klein, studied it with gusto. They encountered an apparent problem in reconciling energy conservation with the new theory and asked Emmy Noether for her help. Her theorem, which we celebrate here, was the result. It resolved the problem of energy conservation in general relativity through the concept that later became known as local gauge invariance (the "second theorem"), and along the way the "first theorem" gave unified insight into the conservation laws of mechanics and electrodynamics.

After publishing the 1918 paper Emmy Noether went on to become a founder of modern abstract algebra. Graduate students in mathematics are familiar with, for example, Noetherian rings. While abstract algebra in the language of ascending chain conditions and unique factorization domains deserves deep and genuine respect, physics students are more at home with Newton's laws and Maxwell's equations, Lorentz transformations, and de Broglie waves, all expressed in the mathematics of analysis. Fortunately for those of us coming to the conversation from a physics background, Emmy Noether's wonderful theorem can be approached and appreciated in the language of calculus and vector spaces. As a beautiful organizing principle of post-introductory physics, Noether's theorem deserves to be widely known among all physics students, novices and senior physicists alike.

As mentioned, of special interest to me are connections between physics concepts. Noether's theorem stands bright and clear like a magnificent summit in an impressive mountain range of ideas. But the peaks and valleys around a prominent peak are part of the landscape too. Likewise, in these pages we explore topics that run alongside Noether's theorem, interesting for their own sake and which, I think, can be more deeply appreciated with Noether's theorem in the background.

This second edition presents the opportunity to delve into the distinction between Noether's "two" theorems (the second theorem extends the first); to add a few more exercises, references, and technical details; to correct errors in the first edition; and to offer a more vivid picture of Emmy Noether's life and influence. This edition, like the first one, is offered as an expression of appreciation. Thank you for joining me in this adventure of making Emmy Noether one of our intellectual companions.

Acknowledgments

To all my teachers I owe a huge debt of gratitude, and this book offers an installment towards repaying it. Instrumental in my Noether's theorem journey were James Gibbons, my undergraduate theoretical mechanics professor at the University of Southern Colorado (now Colorado State University–Pueblo) who introduced Lagrangian dynamics to me and turned me towards a physics major; John David Logan (University of Nebraska), who at Kansas State University taught a graduate course on Noether's theorem with his own textbook [Logan (1977)] and still graciously answers my questions; Larry Weaver of the KSU physics department, who saw my interests and steered me towards Professor Logan's course; and the late Ari Kyrala, for his expressive insights in mathematics-for-physics courses at Arizona State University.

I thank my students for their good cheer and questions in my courses where these notes were developed. Student collaborators in undergraduate research projects featuring Noether's theorem have so far included Nathan Adams, Kevin Cornelius, Jorge Carmona-Reyes, Lucas Dallen, Will Holmes, Curtis McCully, Mohammed Niazi, Reza Niazi, Keith Slinker, Shawna Starkey-York, Geoffrey Taylor, Brian Turner, and Johnnie Renee West.

I appreciate SNU faculty colleagues Brent Eskridge for cheerfully sharing his mastery of LaTeX, Mark Winslow for his enthusiastic, practical encouragement, and Lee Turner for lending his knowledge of math history. With deep respect I thank The Catalysts, an organization of SNU science alumni, for their steadfast support. Thanks extends off campus to Don Lemons of Bethel College (North Newton, Kansas), Don Salsibury of Austin College (Sherman, Texas), and Raghunath Acharya (Arizona State University).

I thank the readers who corresponded with me about the first edition. Deep appreciation also goes to Camilla MacKay and Rachel Appel of the Bryn Mawr College Library; Gareth Peers of Science Photo Library; the Center for History of Physics of the American Institute of Physics; to former Johns Hopkins University Press editor Trevor Lipscombe and his staff for giving the first edition its chance, and Greg Nicholl for turning my rough sketches into crisp line drawings. To Vincent Burke and the current editor Tiffany Gasbarrini and her staff I am grateful for making this second edition possible.

Above all, I thank my family for their support. Book writing projects enter the lives of an author's family too.

Questions

PRIMARY QUESTIONS

What is a functional?

What is "symmetry"? What is "invariance"? What are "conservation laws"?

How are symmetry, invariance, and conservation laws related?

What are Lagrangians and Hamiltonians? Which is more fundamental?

What are generalized coordinates and their velocities?

What are "canonically conjugate" variables?

What are continuous symmetries? What are discrete symmetries?

Who was Noether of Noether's theorem?

What is the difference between Noether's "first" and "second" theorems?

AUXILIARY QUESTIONS

Where does Hamilton's principle come from?

Are Hamilton's principle and Fermat's principle related?

What does traditional notation such as "$\delta\varphi$" mean?

What is the distinction between "stationary" and "extremal" functionals?

Why is a classical mechanics Lagrangian kinetic *minus* potential energy?

Why do some vector components carry superscripts and others have subscripts?

What are tensors?

What is "gauge invariance"? Why distinguish global from local gauge invariance?

What are internal symmetries?

What is "minimal coupling"? What are "covariant derivatives"?

What is the "Jacobian" of a transformation?

What are Legendre transformations and what are they good for?

What is "phase space"? What is it good for?

Why are complex variables used to describe wave functions?

Why do complex scalar field Lagrangian densities lack the $\frac{1}{2}$ of real scalar fields?

Why does Noether's theorem consider *infinitesimal* transformations?

What are "unitarity" and "Hermitian" operators? Why do we need them?

What are "equations of continuity"? How do they describe conservation locally?

What are "proper" and "improper" conservation laws?

What is a group? What does SU(N) mean and what is it for?

What is the role of Hamilton-Jacobi theory?

Part I

WHEN FUNCTIONALS ARE EXTREMAL

Chapter 1

Symmetry

To define the idea of symmetry is certainly not simple. I shall not try to make an all-encompassing or precise definition, against which a contradiction can quickly be brought. ... The idea is still alive and growing. We don't know all that the concept implies. I like best the idea of the seventeenth century philosopher Leibniz. ... For Leibniz, symmetry is related to the indiscernibility of differences. Once you walk into the hall of a Palladian building, you can't quite remember whether you turned left or right.
—Philip Morrison, "On Broken Symmetries," in Judith Wechsler, ed., *On Aesthetics in Science, 1981*

1.1 Symmetry, Invariances, and Conservation Laws

The conservation of energy, linear momentum, angular momentum, and electric charge are among the most fundamental principles of physics. Have you ever wondered why nature cherishes these quantities so much that she conserves them? Asking "why" in this context may be a bottomless question. But we can connect these conservation laws to deeper principles through the elegant theorem published by Emmy Noether in 1918.[1] Noether's theorem relates a huge class of conservation laws to symmetries of space and time and "internal" variables.

Let's see how one of these familiar conservation laws could be connected to a symmetry, using for now only pre-Noether concepts encountered in an introductory physics course. Consider, for example, the conservation of

---

[1]The original: Noether (1918); Mort Tavel's translation: Noether & Tavel (1971); a translation also appears in Kosmann-Schwarzbach (2010).

linear momentum. Let a particle of mass m move in one dimension along an x-axis. Newton's second law says

$$F = \frac{dp}{dt} \qquad (1.1)$$

where F denotes the net force, t the time, and $p = mdx/dt$ the momentum. For the momentum to be conserved—to not change with time—the net force on the particle must vanish. What does this have to do with symmetry?

Let the particle's interactions with the world be expressed through some potential energy function $U(x)$, whose negative gradient equals the force. Write Newton's second law as the differential equation

$$-\frac{\partial U}{\partial x} = \frac{dp}{dt} \qquad (1.2)$$

and approximate the left-hand side as $-\Delta U/\Delta x$. When the particle goes from x to $x + \varepsilon$ then eq. (1.2) may be written

$$-\left(\frac{U(x + \varepsilon) - U(x)}{\varepsilon} \right) = \frac{dp}{dt}. \qquad (1.3)$$

To have $dp/dt = 0$ as $\varepsilon \to 0$, we must have $U(x + \varepsilon) - U(x) \sim \varepsilon^s$, where $s > 1$. For then, as the particle moves from one location to another, the change in U goes to zero faster than ε as $\varepsilon \to 0$, and the particle's momentum shows no measurable change. A tangible example you may recall from your introductory physics course may be a glider moving smoothly over a horizontal air track: "here" is identical to "over there." Evidently, the conservation of linear momentum follows from the system being unchanged by a spatial translation.

To say it another way, it means that if you *do* want the particle's momentum to change, a mere translation through space won't do it; space, itself, is homogeneous. So translational "sameness" signifies a *symmetry* of space. If you want the momentum to change then you have to make the space *over there* different from space *here*, for example by having the glider move against a spring.

Notice that our momentum example forms an "if-then" statement. It does not claim that the particle's environment is translationally symmetric; indeed, whenever a particle's momentum has been changed, the symmetry of space is spoiled, or "broken," at least for one spatial dimension. For instance, the Earth's gravitational field breaks the symmetry between the vertical and horizontal dimensions.

How can "symmetry" be defined in a way that makes it a *quantitative* concept? What *is* symmetry? For the kind of answer we need here, let

Figure 1.1: *A cylinder with axis of symmetry AA. When rotated about this axis, it looks the same before and after the rotation. That's symmetry.*

us look at a cylinder. I invite your attention to the cylinder's axis AA in Figure 1.1. After you rotate the cylinder about this axis, it looks the same as it did before. Unless you paint a dot on the cylinder, or scratch a mark on its surface to break the symmetry, you cannot tell that the cylinder has been rotated. Rotating it *and not being able to tell it's been rotated* captures the essence of what we mean by "symmetry." To make symmetry quantitative, we need to carry out some operation, or transformation, and see if we can detect a difference. If the difference is too small to detect within some infinitesimal tolerance, then we say the system is "invariant under the transformation."

A clockwise rotation of the cylinder by 90° is equivalent, from an observer's perspective, to moving one's location 90° counterclockwise. Thus a transformation can also be a change of reference frame, from an old coordinate system to a new one. If a quantity survives such transformations unchanged, that quantity is said to be invariant. The appearance of the cylinder is invariant under rotation about the AA axis. The existence of invariance reveals an underlying symmetry: the cylinder appears unchanged under the rotation because it is symmetric about the axis of rotation.

"Conservation" as in "conservation of energy" is not the same as "invariant." They are related—and the exploration of that relation forms the substance of our subject—but they are not synonymous. The momentum or energy of a system of particles may be conserved but not necessarily invariant. For example, imagine one billiard ball approaching another. In the reference frame of the billiard table, prior to the collision the cue ball moves and the eight ball sits at rest, and the momentum of the system is nonzero. But in the center-of-mass reference frame, the system's total momentum sums to zero because the cue ball and eight ball approach one another with opposite momentum. In both frames, the collision is analyzed using conservation of momentum within each frame. The table frame sees nonzero momentum, but in the center-of-mass frame sees zero momentum. In this example, momentum is conserved *within* in each frame, but is not invariant *between* them.

"Invariant" means that a quantity's numerical value is not altered by a coordinate transformation. "Conserved," in contrast, means that within a given coordinate system the quantity does not change throughout a process. "Invariance" compares a quantity between reference frames. "Conservation" compares the quantity before and after a process within a reference frame. Noether's theorem relates conservation to invariance, and thus to symmetry.

Coordinate systems or reference frames are not part of nature. They are maps that we introduce into the solution of a problem for our convenience. Therefore, the content of an equation that is supposed to express a truth about nature must transcend the choice of this or that reference frame. One says the equation must be written "covariantly," or the expression is "covariant." In the preceding example of billiard balls, in a given frame the momentum of the two balls adds up to the same total before and after the collision. In the billiard table frame using unprimed vectors, one writes the conservation of momentum as the vector sum before collision equals the vector sum after collision:

$$[\mathbf{p}_{cue} + \mathbf{p}_8]_{before} = [\mathbf{p}_{cue} + \mathbf{p}_8]_{after}. \qquad (1.4)$$

In the center-of-mass frame using primes on the vectors, the same physics content is expressed as

$$[\mathbf{p}'_{cue} + \mathbf{p}'_8]_{before} = [\mathbf{p}'_{cue} + \mathbf{p}'_8]_{after}. \qquad (1.5)$$

Never mind that before the collision the table frame measures $\mathbf{p}_8 = \mathbf{0}$ and the center-of-mass frame measures $\mathbf{p}'_8 \neq \mathbf{0}$; the equation expressing conservation of momentum

$$\left[\sum_n \mathbf{p}_n\right]_{before} = \left[\sum_n \mathbf{p}_n\right]_{after} \qquad (1.6)$$

contains the same content whatever the reference frame; the relationship it describes holds in either frame. The equation is written covariantly. The quantities in the equation transform under a change of coordinates by the same rules as the coordinates themselves.

We will see that conservation of energy, conservation of linear momentum, and conservation of angular momentum are related to invariance under time translations, space translations, and rotations, respectively. These invariances signify underlying spacetime symmetries: the homogeniety of time, the homogeniety of space, and the isotropy of space. The conservation of electric charge emerges from a more abstract symmetry called "gauge invariance." We can go farther: the conservation of more esoteric "charges," such as quark color charge or weak isospin, involve invariances that hold under transformations among so-called internal degrees of freedom. In the first part of this book we consider spacetime symmetries only. Internal degrees of freedom are considered later.

You will have noticed that I have not said *what* are the quantities whose invariance leads to conservation laws. These quantities are called "functionals." In the functional we have a powerful concept that puts almost all of physics into a common language. Everything wonderful that I am going to relate comes through these functionals.

I have organized this book into four parts. The remainder of Part I (chapters 2 and 3) introduces functionals. As a mathematical machine, by definition a functional Γ takes a *function* as input and produces a *real number* as output. That sounds like the task of a definite integral! While a functional as an abstract concept need not be a definite integral, all the functionals considered in this book are expressed as definite integrals. If you pick an input function $x(t)$, stuff it into the input slot of a functional $\Gamma(\ )$ and turn the crank, you get a real number $\Gamma(x)$. But with a different input function $w(t)$ you get a different output number $\Gamma(w)$. Suppose you want to find the function that produces a maximum or minimum value for Γ. One says the functional is to be made an *extremal*, or, as some say, made *stationary*.[2] As chapter 3 describes, the function that makes Γ an extremum is the solution to a differential equation called the Euler-Lagrange equation.

Part II (chapters 4 and 5) studies the conditions for invariance of the functional under transformations of the independent and/or the dependent variables. When we change t to some new t' and x to some new x', so that $x(t) \to x'(t')$, then $\Gamma(x'(t'))$ may or may not be the same number as $\Gamma(x(t))$. We find that Γ meets our formal definition of invariance if and only if a fundamental invariance identity is satisfied.[3] The plot lines of "extremal"

---

[2]The distinction between "extremal" and "stationary" is discussed in section 5.3.

[3]The version of the invariance identity we present here is due to Rund [Rund (1972)] and Trautman [Trautman (1967)], who streamlined the Noether theorem proof and notation.

and "invariance" converge in Noether's theorem (the "first" theorem), as shown in chapter 5. In chapter 5 we also turn the problem around and see how to find transformations that leave a given functional invariant.

Chapter 6 begins Part III by enlarging our program to *fields*, functions of space and time. There we see how charge conservation comes from *global* gauge invariance. Chapter 7 imposes *local* gauge invariance on fields, and we see how far we can push local gauge invariance using Noether's first theorem. As an introduction to an important application of these ideas, we also examine in chapter 7 how local gauge invariance, when applied to "internal" degrees of freedom or generalized "charges," becomes a dynamical principle that leads to an understanding of fundamental forces as the exchange of "gauge bosons."

Chapter 8 extends the program of locally variable (or gauge) transformations to the functional's dependent and independent variables, leading beyond Noether's first theorem to the second theorem. The second theorem contains the first as a special case, provides constraints on differential operators, and expresses conservation laws for such systems in terms of so-called covariant derivatives. In clarifying conservation law issues for the coupled matter-field systems of relativistic gravitation, Emmy Noether helped David Hilbert, Felix Klein, and Albert Einstein put the finishing touches on the general theory of relativity in 1915.

Part IV (chapters 9 and 10) considers other applications of invariance that are not part of Noether's theorem proper, but that share its vocabulary of functionals, transformations, invariances, and conservation laws. Chapter 9 reexamines invariance and conservation in the language of phase space. Whereas Noether's theorem produces conservation laws given equations of motion and an invariance, Hamilton-Jacobi theory uses equations of motion and conservation laws and produces transformations. Through Hamilton-Jacobi theory, the possibilities offered by an indefinite integral version of the functional are developed. From deep within classical mechanics they suggestively point the way to quantum mechanics, as developed in chapter 10.

Because of the central role of conservation laws, one could argue that Noether's Theorem offers a strategic unifying principle for most if not all of physics. Although I never had the honor of meeting her personally, I would be remiss if I did not introduce you to Emmy Noether.

1.2   Meet Emmy Noether

"She was not clay, pressed by the artistic hands of God into a harmonious form, but rather a chunk of human primary rock into which he had blown his creative breath of life."—Hermann Weyl, memorial address for

Emmy Noether, April 26, 1935 in Auguste Dick, *Emmy Noether, 1882–1935*, translated by H. I. Biocher 1981, 112–152

In the judgment of the most competent living mathematicians, Fräulein Noether was the most significant creative mathematical genius thus far produced since the higher education of women began. In the realm of algebra, in which the most gifted mathematicians have been busy for centuries, she discovered methods which have proved of enormous importance in the development of the present-day younger generation of mathematicians.
—Albert Einstein, letter to the *New York Times*, May 5, 1935 in Alice Calaprice, *The Quotable Einstein*, 1996

Amalie Emmy Noether was born on March 23, 1882, the first child of Jewish parents Max (1844–1921) and Amalie Ida Kauffman Noether (1852–1915).[4] Emmy's mother came from a highly educated family that included university scholars in law and history. Emmy's father was descended from iron wholesalers. The first in his line to earn a PhD, Max became a distinguished mathematician, earning his doctorate from the University of Heidelberg, and qualified there in 1870 as a *Privatdozent*.[5] In 1875 Max moved to the University of Elrangen, where Felix Klein (1849–1925)[6] had in 1872 launched the *Erlanger Programm*, which treated geometry as the study of properties of a space that are invariant under a group of transformations. For example, in Euclidean geometry, lengths and angles are unchanged under rotations. This outlook offered a unified approach to classifying the non-Euclidean geometries that had proliferated by the late nineteenth century, pointed to the possibility of new geometries defined from diverse transformation groups [Burton (2011) 602–603], and presumably created the environment that nurtured Emmy's mathematical interests, in which invariance theory played a prominent role. Klein's *Erlanger Programm* brought world prominence to the mathematics department at Erlangen. Max Noether served there the rest of his life. In 1886 Klein accepted a chair at the University of Göttingen, and he brought David Hilbert (1862–1943) to Göttingen in 1895. That move would have effects on Emmy Noether 20 years later, and on readers of this book over a century later.

From 1889 to 1896 Emmy attended the Städtischen Höheren Töchterschule in Erlangen, where she studied languages and piano. She was fond of dancing. As a student she originally planned to teach French and

---

[4]During Max Noether's generation the family name's spelling was changed from Nöther to Noether. Emmy had three younger brothers, Alfred (1883–1918), Fritz (1884–1941), and Gustav Robert (1889–1928) [Brewer & Smith (1982)].

[5]An unsalaried university lecturer with the right to teach independently and advise research students. Lecturers received fees from students rather than a university salary.

[6]Klein is perhaps best known among students for the "Klein bottle."

English, and in 1890 passed the Bavarian State Examination, qualifying her to teach these languages—in schools for women. But her interests had turned increasingly to the eloquent language of mathematics.

This interest was evidently encouraged by her father and his mathematician colleagues, notably Paul Gordan (1837–1912).[7] As Hermann Weyl described in his 1935 memorial address, Emmy developed a close "mathematical kinship" with her father:

> Clebsch has introduced Riemann's ideas into the geometric theory of algebraic curves and [Max] Noether became, after Clebsch had passed away young, his executor in this matter: he succeeded in erecting the whole structure of algebraic geometry of curves on the basis of the so-called Noether residual theorem....
>
> [Max] Noether's residual theorem was later [in the 1920s] fitted by Emmy into her general theory of ideals in arbitrary rings. This scientific kinship of father and daughter—who became in a certain sense his successor in algebra, but stands beside him independent in her fundamental attitude and in her problems— is something extremely beautiful and gratifying. [Dick (1981); Brewer & Smith (1982); James (2002)]

When Gordan passed away in 1912, Max and Emmy Noether wrote his obituary for *Mathematische Annalen.*

After completing grammar school, Emmy wanted to attend university to study mathematics. In the culture of that time and place, women were not allowed to matriculate into German universities,[8] although they could enroll in courses with the professor's permission. As late as 1898, the Academic Senate at the University of Erlangen declared that to admit women as students "would overthrow all academic order." However, in 1900 the spunky Emmy got permission to attend lectures at Erlangen, one of two young women among 986 students during her first semester. There Emmy attended lectures until 1902, and in July 1903 she passed the *matura* examinations, necessary (but not sufficient) for matriculation into university.

That fall she registered at the University of Göttingen,[9] where she attended lectures by distinguished mathematicians who to this day have name recognition among physics students. These included Karl

---

[7]This is the same Gordan of the beloved Clebsch-Gordan coefficients, which grew out of the study of Lie groups and find application to the quantum addition of angular momentum. Rudolf F. A. Clebsch (1833–1872) was a pioneer in invariant theory, and at Giessen collaborated with Paul Gordan.

[8]Women were allowed to enroll in universities in 1861 in France, 1879 in England, and 1885 in Italy. [Brewer & Smith (1982)]

[9]Evidently Emmy was only allowed to audit university classes during this time.

Figure 1.2: *Emmy Noether (1882–1935). This photo was taken sometime before she entered Göttingen University.* (Science Photo Library)

Schwarzschild (1873–1916), Hermann Minkowski (1864–1909), Felix Klein, and David Hilbert.[10] After she had been one semester at Göttingen, Erlangen University relented in their policy against women and allowed female students to matriculate there and take examinations with the same rights as male students. Emmy returned home to Erlangen and entered the university as a degree-earning student in October 1904. She listed mathematics as her program of study. Academic order was *not* overthrown.

At Erlangen, Paul Gordan became Emmy's PhD advisor. Emmy was Gordan's only doctoral student throughout his distinguished career. She completed the PhD in 1907. Dr. Noether's dissertation was titled *On Complete Systems of Invariants for Ternary Biquadratic Forms.*

In his 1935 memorial tribute to Emmy Noether, Hermann Weyl divided her career into three epochs. Epoch 1, 1908–1919, was a time of collaboration with eminent senior mathematicians. The Noether's theorem celebrated in this book was a product of this epoch. In the second epoch, 1920–1926, Emmy Noether became an eminent mathematician in her own right, one whom later generations would call the "mother of abstract algebra" [Tent (2008)], developing the general theory of ideals and Noetherian rings [Moore (1967) 189–196]. By the time of the third epoch, 1927–1935,

---

[10]The Schwarzschild of the Schwarzschild metric in general relativity, the Minkowski of Minkowskian spacetime; most physics majors first hear of Hilbert in the context of "Hilbert space" in quantum mechanics.

Noether was recognized as a leader of the "Noetherian school" of algebraists and continued developing noncommunative algebras such as hypercomplex numbers. During that epoch she also found herself a political refugee.

The Noether theorem discussed in these pages was published in 1918 and thus belongs to the first epoch of Emmy Noether's career, the time of collaboration with top mathematicians such as Paul Gordan, Felix Klein, and David Hilbert. The early part of that era saw her hard at work, as Gordan's doctoral student, on the theory of invariants of binary and higher-order forms. A binary form, for example, is a second-order polynomial $p(x, y)$ in two variables x and y of the form

$$p(x, y) = ax^2 + bxy + cy^2. \tag{1.7}$$

If a, b, and c are constants from a specified set of numbers, $p(x, y)$ is an "algebraic form." If these constant coefficients are replaced by functions that include derivatives of x and y, $p(x, y)$ is a "differential form." A large question in the study of quadratic and higher-order forms asks what quantities remain invariant under a change of variables.[11] In her dissertation, Noether calculated and tabulated 331 invariants among "ternary biquadratic forms"[Brewer & Smith (1982)].

Between 1908 and 1915 Dr. Noether worked without salary at the Erlangen Mathematical Institute. She carried out her own research, attended mathematics conferences, was elected to an Italian mathematical society, and as her father's health declined delivered lectures for him as needed. When Ernst Fischer (1875–1959) joined the Erlangen mathematics faculty after Gordan's retirement, Fischer and Noether collaborated on research. Gordan's methods had been algorithmic and computational—finding invariants by cranking through the possibilities. In contrast, Fischer brought Hilbert's abstract algebra approach to mathematics, along the lines of Hilbert's 1888 paper on basis theory [Byers (1999) 5], methods quickly taken up by Emmy Noether.[12] In 1919 Noether's curriculum vitae credits Fischer—whose specialty was also invariance theory—as influencing her to more abstract ways of mathematical thinking, as she mastered approaches closer to Hilbert's.

Noether became so adept at Hilbert's methods of invariance theory that in April 1915, shortly after her mother died, she moved to Göttingen at the invitation of Hilbert and Klien. Her biographers remark that "already, Emmy Noether was recognized for the extreme generality and abstractness

---

[11]One can see a natural application to differential equations and conservation laws, by setting the $p(x, y)$ of eq. (1.7) equal to zero.

[12]Upon reading one of Hilbert's proofs of a finite basis for specific invariants, Gordan allegedly remarked "Das ist nicht Mathematik; das ist Theologie." It's not certain that Gordan actually said this, or if he did, whether it was in jest, a form of praise, or a criticism [Brewer & Smith (1982)].

of approach which would eventually be seen as her most distinguishing characteristic"[Brewer & Smith (1982) 13].

At Göttingen, women at that time were still not allowed to hold lecturing positions. When Hilbert argued for Noether as a lecturer, prominent nonmathematics members of the faculty senate objected: "What will our soldiers think when they return to the university and find that they are expected to learn at the feet of a woman?" Hilbert famously responded *"Meine Herren,* I do not see that the sex of the candidate is an argument against her admission as a *Privatdozent.* After all, the senate is not a bathhouse!" David Hilbert did not win the argument that day. But determined to keep Emmy Noether at Göttingen, he improvised. Lectures would be announced under his name, but delivered by Fräulein Noether [Reid (1972) 143].

At the end of June and the beginning of July 1915, Albert Einstein spent about a week in Göttingen delivering six lectures on his not-quite-finished general theory of relativity. Noether wrote to Fischer back at Erlangen, "invariant theory is trump here; . . . Hilbert is planning to lecture next week on his Einsteinian differential invariants, and to understand that, the Göttingen people must certainly know something!"[Brewer & Smith (1982) 12]. By "invariant theory," Noether meant differential invariants, in contrast to algebraic invariants. In 1935 Weyl recalled [Kosmann-Schwarzbach (2010) 77],

> Hilbert at that time was over head and ears in the general theory of relativity, and for Klein, too, the theory of relativity and its connections with his old ideas of the Erlangen program[13] brought the last flareup of his mathematical interests and mathematical production.

Even after Einstein unveiled the finished general theory of relativity in November 1915, a problem persisted. Hilbert and Klein encountered a puzzle with energy conservation in Einstein's theory.[14] With her invariant-theoretic knowledge of differential forms, Noether was able to help them. In an exchange of letters, Hilbert wrote to Klein, "Emmy Noether, whose help I sought in clarifying questions concerning my energy law" and Klein wrote to Hilbert, "you know that Fraülein Noether continues to advise me in my work" [Pais (1982) 276]. Noether resolved the problem about energy conservation in general relativity, and along the way proved the theorems we study in these pages. The result was the Noether's theorem we celebrate

---

[13]Since general relativity was by this time treating gravitation as the curvature of spacetime, one can see how general relativity as geometry in a four-dimensional pseudo-Riemannian metric space would be attractive to the authors of the *Erlanger programm.* "Pseudo-Riemannian" means the metric is not positive-definite.

[14]See chapter 8.

here, published in 1918 as *Invariant Variation Problems* [Noether (1918); Noether & Tavel (1971); Kosmann-Schwarzbach (2010)], writing:

> Hilbert enunciates his assertion to the effect that the failure of proper laws of conservation[15] of energy is a characteristic feature of the "general theory of relativity." In order for this assertion to hold good literally, therefore, the term "general relativity" should be taken in a broader sense than usual, and extended also to the foregoing groups depending on N arbitrary functions.

In other words, general relativity is a theory of coupled systems exhibiting local gauge invariance.

Hermann Weyl summarized,

> For two of the most significant sides of the general theory of relativity... she gave at that time the genuine and universal mathematical formulation. [Brewer & Smith (1982) 13; Dick (1981)]

Albert Einstein was one of the first to appreciate Noether's theorem, describing it in a letter to Hilbert as "penetrating mathematical thinking"[Pais (1982)].

Meanwhile, due to the continuing prejudice against women being professors, Hilbert's repeated attempts to gain Noether a permanent appointment were frustrated. In a letter to Hilbert, dated May 24, 1918, Einstein wrote, "It would not have done the Old Guard at Göttingen any harm, had they picked up a thing or two from her. She certainly knows what she is doing"[Calaprice (1996)]. In a letter to Felix Klein dated December 27 of that same year, Einstein objected again to Dr. Noether not being allowed to lecture officially because she happened to be female: "On receiving the new work from Fräulein Noether, I again find it a great injustice that she cannot lecture officially. I would be very much in favor of taking energetic steps in the ministry [to overturn this rule]."

On May 21, 1919, Dr. Noether submitted an application for *Habilatation*, the right to teach at the university as a *Privadozent*. As part of the application she presented a colloquium, and submitted her curriculum vitae and publications. Besides citing her list of papers on abstract algebra and invariants, in her closing paragraph she adds:

> Finally, there are two works on differential invariants and variation problems. These resulted from my assistance to Klein and Hilbert in their work on the Einsteinian general theory

---

[15]By "proper laws" (Hilbert's term) he meant conservation laws that could be cleanly expressed by an equation of continuity; see section 6.4.

of relativity.... The second of these, *Invariant Variation Problems,* which is my Habilitation paper, treats arbitrary continuous Lie groups, finite or infinite, and draws conclusions from a special case of invariance relative to such a group. These general results contain as special cases some known results concerning proper intergrals from mechanics, stability theorems, and certain dependencies among field equations arising in the theory of relativity, while, on the other hand, the converses of these theorems are also given. [Brewer & Smith (1982) 15–16]

As a result of some slight liberalization of policies following World War I, the university granted Dr. Noether the right to lecture under her own name in 1919. In 1922 she was finally named an "unofficial associate professor," a purely honorary position. A subsequent teaching appointment in abstract algebra provided her with a modest salary. She remained at Göttingen for the next decade, except for visiting professorships, where she was warmly received, at Moscow in 1928–1929 and Frankfurt in the summer of 1930. By 1931 she was an associate professor at Göttingen [Brewer & Smith (1982) 75].

The second and third epochs of Emmy Noether's career lie outside the scope of this book. But her work during those years wrote new chapters in abstract algebra that made her name famous among mathematicians everywhere. Those contributions include the origination in 1920 of the *left ideals* and *right ideals,* followed by *Noetherian rings* with their "ascending chain condition" in 1921. Weyl wrote that "she changed the face of algebra by this work." Her influence extended well beyond her students. When mathematician B. L. Van der Waerden spent a year at Göttingen studying with her, he returned to Amsterdam inspired, and wrote a two-volume treatise, *Moderne Algebra,* of which Garrett Birkhoff wrote decades later with the benefit of historical perspective,

> both the axiomatic approach and much of the content of "modern" algebra dates back to before 1914. However, even in 1929, its concepts and methods were still considered to have marginal interest compared with those of analysis.... By exhibiting their mathematical and philosophical unity, and by showing their power as developed by Emmy Noether and her younger colleagues,... Van der Warden made "modern algebra" suddenly seem central in mathematics. [Brewer & Smith (1982) 19]

During the 1927–1935 epoch Dr. Noether's major publications included ideal theory of hypercomplex number systems and their applications to group representations [Brewer & Smith (1982) 146] and noncommuncative algebras.

Noether was a coeditor of the collected mathematical papers of Richard Dedekind (1831–1916), edited the correspondence of Dedekind and Georg Cantor (1845–1918), and did some editing for *Mathematische Annalen* [Dick (1981) 42]. At Göttingen she had an eager group of loyal students, and many of their mathematics discussions took place in the welcoming quarters of Emmy's modest apartment. She was "so amazingly lively!" recalled one of her students. Another remembered, "she lived in close communion with her pupils; she loved them, and took an interest in their personal affairs." The mathematician and philosopher Norbert Wiener (1894–1964)[16] recalled encountering Emmy Noether and a group of her students on a train, "probably the best woman mathematician there has ever been... looking like an energetic and very near-sighted washerswoman... and her many students flocked around her like a clutch of ducklings about a kind, motherly hen"[Brewer & Smith (1982) 40]. She was blessed with seemingly boundless energy. In her lectures she talked loudly, with large gestures, and frequently thought her way through new ideas while in front of the class—which could make her difficult to follow for listeners not used to it—but her students adored her. In middle age, with her round spectacles and ample lap, she looked like she could have been your favorite auntie—which she was to the children of her brothers. She was not sentimental, but expressed her affection in an almost rough, jolly way [Dick (1981) 46]. She was always generous in sharing ideas with her students, and in talks and publications gave them ample credit for their contributions.

The dark clouds of arrogant intolerance that gathered over Fascist Europe in the 1930s did not spare Emmy Noether. With the appointment of Adolf Hitler as chancellor of Germany in January 1933, swiftly followed by brutal consolidation of Nazi control of German institutions, Dr. Noether, like other Jewish professors in German universities, was abruptly dismissed. Herman Weyl recalled,

> A stormy time of struggle like this one we spent in Göttingen in the summer of 1933 draws people closely together; thus I have a particularly vivid recollection of these months. Emmy Noether— her courage, her frankness, her unconcern about her own fate, her conciliatory spirit—was in the midst of all the hatred and meanness, despair and sorrow surrounding us, a moral solace.... Her heart knew no malice.

With the assistance of the Rockefeller Foundation's Emergency Committee to Aid Displaced German Scholars, in 1933 Bryn Mawr College in Pennsylvania offered Dr. Noether a faculty position. There she was treated

---

[16]Among his many accomplishments, Wiener was the founder of cybernetics, the study of feedback mechanisms.

Figure 1.3: *Emmy Noether, "probably the best woman mathematician there has ever been."* (Bryn Mawr College)

with the respect she had always deserved, but which had too often been denied her before. The president of Bryn Mawr College, Marion Park, wrote to a friend in November 1933, "I am venturing to ask you whether by any lucky chance you can come down to Bryn Mawr in December and see Dr. Noether in action!" In 1934 President Park reported to the Rockefeller Foundation that the newly created Emmy Noether Fellowship would be awarded to distinguished students. In a letter of January 1935, Norbert Wiener wrote,

> Miss Noether is a great personality; the greatest woman mathematician who has ever lived; and the greatest woman scientist of any sort now living, and a scholar at least on the plane of Madame Curie. Leaving all questions of sex aside, she is one of the ten or twelve leading mathematicians of the present generation in the entire world and has founded what is certain to be the most important close-knit group of mathematicians in Germany—the Modern School of Algebraists. ... In all the cases of German refugees, whether in this country or elsewhere, that of Miss Noether is without doubt the first to be considered. [Brewer & Smith (1982) 33–34]

In March 1935, the newly formed Institute for Advanced Study in nearby Princeton, New Jersey, offered Dr. Noether a stipend to visit the institute and conduct a weekly seminar. One wonders what might have developed in unified field theory had she and Albert Einstein, who was newly installed there, been able to collaborate...

Alas! following complications from surgery to remove a tumor on April 10, 1935, in which her initial recovery seemed to be going fine, Emmy Noether went into a coma. She passed from this life on April 14, at the age of 57. Over the next few weeks, tributes to her appeared in several languages. At the Bryn Mawr memorial service where Hermann Weyl spoke with such eloquence on April 26, 1935, he summarized her personality as "warm as a loaf of bread"[Dick (1981)]. Emmy Noether's ashes were interred on the campus of Bryn Mawr College, in an honored place, under the cloister walkway of the Martha Carey Thomas Library.[17]

With Emmy Noether's premature death, besides the personal loss that befell her friends and extended family, and the loss to mathematics and physics, one also mourns her loss to the culture of science. She was and still is an icon for women in the sciences, a model of integrity for not allowing injustice to turn a victim towards bitterness, and an inspiring intellectual companion for anyone who appreciates a life of the mind.[18]

Emmy Noether's achievements live beyond her physical presence. She left a great legacy to mathematics, as in her studies of invariants, abstract algebra (notably the theory of rings and ideals), hypercomplex numbers, and applications of group theory to combinatorial topology [Brewer & Smith (1982) 22], to name a few of her important contributions. To mathematical physics, she also left a splendid legacy through her powerful theorem celebrated in this book. She advised a dozen doctoral students, authored over 40 publications,[19] and inspired the "Noether school" of algebraists. Today the Emmy Noether Gymnasium in Berlin and the Emmy Noether Gymnasium in Erlangen honor her memory. In 1980 the Association for Women in Mathematics established the annual Emmy Noether Lecture, to "honor women who have made fundamental and sustained contributions to the mathematical sciences." From a place of honor among the mathematical sciences, Emmy Noether's theorem of 1918 forms a central organizing principle for the great range of physics.

---

[17]Martha Carey Thomas (1857–1935) was a first dean and second president of Bryn Mawr College. She was an active suffragist, and set a tone of excellence through her vision, tenacity, and independent spirit. When Thomas was denied admission to American universities because she happened to be female, she earned her PhD in linguistics at the University of Zürich. She and Emmy Noether must have been kindred spirits, and it is fitting that their ashes share the same hollowed ground.

[18]For a list of numerous sources on Noether's theorem (first and second), including biographical and historical resources, see Neuenschwander (2014b).

[19]See Brewer & Smith (1982) for a list of Noether's publications and descriptions of her doctoral students and their accomplishments; see Dick (1981) for a publications list and the dissertation titles of Noether's 13 doctoral students.

Questions for Reflection and Discussion

Q 1.a. In the argument that leads from translational invariance to the conservation of linear momentum, why was it necessary that the force be derivable from a potential energy function?

Q 1.b. List all the invariants you can think of for the following settings:
a. Newtonian mechanics, with its Galilean transformation between inertial frames;
b. Special relativity, for inertial frames related by a Lorentz transformation;
c. Classical mechanics, for the transformation from an inertial frame to a rotating frame.

Q1.c. A quantity may be conserved but not invariant. Can a quantity be invariant but not conserved?

Q 1.d. Show that if space is isotropic about two points then it is also homogeneous.

Q 1.e. Obtain a translation of Emmy Noether's 1918 paper (or the 1918 original, in German!) [Noether (1918); Noether & Tavel (1971); Kosmann-Schwarzbach (2010)]. As you progress through this text, consult Noether's paper and notice the similarities and differences between her approach and the approach followed here.

Q1.f. If conservation laws follow from symmetries, what symmetry accounts for the conservation of electric charge?

Q1.g. In this book we celebrate symmetry. However, we realize that lack of symmetry, or "symmetry breaking," is also essential to life as we know it.
a. Describe instances where symmetry breaking makes life possible and/or interesting. For specific instances to consider, contemplate asymmetry in architecture; machine design; face recognition; chemical reactions that depend on "handedness" of some molecules; traffic dynamics; and art.
b. Does the second law of thermodynamics have anything to say about the invariance (or lack thereof) of a system's evolution in time?

Exercises

1.1. Show how to express Newton's second law, $\mathbf{F} = d\mathbf{p}/dt$, in terms of the change in
a. mechanical energy, and
b. angular momentum.

c. In each case identify the criteria for the corresponding conservation law, and suggest for it an underlying symmetry of space or time.

1.2. An example from special relativity (see appendix B) illustrates the distinct roles of invariance and conservation in solving a problem. A proton of mass m moves with momentum $\mathbf{p}$ and kinetic energy K through the lab frame and collides with another proton initially at rest. Find in this frame the minimum value of K necessary for the production of a new proton and antiproton pair,

$$p + p \rightarrow 3p + \bar{p}, \tag{1.8}$$

where $\bar{p}$ denotes an antiproton. Hints: use the invariance of $E^2 - (pc)^2 = (mc^2)^2$ between the laboratory and the center-of-mass frames (c = speed of light), and the covariance of expressions for energy and momentum conservation.

Chapter 2

Functionals

Variable quantities called functionals play an important role in many problems arising in analysis, mechanics, geometry, etc. By a functional, we mean a correspondence which assigns a definite (real) number to each function (or curve) belonging to some class. —I. M. Gelfand and S. V. Fomin, *Calculus of Variations*, translated by Richard A. Silverman, 1963

2.1   Single-Integral Functionals

At the foundation of Noether's theorem stand mathematical objects called *functionals*. To invoke Noether's (first) theorem, we make two distinct demands on a functional: (1) that it be an *extremal* (or *stationary*), and (2) that it be *invariant* under a continuous transformation. Before going there, let's get acquainted with these functionials.

Generically, a functional is a mapping from a well-defined set of functions to the real numbers (Figure 2.1). A functional is like a vending machine. Into the input slot you insert a function selected from a set of allowed possibilities. The machine clanks and grinds, and out pops a real number as the output. Definite integrals map a function to a real number, and the functionals of interest in this book are expressed as definite integrals. A few examples provide illustrations.

Example (Distance Functional): I would offend tradition and good sense to not include a standard illustrative example, the "distance functional." In the xy plane, a function $y = y(x)$ describes a path. The distance on the path from point $(x, y) = (a, y(a))$ to point $(x, y) = (b, y(b))$ is found

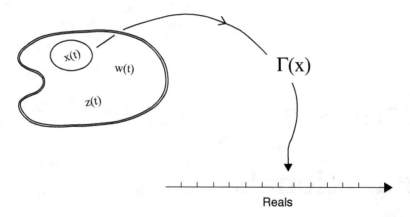

Figure 2.1: *Definition of a functional: a mapping from a set of functions* $\{x(t)\}$ *to the set of real numbers.*

by summing up little increments of length ds. With dx^2 denoting $(dx)^2$, we have

$$\text{distance} = \int_a^b ds$$

$$= \int_a^b \sqrt{dx^2 + dy^2} \tag{2.1}$$

$$= \int_a^b \sqrt{1 + y'^2}\, dx,$$

where y' denotes dy/dx, which we assume exists on the interval $[a, b]$. This requirement defines the set of all permissible functions $\{y(x)\}$. The shortest path constrained to a surface, that connects two points on that surface, is called the *geodesic*.

Example (another Distance Functional): The distance traveled on a journey from Baltimore to Erlangen depends on the path taken. To specify a path one could chart the trajectory in terms of longitude as a function of latitude. For a journey over the surface of the Earth, in spherical coordinates (r, θ, φ) the distance traveled may be computed from the definite integral

$$\text{distance} = \int_{\text{Baltimore}}^{\text{Erlangen}} ds \tag{2.2}$$

where

$$ds^2 = dr^2 + r^2(d\theta^2 + \sin^2\theta d\varphi^2). \tag{2.3}$$

At constant radius R from the Earth's center, the distance is given by

$$\text{distance} = R \int_{\text{Baltimore}}^{\text{Erlangen}} \sqrt{1 + \varphi'^2 \sin^2 \theta} \, d\theta \tag{2.4}$$

and $\varphi' \equiv d\varphi/d\theta$. If the trajectory is expressed in terms of longitude as a given function of latitude, $\varphi = \varphi(\theta)$, then we can evaluate the integral and find the distance along that path between the two points on the globe. A sensible airline will find the path that makes the distance traveled a minimum, the geodesic.

Example (Fermat's Principle): A light ray passes through a medium of refractive index n. The light's speed through the medium is $v = ds/dt = c/n$, where c denotes the speed of light in vacuum. Let the ray move in an xy plane. The refractive index may vary with position so that $n = n(x, y)$. When the light goes from a fixed initial point a to a fixed final point b, the elapsed time is

$$\Delta t = \int_{t(a)}^{t(b)} dt. \tag{2.5}$$

The time increment may be written $dt = ds/v = n(x,y)ds/c$:

$$\Delta t = \frac{1}{c} \int_{x(a)}^{x(b)} n(x,y) \sqrt{1 + y'^2} \, dx. \tag{2.6}$$

For any given path $y = y(x)$, one does the integral and computes the elapsed time. Fermat's principle of geometrical optics postulates that the ray's actual trajectory between fixed points will be the one for which Δt is a minimum.[1]

You may notice in functionals that treat one spatial variable as a function of another, as with curves in the xy plane, the integral may be rewritten in terms of a parameter t. This t could be elapsed time from a point of departure, arc length along a curve, and so on. In the first distance functional example, suppose a parameterization $x \to x(t)$ and $y \to y(t)$ is introduced. The particle sets off from $(x, y) = (a, y(a))$ at time $t(a)$ and follows some curve $y = y(x)$ to arrive at $(b, y(b))$ at time $t(b)$. Denoting $\dot{x} \equiv dx/dt$, it follows that $dx = \dot{x}dt$ and similarly for dy. Now the functional may be written parametrically in terms of t as

$$\text{distance} = \int_{t(a)}^{t(b)} \sqrt{\dot{x}^2 + \dot{y}^2} \, dt. \tag{2.7}$$

---

[1]The "minimum" was stated by Fermat and continues in most introductory optics treatments; see the comment by Jenkins and White in section 5.3.

In the distance functional, the integrand had changed from $\sqrt{1 + y'^2}$ integrated over x, to $\sqrt{\dot{x}^2 + \dot{y}^2}$ integrated over t. Same information, different parametrization.

The same procedure can be done with the Fermat's principle functional:[2]

$$\Delta t = \frac{1}{c} \int_a^b n(x, y) \sqrt{dx^2 + dy^2}$$

$$= \frac{1}{c} \int_{t(a)}^{t(b)} n(x(t), y(t)) \sqrt{\dot{x}^2 + \dot{y}^2} \; dt. \tag{2.8}$$

A system may have more than one dependent variable; for example, in the blue space that birds fly through, a bird's location may be expressed by three spatial coordinates, which together form the components of a position vector. Let the vector components be denoted x^μ, where $\mu = 1, 2, 3$ in ordinary three-dimensional space; in special and general relativity, $\mu = 0$, $1, 2, 3$ for the four dimensions of spacetime.[3] Denote the functional as Γ, and let its integrand L be a function of the independent variable t, the dependent variables $x^\mu(t)$, and their first derivatives with respect to t. For example, in three-dimensional space, where the coordinates of a particle in motion depend upon a parameter t such as time, Γ takes the form

$$\Gamma = \int_a^b L(t, x^\mu, \dot{x}^\mu) dt \tag{2.9}$$

where $\dot{x}^\mu \equiv dx^\mu / dt$.

Example (Hamilton's Principle): In mechanics, the functional of cardinal importance is the time integral of the difference between a system's kinetic and potential energies:

$$\Gamma = \int_a^b (K - U) dt. \tag{2.10}$$

For a single particle, in classical mechanics the kinetic energy is given by $K = \frac{1}{2} m(\dot{x}^2 + \dot{y}^2 + \dot{z}^2)$, or in terms of so-called generalized coordinates q^μ,

---

[2]Note that in parametric representation of the Fermat's principle functional, the coefficient of dt is $nv/c = 1$. This often occurs in parametric representations, where the integrand is a *function*, but it's *numercial value* is a constant. This is not unusual: for an example in elementary mechanics, the conservation of energy can be written $1 = (mv^2)/2E + U(x)/E$, even though v and x are variables.

[3]See appendixes A and B for a discussion of coordinates with upper and lower indices. In Euclidean spaces the distinction between x^μ and x_μ makes no difference, but the distinction becomes important in special and general relativity, and in the geometries of curved spaces.

$K = \frac{1}{2}mg_{\mu\nu}\dot{q}^\mu\dot{q}^\nu$, where the metric tensor components $g_{\mu\nu}$ (see appendix A) turn coordinate displacements into distances[4] and repeated indices are summed. For example, in spherical coordinates,

$$K = \frac{1}{2}m\left(\dot{r}^2 + r^2\dot{\theta}^2 + r^2(\sin^2\theta)\dot{\varphi}^2\right), \tag{2.11}$$

so that the nonzero $g_{\mu\nu}$ are $g_{rr} = 1, g_{\theta\theta} = r^2, g_{\varphi\varphi} = r^2\sin^2\theta$. Through the $g_{\mu\nu}$ the kinetic energy may be a function of the coordinates themselves, in addition to being a function of coordinate velocities. The potential energy is a function of the coordinates and possibly the time, $U = U(t, q^\mu)$. Given a trajectory that specifies a particle's coordinates as functions of time, $q^\mu(t)$, one evaluates Γ for that trajectory. According to Hamilton's principle, a particle's *actual* trajectory between fixed times is one for which Γ is a minimum.[5] More specifically, consider a projectile falling, without air resistance, under the control of a uniform gravitational field directed vertically down, the direction opposite the $+z$ axis. Therefore the potential energy is mgz. The trajectory followed, according to Hamilton's principle, is precisely the path for which this integral is a minimum:

$$\Gamma = \int_a^b \left(\frac{1}{2}m(\dot{x}^2 + \dot{y}^2 + \dot{z}^2) - mgz\right) dt. \tag{2.12}$$

Example (Special Relativity and Free Fall in Gravity-Free Spacetime): No matter which reference frame measures proper time between any two events,[6] all inertial observers can use their reference frame's time and space measurements (dt, dx, dy, dz) to compute the proper time[7] increment $d\tau$ with the aid of the invariant spacetime interval, according to

$$c^2 d\tau^2 = c^2 dt^2 - dx^2 - dy^2 - dx^2. \tag{2.13}$$

One way of stating the postulates of special relativity asserts a relativistic analog to Fermat's principle: Of all world lines through spacetime that

---

[4]In Euclidean space mapped with Cartesian coordinates, $g_{\mu\nu} = \delta_{\mu\nu}$, elements of the unit matrix, or Kronecker delta. If you are not familiar with tensors yet, for now it suffices to think of a two-index tensor as a matrix. However, under coordinate transformations there are additional requirements put on tensors that do not necessarily apply to matrices; see appendix A.

[5]Some authors say that the trajectory actually followed is the one for which Γ is *extremal*, others say *stationary*. See section 5.3.

[6]The distinction between two *places* and two *events* should be noted.

[7]As a particle's world line takes it from spacetime event a to event b, the proper time is the "wristwatch time" recorded by the particle itself; in other words, the time between the two events as measured in the reference frame where they occur at the same place. See Taylor & Wheeler (2000).

a freely falling particle could be imagined to follow from event a to event b, the world line actually followed between those two events is the one that maximizes the elapsed proper time,

$$\Delta\tau = \int_a^b d\tau$$

$$= \int_{t(a)}^{t(b)} \sqrt{1 - \frac{1}{c^2}(\dot{x}^2 + \dot{y}^2 + \dot{z}^2)}\, dt \qquad (2.14)$$

$$= \int_{t(a)}^{t(b)} \gamma^{-1}\, dt$$

where $\dot{x} \equiv dx/dt$ and $\gamma \equiv 1/\sqrt{1 - \frac{v^2}{c^2}}$.

Example (from finance): An investor plans to invest annually N_B dollars in bonds and N_S dollars in stocks, where these numbers may vary from year to year. Let Y_B be the annual yield for the bonds, and Y_S the annual yield for stocks.[8] The total amount of earnings made by the investments from year a to year b will be

$$\text{earnings} = \int_a^b [Y_B(t)N_B(t) + Y_S(t)N_S(t)]\, dt. \qquad (2.15)$$

Suppose Y_B and Y_S, as functions of time, are estimated in advance. If one chooses different scenarios for the investor's annual contributions N_B and N_S as functions of time from year a to year b, then one can compute the total earnings that will follow from each scenario. To choose functions $N_S(t)$ and $N_B(t)$ each year is to choose a "path" through a two-dimensional "space" of stocks and bonds. Of course, a wise investor will seek a trajectory that maximizes the accumulated earnings.

These examples offer a sampling of functionals that are definite integrals. The formal definition of a functional may now be given.

2.2   Formal Definition of a Functional

The functionals we consider in this book are defined in terms of definite integrals.

Definition: A *functional*, Γ, is a mapping from a well-defined set of functions to the real numbers. The domain of the mapping is the set of

---

[8]E.g., an 8% annual increase means $Y = 1.08$.

twice differentiable functions $\{q^\mu(t)\}$ on the closed interval $[a, b]$, where the label μ distinguishes N dependent variables. The mapping is given by a definite integral of the form

$$\Gamma = \int_a^b L(t, q^\mu, \dot{q}^\mu) \, dt, \tag{2.16}$$

where $\dot{q}^\mu \equiv dq^\mu/dt$. The function $L(t, q^\mu, \dot{q}^\mu)$ is called the *Lagrangian* of the functional. Notice that L may depend explicitly on the independent variable, the dependent variables, and the first derivatives of the dependent variables.

Example: Consider the distance functional in the xy plane from the origin to (1,1):

$$\Gamma = \int_0^1 \sqrt{1 + y'^2} \, dx \tag{2.17}$$

where $y = y(x)$ and $y' = dy/dx$. If we choose $y = x$, then $\Gamma = \sqrt{2} \approx 1.414$; if we choose $y = x^2$, then $\Gamma = \frac{1}{4}[2\sqrt{5} + \ln(2 + \sqrt{5})] \approx 1.479$.

Generalizations to this definition of the functional are readily made. The Lagrangian can depend on higher derivatives. As we shall see in chapter 3, the requirement that Γ be a maximum or minimum leads to a differential equation of one order higher than the highest-order derivative in the Lagrangian. Because most differential equations in physics are of second order, for them we need only Lagrangians that include the first derivatives of the dependent variables. Extensions to higher-order derivatives have their uses, but accommodating them here is not essential to fluency in the principles of Noether's theorem.

In later chapters we generalize the definition of a functional to multiple-integral problems, when two or more independent variables are necessary. For example, in wave physics the wave function depends on space and time; in relativity, four-vectors and other tensors are functions of (t, x, y, z). Again, the fundamentals of our subject are not essentially changed; the changes are in important details—that is, total derivatives such as $\dot{q}^\mu = dq^\mu/dt$ get replaced with partial derivatives. These considerations occupy us in Part III.

These examples illustrate how interest in functionals may include making them a maximum (as in the stocks and bonds investment example) or a minimum (as in the airline flight paths). In either case, one speaks of the functional as being made an extremal (or stationary). How to systematically find dependent variables $q^\mu(t)$ that makes the functional an extremal forms the subject of chapter 3.

Questions for Reflection and Discussion

Q2.a. Under what circumstances are the following definite integrals functionals?

a. Mechanical work as a particle moves from position **a** to position **b**, while acted upon by the force **F**:

$$W = \int_{\mathbf{a}}^{\mathbf{b}} \mathbf{F} \cdot d\mathbf{r}. \qquad (2.18)$$

b. The entropy change ΔS, in terms of heat dQ added to a system at absolute temperature T, for a change of thermodynamic state from a to b:

$$\Delta S = \int_{a}^{b} \frac{dQ}{T}. \qquad (2.19)$$

c. The voltage (or emf) induced by Faraday's law, $\oint_{\mathcal{C}} \mathbf{E} \cdot d\mathbf{r}$, where $\mathbf{E}$ denotes the electric field and $\mathcal{C}$ a closed path.

d. The flux of an electric, magnetic, or gravitational vector field through a specific surface.

Q2.b. For the functionals illustrated in this chapter, list the independent and dependent variables, and identify the Lagrangian. Propose simple special-case trajectories that produce an extremum for:

a. Distance functional in the xy plane.

b. Distance functional on a spherical surface.

c. Time functional for light in a refractive medium.

d. Proper time functional for a particle falling freely in spacetime.

e. Projectile motion functional.

f. Earnings functional.

Q2.c. Construct one or more functionals that are *not* definite integrals.

Exercises

2.1. Imagine a light ray traveling at the angle α above the x-axis when it encounters at $x = 0$ a medium with refractive index $n = n_0(1 + x/a)$, where n_0 and a are constants and $x \geq 0$. Let the location of the encounter be the origin of an xy coordinate system, and evaluate the time required for a light ray to travel from the origin to $(x, y) = (a, y(a))$ along these paths:

a. $y = x$.

b. $y = a[\cosh\left(\frac{x}{a}\right) - 1]$.

2.2. According to Hamilton's principle, in mechanics the "action" functional,

$$\Gamma = \int_0^T (K - U)\, dt \tag{2.20}$$

when minimized determines the particle's trajectory. Thus Hamilton's principle is often called the principle of least action. The term "action" can sometimes be ambiguous; this exercise addresses the ambiguity. Define the total mechanical energy E, kinetic plus potential energy, $E = K + U$, where $K = \frac{1}{2}m(dx/dt)^2$.

a. Show that

$$\Gamma = \int_0^T 2K\, dt - \int_0^T E\, dt. \tag{2.21}$$

b. Show that Γ can be further developed into

$$\Gamma = \int_{x(0)}^{x(T)} p\, dx - \int_0^T E\, dt, \tag{2.22}$$

where p denotes the momentum mv.[9]
c. Show that

$$p = \frac{\partial \Gamma}{\partial x} \tag{2.23}$$

and

$$E = -\frac{\partial \Gamma}{\partial t}. \tag{2.24}$$

These results foreshadow similar ones in Hamilton-Jacobi theory (chapter 9) and quantum mechanics (chapter 10).
d. If we already know that $E = \text{const.}$, show that Hamilton's principle becomes the statement

$$\int_{x(0)}^{x(T)} p\, dx = \text{min.} \tag{2.25}$$

In contrast, Hamilton's principle, $\int_a^b (K - U)dt = \text{min.}$, does not impose the conservation of energy as a supplementary condition. Some authors call $\int_a^b p\,dx$ the "abbreviated action" to distinguish it from "the" action, $\int_a^b (K - U)dt$. Minimizing either action says the same thing if the energy

---

[9]The conceptually distinct and sometimes numerically identical "canonical momentum" is introduced in chapter 3.

happens to be conserved. Because Hamilton's principle does not require E to be constant, it is the more general of the two.

2.3. Consider a particle of mass m falling vertically in a uniform gravitational field of magnitude g. The potential energy is mgy, where the y-axis is vertical and increases upward. Set up the functional

$$\Gamma = \int_0^{t_0} \left(\frac{1}{2}m\dot{y}^2 - mgy\right) dt. \tag{2.26}$$

Compute Γ explicitly by evaluating the definite integral in the following cases: a. Suppose that $y(t) = A\left(\frac{t}{t_0}\right)^2$, where A and t_0 are constants;
b. $y(t) = A(1 - e^{-t/t_0})$.
c. Compare the results of parts (a) and (b). Which is least?

2.4. Consider a particle undergoing periodic motion with period T, as described by the functional

$$\Gamma = \int_0^T \left(\tfrac{1}{2}m\dot{x}^2 - \tfrac{1}{2}m\omega^2 x^2\right) dt. \tag{2.27}$$

a. Suppose someone tries the solution $dx/dt = +v_0 = $ const. during the first half of a cycle and $dx/dt = -v_0$ during the second half. Compute Γ explicitly by evaluating the definite integral.
b. Compute Γ if $x(t) = A\cos(\omega t)$, where $\omega = 2\pi/T$.
c. Compare the Γ of part (a) to the Γ of part (b). Which is least?

2.5. Consider the investment functional

$$\text{earnings} = \int_0^T [Y_B(t)N_B(t) + Y_S(t)N_S(t)] \, dt. \tag{2.28}$$

Evaluate the earnings, assuming the annual investment N_0 is partitioned equally between stocks and bonds, and the yields vary as follows:
a. The yields are identical and constant at Y_0.
b. The bond yield is a constant $2Y_0$ and the stock yield declines as $Y_S = Y_0 e^{-t/T}$.
c. The bond yield is a constant $2Y_0$ and the stock yield is periodic with period T, where $Y_S = Y_0(1 + \cos\omega t)$ with $\omega = 2\pi/T$.
d. How to the earnings of scenarios (a), (b), and (c) compare?

Chapter 3

Extremals

In the Newtonian formulation, a certain force on a body is considered to produce a definite motion; that is, a definite effect is always associated with a certain cause. According to Hamilton's Principle, however, the motion of a body may be considered to result from the attempt of Nature to achieve a certain purpose, namely, to minimize the time integral of the difference between the kinetic and potential energies. —Jerry B. Marion, *Classical Dynamics of Particles and Systems*, 1970, 198

3.1   The Euler-Lagrange Equation

The distance traveled by an airplane flying from Baltimore to Erlangen is a functional: the number you get for the distance depends on the route. There is no unique route that makes the distance a maximum. But only one route exists between Baltimore and Erlangen, at a fixed distance from the center of the Earth, that makes the distance a minimum. That route is called the "geodesic" between the initial and final points, and requires

$$\text{distance} = R \int_{\text{Baltimore}}^{\text{Erlangen}} \sqrt{1 + \varphi'^2 \sin^2 \theta}\, d\theta = \min. \tag{3.1}$$

where $\varphi' \equiv d\varphi/d\theta$.

Geometrical optics may be derived from Fermat's principle. Of all possible paths connecting two fixed points in a refractive medium, the path actually followed by a light ray is the one that minimizes the elapsed time:

$$\Delta t = \frac{1}{c} \int_{x(a)}^{x(b)} n(x, y) \sqrt{1 + y'^2}\, dx = \min. \tag{3.2}$$

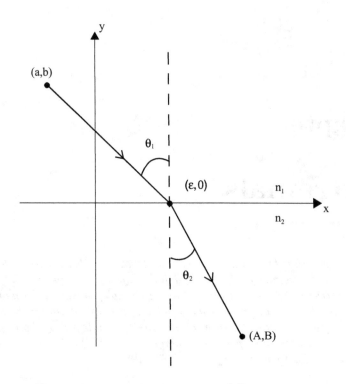

Figure 3.1: *The geometry in the derivation of Snell's law and the law of reflection, using Fermat's principle.*

where $y' \equiv dy/dx$. If the refractive index n happens to be constant throughout a region, then the shortest *time* of Fermat's principle reduces to the path of shortest *distance*, the geodesic.

Example: Let's apply Fermat's principle to a light ray that travels from one fixed point to another, encountering along the way a piecewise constant index of refraction. Let the boundary between the two refractive media be the $y = 0$ plane, with uniform index of refraction n_1 above this plane, and uniform index $n_2 > n_1$ below it. Thus the rays above and below the $y = 0$ plane are straight lines (Figure 3.1).

Let a light ray move from (a, b) above the plane, to (A, B) below it, crossing the x-axis at some location ε, which is to be determined from Fermat's principle. The elapsed time functional T can be written as a function of ε. Letting s denote the lengths,

$$
\begin{aligned}
T(\varepsilon) &= \frac{n_1}{c} s_1 + \frac{n_2}{c} s_2 \\
&= \frac{n_1}{c} \sqrt{(\varepsilon - a)^2 + b^2} + \frac{n_2}{c} \sqrt{(A - \varepsilon)^2 + B^2}.
\end{aligned}
\tag{3.3}
$$

According to Fermat's principle, on the extremal path T must be a minimum. Setting $dT/d\varepsilon = 0$, we obtain

$$0 = \frac{n_1(\varepsilon - a)}{s_1} - \frac{n_2(A - \varepsilon)}{s_2},\qquad(3.4)$$

which becomes Snell's law,

$$n_1 \sin\theta_1 = n_2 \sin\theta_2.\qquad(3.5)$$

As a bonus, if the ray is reflected so that $B \to -B$ and $n_1 = n_2$, then we obtain the law of reflection, $\theta_1 = \theta_2$.

This example illustrates an important point: Had Fermat's principle stated the time T was to be made a maximum, or sought an inflection point on a graph of $T(\varepsilon)$ vs. ε, the same procedure of setting the derivative equal to zero would have been followed, and the outcome—Snell's law—would have been the same. In a calculus problem, setting $df(x)/dx = 0$ only tells us where the function $f(x)$ has zero slope; the zero first derivative does not, by itself, distinguish between maxima, minima, or inflection points. That Fermat's principle requires the time functional to be a minimum, that an airline route functional should be a minimum, or that an earnings functional seeks a maximum, arise from reasons of physics, geometry, or finance that lie outside the mathematics of the calculus of variations itself.

Example: Imagine a baseball released from rest and allowed to fall with negligible air resistance in a uniform gravitational field $\mathbf{g}$ directed vertically downward. Let the y-axis point upward. With potential energy $U(y) = mgy$, and the particle released from rest at $y = 0$ at time $t = 0$, for $t > 0$ conservation of mechanical energy $K + U = $ const. gives $K = -mgy > 0$. From Hamilton's principle, the trajectory $y = y(t)$ follows by minimizing the time integral of $K - U$. Invoking conservation of energy, this means the Lagrangian $K - U$ equals $2K > 0$ for any time after the baseball is dropped. Evidently we may say[1]

$$\Gamma = \int_0^t (K - U)\, dt' \geq 0.\qquad(3.6)$$

Therefore the minimum value that Γ can possibly have is 0. Suppose we set $\Gamma = 0$ for all t after the baseball is released. Let's take Hamilton's principle seriously and see how far it goes towards giving the trajectory function $y(t)$. Parameterize the solution as $y = -At^n$, consistent with the

---

[1]In eqs. (3.6) and (3.7), t' is a dummy variable of integration and $y = y(t')$ because t appears in the upper limit.

initial condition that $y(0) = 0$ and taking into account that $y < 0$ as the ball drops. Determine n by setting $\Gamma = 0$. Our functional[2]

$$\Gamma = \int_0^t \left(\frac{1}{2} m \dot{y}^2 - mgy \right) dt' \tag{3.7}$$

integrates to

$$\Gamma = \frac{1}{2} m A^2 \frac{n^2 t^{2n-1}}{2n-1} - mgA \frac{t^{n+1}}{n+1}. \tag{3.8}$$

Invoking Hamilton's principle, in this case we set $\Gamma = 0$, which yields

$$t^{2-n} = \frac{An^2}{2g} \left(\frac{n+1}{2n-1} \right) \tag{3.9}$$

and we notice the right-hand side is a constant. For the left-hand side to be constant for all t, we must set $n = 2$. With $n = 2$, we also get $A = g/2$. Therefore we conclude that $y = -\frac{1}{2} gt^2$.

This example works out neatly enough to the expected answer, and it illustrates an important limitation: Hamilton's principle cannot, by itself, give us the *unique* solution; we had to build initial conditions into the solution by imposing $y(t) \leq 0$ for $t \geq 0$, as a *supplement* to Hamilton's principle. Of course, this should not be surprising. In evaluating integrals or solving differential equations, initial conditions (or boundary conditions)[3] must be specified in order to obtain unique solutions. This similarity between functionals and differential equations both needing initial or boundary conditions is not a coincidence. The general strategy for finding a function that makes the functional an extremal requires us to solve a differential equation, the Euler-Lagrange equation. As with any differential equation, to obtain a unique solution, initial or boundary conditions must be given.

An ambiguity in the Lagrangian always exists in any definite integral functional: a derivative dS/dt can always be added to the Lagrangian without changing the numerical value of Γ, provided $S(b) = S(a)$:

$$\int_a^b \left(L + \frac{dS}{dt} \right) dt = \int_a^b L \, dt + [S(b) - S(a)]. \tag{3.10}$$

This feature will recur in our deliberations, including something called "divergence invariance" (section 4.3) and in Hamilton-Jacobi theory (section 9.4).

---

[2]Of course, if we had never encountered the free-fall problem before we would have no reason to assume the correct trajectory is a power law in t. But we can think of At^n as the leading term in a Taylor series expansion of $y(t)$.

[3]This is the role of limits on definite integrals. In our example, the upper limit t was arbitrary.

Let us now develop the Euler-Lagrange equation. Consider the functional

$$\Gamma = \int_a^b L(t, q^\mu, \dot{q}^\mu) \, dt \tag{3.11}$$

whose Lagrangian L depends on one independent variable t, N dependent variables $q^\mu(t)$, and their first derivatives $\dot{q}^\mu$, where $\mu = 1, 2, \ldots, N$. When all N coordinates are independent of each other,[4] we say the system has "N degrees of freedom." The set of generalized coordinates q^μ form a vector in an N-dimensional vector space.[5] We also assume that all the derivatives of q^μ exist that may be needed. We will show that the set $\{q^\mu(t)\}$, in other words the components of the vector $\mathbf{q}$ that makes Γ an extremal, are solutions of the N Euler-Lagrange equations

$$\frac{\partial L}{\partial q^\mu} = \frac{d}{dt} \frac{\partial L}{\partial \dot{q}^\mu}, \qquad \mu = 1, 2, \ldots, N. \tag{3.12}$$

The existence of solutions to the Euler-Lagrange equation forms a necessary condition for the functional to be an extremal. Demonstrating this claim is a two-step process.

Step 1: Let $q^\mu(t)$ be a generalized coordinate on the trajectory that makes Γ an extremal. Embed this path in a family of paths (denoted with primes, see Figure 3.2) obtained from the extremal path through a continuously variable parameter ε,

$$q'^\mu = q^\mu + \varepsilon \zeta^\mu \tag{3.13}$$

where $\zeta^\mu(a) = \zeta^\mu(b) = 0$. In addition, we require that the ζ^μ belong to the same vector space as the q^μ and have the same number of derivatives. Beyond these requirements, $\zeta^\mu(t)$ is otherwise arbitrary. Now the functional Γ has become a function of ε,

$$\Gamma(\varepsilon) = \int_a^b L(t, q'^\mu, \dot{q}'^\mu) \, dt. \tag{3.14}$$

As ε is varied, different q'^μs result, and with them different values of $\Gamma(\varepsilon)$. But Γ becomes an extremum when $d\Gamma/d\varepsilon = 0$, where $d\Gamma/d\varepsilon$ means[6]

$$\frac{d\Gamma}{d\varepsilon} = \lim_{\varepsilon \to 0} \frac{\Gamma(\varepsilon) - \Gamma(0)}{\varepsilon}. \tag{3.15}$$

---

[4] For handling constraints between them, see section 3.6.

[5] The vector space can, in general, be familiar or quite abstract. Examples range from the Euclidean space of classical mechanics or the Minkowskian spacetime of relativity, to the "stock and bonds space" of the financial example, or the abstract complex-valued state spaces of quantum mechanics.

[6] I mention this because the elegant and traditational notation $\delta\Gamma = 0$, used in much of the literature (and discussed in this section), potentially creates the misconception that making the functional an extremal involves some new kind of weird derivative. It doesn't.

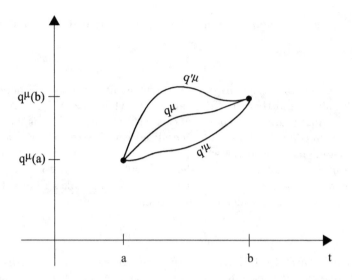

Figure 3.2: *The extremal path q^μ and some varied paths $q'^\mu = q^\mu + \varepsilon\zeta^\mu$, where $\zeta^\mu(a) = \zeta^\mu(b) = 0$.*

This is to occur on the path for which $\varepsilon = 0$. We therefore evaluate

$$\left[\frac{d\Gamma}{d\varepsilon}\right]_{\varepsilon=0} = 0. \tag{3.16}$$

Assuming the continuity of the partial derivatives, by Leibniz's rule the derivative can be brought under the integral sign. With the chain rule and summing over repeated indices we obtain:

$$\frac{d\Gamma}{d\varepsilon} = \int_a^b \left[\frac{\partial L}{\partial q'^\mu}\frac{\partial q'^\mu}{\partial\varepsilon} + \frac{\partial L}{\partial\dot{q}'^\mu}\frac{\partial\dot{q}'^\mu}{\partial\varepsilon}\right] dt. \tag{3.17}$$

Setting $\varepsilon = 0$, this becomes

$$\left[\frac{d\Gamma}{d\varepsilon}\right]_0 = \int_a^b \left[\frac{\partial L}{\partial q^\mu}\zeta^\mu + \frac{\partial L}{\partial\dot{q}^\mu}\dot{\zeta}^\mu\right] dt = 0. \tag{3.18}$$

To turn the $\dot{\zeta}^\mu$ into a ζ^μ so it can be factored out, integrate the second term by parts, using

$$\frac{d}{dt}\left[\frac{\partial L}{\partial\dot{q}^\mu}\zeta^\mu\right] = \frac{\partial L}{\partial\dot{q}^\mu}\dot{\zeta}^\mu + \zeta^\mu\frac{d}{dt}\frac{\partial L}{\partial\dot{q}^\mu}, \tag{3.19}$$

which yields

$$\int_a^b \left[\frac{\partial L}{\partial q^\mu} - \frac{d}{dt} \frac{\partial L}{\partial \dot{q}^\mu} \right] \zeta^\mu dt + \left[\zeta^\mu \frac{\partial L}{\partial \dot{q}^\mu} \right]_a^b = 0. \tag{3.20}$$

The integrated terms vanish because $\zeta^\mu(b) = \zeta^\mu(a) = 0$, leaving

$$\int_a^b \left[\frac{\partial L}{\partial q^\mu} - \frac{d}{dt} \frac{\partial L}{\partial \dot{q}^\mu} \right] \zeta^\mu dt = 0. \tag{3.21}$$

A traditional and elegant δ-notation for these steps, attributed to Lagrange and used by Emmy Noether in her 1918 paper and by a host of authors across the years, rewrites $\varepsilon \zeta^\mu$ as

$$q'^\mu - q^\mu \equiv \delta q^\mu \tag{3.22}$$

for the variation of the path, and writes

$$\delta \int_a^b L \, dt = 0 \tag{3.23}$$

for the procedure denoted by eq. (3.16). Thus eq. (3.18) could be written in Lagrange's notation as

$$\delta \Gamma = \int_a^b \left[\frac{\partial L}{\partial q^\mu} \delta q^\mu + \frac{\partial L}{\partial \dot{q}^\mu} \delta \dot{q}^\mu \right] dt = 0. \tag{3.24}$$

Another way to look at the δ-notation is to let $\delta \Gamma \to d\Gamma$, let $\varepsilon \to d\varepsilon$, then set $\delta q^\mu = (d\varepsilon) \zeta^\mu$, and write

$$d\Gamma = \int_a^b \left[\frac{\partial L}{\partial q^\mu} \zeta^\mu + \frac{\partial L}{\partial \dot{q}^\mu} \dot{\zeta}^\mu \right] d\varepsilon \, dt. \tag{3.25}$$

This sets up the derivative with respect to ε as outlined above. The Lagrange notation, while elegant and traditional, takes getting used to. Essentially, something with a δ on it is a differential, although δq^μ is made to vanish at the endpoints. In this book I prefer to make explicit the variational parameter ε.

Our objective is to find the set of functions $\{q^\mu(t)\}$ that make Γ an extremal. Since ζ^μ is arbitrary (other than the conditions stated),[7] for the integral in eq. (3.21) to vanish *whatever* ζ^μ happens to be, evidently the

---

[7]Recall that the derivatives of the arbitrary function ζ^μ, like the derivatives of the dependent variables q^μ, must exist in the interval of integration.

factor in the square brackets must vanish identically. This conclusion, while intuitively plausible, must be proved, bringing us to the second step.

Step 2: **Theorem:** If $A(t)$ is continuous on the open interval (a, b), and if $h(t)$ vanishes at both endpoints but is otherwise arbitrary on the closed interval $[a, b]$, and if

$$\int_a^b A(t)h(t)dt = 0, \tag{3.26}$$

then $A(t) = 0$ throughout the region of integration.

At first glance the conclusion that $A(t)$ must vanish on $[a, b]$ might seem so obvious that no proof would be necessary. After all, $h(t)$ can be anything we like so long at it belongs to the same space as $A(t)$ and vanishes at the endpoints—what else could $A(t)$ be other than zero? However, one can imagine scenarios where A and h are orthogonal functions on $[a, b]$, as in the case of $\sin(n\pi x/L)$ with $n = 1, 2, 3, \ldots$ on the interval $[0, L]$. For these functions,

$$\int_0^L \sin\left(\frac{n\pi x}{L}\right) \sin\left(\frac{n'\pi x}{L}\right) \, dx = \frac{L}{2}\delta_{nn'}, \tag{3.27}$$

so the integral could vanish but neither function be indentically zero throughout the interval—even though both of them vanish at the endpoints. Intuitive arguments, while suggestive, are not proofs.

For a proof by contradiction, suppose there exists a point t_0 between a and b for which $A(t_0) > 0$. Because A is continuous, there exists an interval from t_1 to t_2 about t_0, where $a < t_1 < t_0 < t_2 < b$, such that $A(t) > 0$ in this interval.[8] If we can find at least one function h that satisfies the stated conditions on it, then the integral will not vanish. For example, suppose

$$h(t) = \sin\left(\frac{\pi(t - a)}{b - a}\right). \tag{3.28}$$

This describes half a sine wave of wavelength $2(b - a)$, with nodes at the endpoints, so that $h(b) = h(a) = 0$. Furthermore, $h(t)$ is nicely continuous on $[a, b]$, and $h(t) > 0$ within (a, b). Now for the interval from t_1 to t_2 it follows that

$$\int_{t_1}^{t_2} A(t)h(t) \, dt > 0, \tag{3.29}$$

with no guarantee that the integral over the rest of the interval will cancel this out, contrary to the postulate that the integral of $A(t)h(t)$ from a to b vanishes. Therefore, to guarantee the integral vanishes, $A(t)$ must be zero throughout $[a, b]$. This result is called the fundamental lemma of the calculus of variations [Logan (1977) 8].

---

[8]Thus $A(t)$ cannot be, for example, a Dirac delta function.

The lemma justifies our returning to eq. (3.21) and setting the term in square brackets equal to zero, to obtain the Euler-Lagrange equation (ELE),[9]

$$\frac{\partial L}{\partial q^\mu} - \frac{d}{dt}\frac{\partial L}{\partial \dot{q}^\mu} = 0. \tag{3.30}$$

To find a $\mathbf{q}(t)$ that makes Γ an extremal, one solves the N ELEs for the q^μ. Thus the ELE forms a necessary condition for the functional to be an extremal.

Is the ELE also sufficient for Γ to be an extremal? Yes, it is: starting from the ELE, one can multiply it by $\zeta^\mu dt$ and add the boundary terms back in, and integrate from a to b; the steps are reversible (see discussion question Q.3.j).

Whenever the Lagrangian includes *first* derivatives of the generalized coordinates, the Euler-Lagrange equation is a *second*-order partial differential equation, as can be demonstrated explicitly. Remembering the argument of L depends on t, on the generalized velocities and on their first derivatives, by the chain rule the ELE takes the expanded form

$$\frac{\partial L}{\partial q^\mu} = \frac{\partial^2 L}{\partial t \partial \dot{q}^\mu} + \frac{\partial^2 L}{\partial q^\mu \partial \dot{q}^\nu}\dot{q}^\nu + \frac{\partial^2 L}{\partial \dot{q}^\mu \partial \dot{q}^\nu}\ddot{q}^\nu \tag{3.31}$$

(sum repeated indices). Therefore this condition, the ELE, that must be met for the functional to be an extremal, will make sense only if the second derivatives of the functions q^μ exist.[10] If one considers functionals whose Lagrangians contain second derivatives, then the ELE will be third order in the derivatives of the q^μ, and of course these derivatives must exist for the result to be meaningful—and so on to higher orders. Most physics differential equations are of second order. Therefore they require Lagrangians that contain only first-order derivatives of the dependent variables.[11]

Let's examine a couple of elementary examples to illustrate how the ELE is used. Consider a simple harmonic oscillator of mass m moving

---

[9]In Noether's 1918 paper, she defines $\psi_k \equiv \frac{\partial L}{\partial q^k} - \frac{d}{dt}\frac{\partial L}{\partial \dot{q}^k}$. When the functional is an extremal, then $\psi_k = 0$. Thus the ψ_k can be carried around in equations when the functional has not yet been made an extremal. We will do so in chapters 4, 5, and 8.

[10]Thus the second derivatives of the ζ^μ must also exist for the q'^μ to remain among the set of functions allowed in the functional.

[11]For a counterexample, suppose a uniform cantilevered beam of length ℓ is oriented along a horizontal x-axis and subjected to a vertical load $W(x)$. The deflection y of the beam is described by

$$YI\frac{d^4y}{dx^4} = W(x) \tag{3.32}$$

for $0 \le x \le \ell$, where Y is the beam's modulus of elasticity and I the moment of inertia about the supporting point [Rainville & Bedient (1974) 219–220]. Its Lagrangian would depend on d^3x/dt^3 (see ex. 3.22).

along the x-axis, acted upon by a linear spring of spring constant k. The Lagrangian is[12]

$$L(x, \dot{x}) = \frac{1}{2}m\dot{x}^2 - \frac{1}{2}kx^2 \tag{3.33}$$

and the ELE,

$$\frac{\partial L}{\partial x} - \frac{d}{dt}\left(\frac{\partial L}{\partial \dot{x}}\right) = 0 \tag{3.34}$$

gives at once

$$-kx = m\ddot{x}, \tag{3.35}$$

the familiar equation of motion for a simple harmonic oscillator.

In this example notice that the right-hand side of the ELE gives the rate of change of momentum: $\frac{d}{dt}\frac{\partial L}{\partial \dot{x}} = \frac{d(m\dot{x})}{dt}$. This suggests going back to the ELE and defining, for generalized coordinate q^μ, a corresponding generalized momentum p_μ according to

$$p_\mu \equiv \frac{\partial L}{\partial \dot{q}^\mu}. \tag{3.36}$$

The momentum p_μ is said to be canonically conjugate to q^μ, the "canonical momentum." It is not necessarily the same as the mv momentum. For example, consider a pendulum oscillating in a vertical plane under the influence of gravity. With a fixed length l, in the presence of a uniform gravitational field of magnitude g, and taking the potential energy to be zero at the bottom of the swing where $\theta = 0$, the Lagrangian is

$$L(\theta, \dot{\theta}) = \frac{1}{2}ml^2\dot{\theta}^2 - mgl(1 - \cos\theta). \tag{3.37}$$

The momentum conjugate to θ is

$$\frac{\partial L}{\partial \dot{\theta}} = mr^2\dot{\theta}, \tag{3.38}$$

which will be recognized as the pendulum's angular momentum about the point of suspension. While the ELE for the simple harmonic oscillator on the x-axis gave the same result as $F = ma$, the ELE for the pendulum

---

[12]This particular Lagrangian has no explicit time dependence, but it could be included if, for example, the spring were to weaken with metal fatigue, making k a function of time.

gives the same result as the rotational version of Newton's second law, $\tau = d\ell/dt$, where τ is the torque and ℓ the angular momentum.

One of the many beauties of doing physics with Lagrangians is that linear and angular momentum are treated the same. If q^μ is a linear coordinate such as x or r, then the canonical momentum is a linear momentum component, but if q^μ is an angle then its canonical momentum is a component of angular momentum. Thus $\partial L/\partial q^\mu$ can, in a mechanics application, be either a force or a torque, a "generalized force." When written in terms of canonical momentum, the ELE takes on a simple appearance:

$$\frac{\partial L}{\partial q^\mu} = \dot{p}_\mu. \tag{3.39}$$

Mechanics examples illustrate the practical beauty of the ELE: vector equations of motion that cannot be easily handled with elementary techniques of resolving $\mathbf{F} = m\mathbf{a}$ into components, can be handled readily with the ELE, which uses energies. The various ELE's, one for each q^μ, does the splitting into components automatically. For instance, let the pendulum's string of fixed length l be replaced by a spring of variable length $l + r$ and spring constant k. The Lagrangian becomes

$$L(r, \dot{r}, \theta, \dot{\theta}) = \tfrac{1}{2}m\dot{r}^2 + \tfrac{1}{2}m(l+r)^2\dot{\theta}^2 - mg(l+r)(1 - \cos\theta) - \tfrac{1}{2}kr^2. \tag{3.40}$$

Now there are two ELEs, one for r and one for θ. They are coupled and nonlinear, but they are the correct equations of motion. Whether I can solve the ELEs is another issue, but Hamilton's principle, through the ELE, gives the correct equations of motion in a straightforward manner.

For a nonmechanics example, let us try our Euler-Lagrange formalism on Fermat's principle, in the simple instance of light traveling in a medium with uniform refractive index. We already know the answer—light travels in a straight line under such circumstances—so this exercise demonstrates ELE technique. With constant refractive index n, the functional of Fermat's principle becomes

$$s = \frac{n}{c} \int_a^b \sqrt{1 + y'^2}\, dx \tag{3.41}$$

where $y' = dy/dx$. Since $\partial L/\partial y = 0$, the ELE gives $y' = \text{const.}$, which integrates at once to the equation of a straight line, $y = Ax + B$, for some constants A and B. Notice that to get a unique straight line, supplementary boundary conditions must be supplied to nail down values for A and B.

The ELE can also be written another way, where instead of N separate ELEs, one for each generalized coordinate q^μ, all these components are summed out in one equation. To see this alternate form of the ELE, evaluate

dL/dt, recalling that the Lagrangian depends on t, q^μ, and $\dot{q}^\mu$. From the chain rule, we have

$$\frac{dL}{dt} = \frac{\partial L}{\partial t} + \frac{\partial L}{\partial q^\mu}\dot{q}^\mu + \frac{\partial L}{\partial \dot{q}^\mu}\ddot{q}^\mu$$

$$= \frac{\partial L}{\partial t} + \dot{p}_\mu \dot{q}^\mu + p_\mu \ddot{q}^\mu.$$
$$(3.42)$$

In going from the first to the second line, the ELE has been used in the second term, and the definition of canonical momentum in the third term. Regrouping and recognizing a total derivative among three of the terms, this may be rearranged as follows:

$$\frac{\partial L}{\partial t} = -\frac{d}{dt}[p_\mu \dot{q}^\mu - L]. \qquad (3.43)$$

The quantity under the total derivative is called the Hamiltonian, H:

$$H \equiv p_\mu \dot{q}^\mu - L. \qquad (3.44)$$

Formally, in the developments to follow,[13] the Hamiltonian is a function of the independent variable t, the dependent variables q^μ, and their conjugate momenta p_μ, so that $H = H(t, q^\mu, p_\mu)$. In other words, eq. (3.44) defines a Legendre transformation (see appendix D) from $L(t, q^\mu, \dot{q}^\mu)$ to $H(t, q^\mu, p_\mu)$.

In terms of the Hamiltonian, the ELE takes the abbreviated form of eq. (3.43),

$$\frac{\partial L}{\partial t} = -\dot{H}. \qquad (3.45)$$

Let us compute the Hamiltonian for our elementary examples above. In the case of the simple harmonic oscillator, $p = \partial L/\partial \dot{x} = m\dot{x}$, so that $p = \dot{x}/m$, and thus

$$H(x,p) = m\dot{x}^2 - \tfrac{1}{2}m\dot{x}^2 + \tfrac{1}{2}kx^2$$

$$= \frac{p^2}{m} - \frac{p^2}{2m} + \tfrac{1}{2}kx^2$$

$$= \frac{p^2}{2m} + \tfrac{1}{2}kx^2$$
$$(3.46)$$

$$= E$$

---

[13]We are thinking here especially of Hamilton's equations, where the second-order Euler-Lagrange equation is effectively split into a pair of first-order equations; see chapter 9.

where $E = K + U$ is the oscillator's total mechanical energy. For the pendulum of variable length, $p_r = \partial L/\partial \dot{r} = m\dot{r}$ and $p_\theta = \partial L/\partial \dot{\theta} = m(l + r)^2\dot{\theta}$, so that $\dot{r} = p_r/m$ and $\dot{\theta} = p_\theta/m(l + r)^2$. Therefore,

$$
\begin{aligned}
H &= \dot{r}p_r + \dot{\theta}p_\theta - L \\
&= \frac{p_r^2}{2m} + \frac{p_\theta^2}{2m(l + r)^2} + mg(l + r)(1 - \cos\theta) \qquad (3.47) \\
&= E
\end{aligned}
$$

where, again, H equals the total mechanical energy E. In many physics applications, the numerical value of the Hamiltonian is typically but not necessarily equal to the mechanical energy. In other words, the definition of the Hamiltonian is *not* $H = K + U$, but rather

$$
H(t, q^\mu, p_\mu) \equiv p_\nu \dot{q}^\nu - L(t, q^\mu, \dot{q}^\mu) \qquad (3.48)
$$

where, in going from L to H, the generalized velocities $\dot{q}^\nu$ are swapped out for their corresponding momenta p_ν. But in physics applications, where the Lagrangian is an energy, because L and H have the same dimensions, the Hamiltonian is always the energy of *something*—for instance, the total energy of a particle and a field in a coupled system. Whether or not H numerically equals $K + U$, a distinction exists between E the *number* and H the *function*.

3.2 Conservation Laws as Corollaries to the Euler-Lagrange Equation

We have seen two ways of writing the ELE:

$$
\frac{\partial L}{\partial q^\mu} = \dot{p}_\mu \qquad (3.49)
$$

where

$$
p_\mu \equiv \frac{\partial L}{\partial \dot{q}^\mu} \qquad (3.50)
$$

denotes the momentum conjugate to q^μ; and

$$
\frac{\partial L}{\partial t} = -\dot{H} \qquad (3.51)
$$

where

$$H \equiv \dot{q}^{\mu} p_{\mu} - L \qquad (3.52)$$

denotes the Hamiltonian. For future reference let us call these differential equations respectively the "p-dot" and "H-dot" versions of the Euler-Lagrange equation. Each of them offers distinct conservation laws.

Conservation of Canonical Momentum: From the p-dot equation: $p_{\mu} = $ const. if and only if $\partial L/\partial q^{\mu} = 0$. In words, the canonical momentum conjugate to the generalized coordinate q^{μ} is constant if and only if the Lagrangian does not depend *explicitly* on q^{μ}.

This conservation law corresponds to a symmetry! If L does not depend explicitly on q^{μ}—in other words if L remains unaffected by the translation $q^{\mu} \to q^{\mu} + \Delta q^{\mu}$—then and only then will $\partial L/\partial q^{\mu} = 0$. If q^{μ} denotes a coordinate with dimensions of length, like x or r, then p_{μ} is a component of linear momentum. If q^{μ} denotes an angle θ, then the translation is a rotation and p_{θ} is a component of angular momentum. Thus the conservation law inherent in the p-dot equation covers the conservation of both linear and angular momentum.

When this conservation law holds, if you want to change the system's linear momentum, a mere translation through space will not do it; if you want to change the system's angular momentum, a mere rotation will not do it. For linear momentum to be conserved, space must be homogeneous. For angular momentum to be conserved, space must be isotropic.

Conservation of the Hamiltonian: From the H-dot equation, the Hamiltonian is constant if and only if the Lagrangian does not depend *explicitly* on the independent variable t, so that $\partial L/\partial t = 0$. Again, this signifies a symmetry. In a physics application where t means time, the Hamiltonian is constant if and only if the system is symmetric under a time translation $t \to t + \Delta t$. Then and only then will $\partial L/\partial t = 0$ and H be conserved. In physics applications, the Hamiltonian is an energy, and conservation of energy occurs because of invariance under a time translation.

Example, Central Force Motion: Let a particle of mass m move under the influence of a central potential energy $U(r)$, where r denotes the radial distance from the origin to the particle (Figure 3.3).[14] In the familiar

---

[14] If the force center is sufficiently massive to be considered immovable (e.g., a communications satellite orbiting the Earth), then r is the distance from the immovable force center to the particle. If the central body recoils (e.g., binary stars), then by putting the origin at the center of mass of the two-body system, the two-body central force problem becomes effectively a one-body problem by (a) defining r as the distance between the two bodies, and (b) replacing m with the "reduced mass" $\mu \equiv m_1 m_2/(m_1 + m_2)$. See any mechanics textbook, such as [Marion (1970); Taylor (2005)].

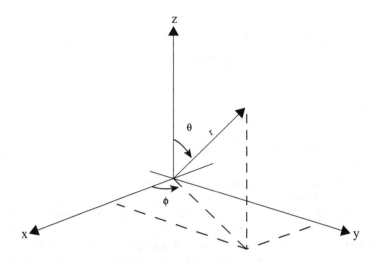

Figure 3.3: *Spherical coordinates superimposed on rectangular coordinates.*

spherical coordinates (r, θ, φ) the particle's velocity is[15]

$$\mathbf{v} = \dot{r}\hat{\mathbf{r}} + r\dot{\theta}\hat{\boldsymbol{\theta}} + (r\dot{\varphi}\sin\theta)\hat{\boldsymbol{\varphi}}, \qquad (3.53)$$

where in spherical coordinates θ measures latitude ($\theta = 0$ at the "north pole" and $180°$ at the "south pole"), and φ measures longitude. Therefore,

$$\mathbf{v} \cdot \mathbf{v} = \dot{r}^2 + r^2\dot{\theta}^2 + r^2\dot{\varphi}^2\sin^2\theta. \qquad (3.54)$$

From Hamilton's principle, the Lagrangian is

$$L(r, \theta, \dot{r}, \dot{\theta}, \dot{\varphi}) = \tfrac{1}{2}m\mathbf{v} \cdot \mathbf{v} - U(r). \qquad (3.55)$$

From the ELE for r we notice that $\partial L/\partial r \neq 0$, so p_r is *not* constant—a force acts in the r direction. But $\partial L/\partial \varphi = 0$, which means $p_\varphi = mr^2\dot{\varphi}\sin^2\theta \equiv I\dot{\varphi}$, where $I \equiv mr^2\sin^2\theta$ denotes the moment of inertia about the z-axis, and $I\dot{\varphi}$ is the z-component of angular momentum. Without loss of generality we can make the orbit about the z-axis be the $\theta = 90°$ plane, so that $p_\varphi = mr^2\dot{\varphi} = \text{const}$. Now $\partial L/\partial\theta = 0$ so $p_\theta = \text{const}$. but this vanishes since the particle orbits in the $\theta = 90°$ plane. Thus the vector $\boldsymbol{\ell} = \mathbf{r} \times \mathbf{p}$ remains constant in time, and reduces to

$$\boldsymbol{\ell} = mr^2\dot{\varphi}\,\hat{\mathbf{z}}, \qquad (3.56)$$

where $\hat{\mathbf{z}}$ denotes the unit vector along the positive z-axis.

---

[15]From elementary considerations of central forces, we know the torque on mass m vanishes, so that its angular momentum is constant, which requires the particle to move in a plane. Typically this is taken to be the $\theta = 90°$ plane from the outset, but here we want to see the conservation laws emerge from the ELE.

If the central potential's Lagrangian contains no explicit time dependence, so that $\partial L/\partial t = 0$, then by the H-dot version of the ELE the Hamiltonian is constant. In this example the Hamiltonian is

$$H = \dot{r}p_r + \dot{\theta}p_\theta + \dot{\varphi}p_\varphi - L$$
$$= \frac{p_r^2}{2m} + \frac{\ell^2}{2mr^2} + U(r). \tag{3.57}$$

This H numerically equals the particle's mechanical energy E, and is conserved.

Conservation theorems for the canonical momenta and the Hamiltonian emerge whenever the Lagrangian is invariant under the corresponding transformations. These conservation laws will also emerge as special cases of Noether's theorem, which in a broader context considers invariance of the *functional*, not just the *Lagrangian*. Noether's theorem offers as its generally conserved quantity a linear combination of H and the p_μ.

3.3   On the Equivalence of Hamilton's Principle and Newton's Second Law

This is the celebrated principle of Hamilton. It will serve very usefully as an introduction to what is usually referred to as advanced mechanics and about which the reader of this book may well expect to learn a little since it is basic for theoretical physics in general.—R. B. Lindsay, *Physical Mechanics*, 1950

Hamilton's principle for classical mechanics postulates that, of all the trajectories a particle might conceivably follow between any two fixed times a and b, the path actually followed is the one for which the time integral of the difference between the kinetic and potential energies is minimized. In a one-spatial-dimension example, this means

$$\int_a^b \left(\frac{1}{2}m\dot{x}^2 - U(t, x) \right) dt = min. \tag{3.58}$$

Its Euler-Lagrange equation (p-dot version) gives

$$-\frac{\partial U}{\partial x} = m\ddot{x}. \tag{3.59}$$

Therefore, Hamilton's principle implies Newton's second law.

Conversely, Newton's second law implies Hamilton's principle, with Lagrangian $L = K - U$ (for nondissipative systems), as we will show. When statement A implies statement B and statement B implies statement A, the two statements are equivalent. Hamilton's principle and Newton's second law are equivalent principles for classical mechanics, provided each interaction has a potential energy function.[16]

To demonstrate that $\mathbf{F} = m\mathbf{a}$ implies $\delta \int_a^b (K - U)dt = 0$, we start with a component of $F = ma$ written in rectangular coordinates x^μ:[17]

$$-\frac{\partial U}{\partial x^\mu} = m\frac{d^2 x^\mu}{dt^2}. \tag{3.60}$$

We will show that by changing coordinates from the x^μ to some generalized coordinates q^μ, eq. (3.60) becomes[18]

$$\frac{\partial(K - U)}{\partial q^\mu} = \frac{d}{dt}\frac{\partial(K - U)}{\partial \dot{q}^\mu}. \tag{3.61}$$

This we recognize as the Euler-Lagrange equation, with Lagrangian

$$L = K - U, \tag{3.62}$$

which implies Hamilton's principle. Let us see how this comes about.

To convert from rectangular to generalized coordinates we need to do some preliminary spadework. Each x^μ can be written as a function of all the q^ν,

$$x^\mu = x^\mu(q^\nu), \tag{3.63}$$

where $\mu, \nu = 1, 2 \ldots, N$. From the chain rule (as usual, sum over repeated indices), the time derivative of x^μ becomes

$$\dot{x}^\mu = \frac{\partial x^\mu}{\partial q^\nu}\dot{q}^\nu. \tag{3.64}$$

If we differentiate this with respect to $\dot{q}^\rho$ and note

$$\frac{\partial \dot{q}^\nu}{\partial \dot{q}^\rho} = \delta^\nu{}_\rho, \tag{3.65}$$

where $\delta^\nu{}_\rho$ is the Kronecker delta (1 if $\nu = \rho$ and 0 if $\nu \neq \rho$), it follows that

$$\frac{\partial \dot{x}^\mu}{\partial \dot{q}^\rho} = \frac{\partial x^\mu}{\partial q^\rho}. \tag{3.66}$$

Now return to eq. (3.60). Multiply it by $\partial x^\mu/\partial q^\nu$ and sum over μ. Thanks to the chain rule, the left-hand side will be recognized as $-\partial U/\partial q^\nu$. So far our way of writing a component of $\mathbf{F} = m\mathbf{a}$ has become

$$-\frac{\partial U}{\partial q^\nu} = m\ddot{x}_\mu \frac{\partial x^\mu}{\partial q^\nu}. \tag{3.67}$$

---

[16]Nonconservative forces can be handled as well, as we shall see, provided we enlarge what we mean by the Lagrangian.

[17]In rectangular Cartesian coordinates, upper and lower indices make no difference.

[18]See also Campbell (1947) 305–307 and Symon (1953) 309–311.

By using the product rule for derivatives on the right-hand side, eq. (3.67) may be rewritten

$$-\frac{\partial U}{\partial q^\nu} = \frac{d}{dt}\left[m\dot{x}_\mu \frac{\partial x^\mu}{\partial q^\nu}\right] - m\dot{x}_\mu \frac{d}{dt}\frac{\partial x^\mu}{\partial q^\nu}. \tag{3.68}$$

In the first term on the right-hand side, replace $\partial x^\mu/\partial q^\nu$ with $\partial \dot{x}^\mu/\partial \dot{q}^\nu$, so this first term becomes

$$\begin{aligned}
\frac{d}{dt}\left[m\dot{x}_\mu \frac{\partial x^\mu}{\partial q^\nu}\right] &= \frac{d}{dt}\left[m\dot{x}_\mu \frac{\partial \dot{x}^\mu}{\partial \dot{q}^\nu}\right] \\
&= \frac{d}{dt}\frac{\partial}{\partial \dot{q}^\nu}\left(\frac{1}{2}m\dot{x}_\mu \dot{x}^\mu\right) \\
&= \frac{d}{dt}\left(\frac{\partial K}{\partial \dot{q}^\nu}\right).
\end{aligned} \tag{3.69}$$

In the second term of eq. (3.68), use the chain rule on $x^\mu = x^\mu(q^\sigma)$ to write it as

$$\begin{aligned}
-m\dot{x}_\mu \frac{d}{dt}\frac{\partial x^\mu}{\partial q^\nu} &= -m\dot{x}_\mu \frac{\partial^2 x^\mu}{\partial q^\nu \partial q^\sigma}\dot{q}^\sigma \\
&= -m\dot{x}_\mu \frac{\partial}{\partial q^\nu}\left(\frac{\partial x^\mu}{\partial q^\sigma}\dot{q}^\sigma\right) \\
&= -m\dot{x}_\mu \frac{\partial \dot{x}^\mu}{\partial q^\nu} \\
&= -\frac{\partial}{\partial q^\nu}\left(\frac{1}{2}m\dot{x}_\mu \dot{x}^\mu\right) \\
&= -\frac{\partial K}{\partial q^\nu}
\end{aligned} \tag{3.70}$$

where in going from the first to the second line we have used $\partial \dot{q}^\sigma/\partial q^\nu = 0$. Now eq. (3.68) reads

$$-\frac{\partial U}{\partial q^\nu} = \frac{d}{dt}\left(\frac{\partial K}{\partial \dot{q}^\nu}\right) - \frac{\partial K}{\partial q^\nu}. \tag{3.71}$$

Transpose $\partial K/\partial q^\nu$ to the left-hand side, and note that, because the potential energy function contains no velocities, we may add $-U$ to the K under the derivative with respect to $\dot{q}^\nu$. We obtain at last

$$\frac{\partial(K-U)}{\partial q^\nu} = \frac{d}{dt}\frac{\partial(K-U)}{\partial \dot{q}^\nu}, \tag{3.72}$$

the Euler-Lagrange equation for the Lagrangian

$$L(t, q^\mu, \dot{q}^\nu) = K(q^\mu, \dot{q}^\mu) - U(t, q^\mu), \qquad (3.73)$$

showing that Newton's second law leads to the Euler-Lagrange equation that would follow from making extremal the time integral of $K - U$, expressed in generalized coordinates.

We have demonstrated that Hamilton's principle and Newton's second law are equivalent for classical mechanics: each implies the other. However, when we look beyond classical mechanics, Hamilton's principle wins more fundamental accolades hands down, for it can be generalized to systems about which $\mathbf{F} = m\mathbf{a}$ says little or nothing, such as the electromagnetic field, general relativity, and quantum theory. In the broader view, Hamilton's principle is the more comprehensive, with Newton's second law as one of its special cases.

3.4   Where Do Functional Extremal Principles Come From?

Following in the footsteps of Hero and Fermat, he [Maupertuis] then proclaimed that this simplicity causes nature to act in such a way as to render a certain quantity, which he named the "action," a minimum.

—Wolfgang Yourgrau and Stanley Mandelstam, *Variational Principles in Dynamics and Quantum Theory*, 1968, 19

All of this is splendid enough, but still we may ask: where do functional extremal principles, such those of Fermat and Hamilton, come from? Hamilton's principle is often introduced in classical mechanics textbooks by merely stating it. Presented as a gift from above, it sounds mysterious. But then we might reflect, where did Newton's laws of motion come from? They were not derived deductively from anything deeper. If you want Newton's laws or Hamilton's principle or Fermat's principle, you have to postulate them. They are informed conjectures, elevated to the status of axioms, from which the development of the theory flows. However, they were not proposed arbitrarily or by guesswork. What was the motivation?

Variational problems in geometry appeared in antiquity, as "Dido's problem" illustrates. The story goes that when Queen Dido migrated about 814 BCE as a refugee from Tyre in Lebanon, and landed in Tunisia on the coast of north Africa, she negotiated with the locals for a site on which to settle. She offered them a tidy sum for the amount of land that could be enclosed with a ox hide, to which the sellers readily agreed. But Dido, being clever, cut the hide into thin strips and tied them together. Her mathematics problem was to determine the shape of the perimeter that maximized the enclosed area.

The city of Carthage was founded upon a variational principle, centuries before the calculus was invented!

Around the second century BCE, commerce began moving from Samarkand to Chang'an along the Silk Road. The Silk Road travelers sought routes that minimized time, danger, and expense—the closest they could come to a geodesic. It was

> a considerable undertaking not only in terms of distance (over 3,000 miles), but also because of the the geographical obstacles along the way. From Samarkand, the easternmost edge of the Eurasian plains, there were several roads east, but all had to negotiate the great mountain ranges which divided Transoxania from the Tarim basin. One route traversed the foothills of the Pamirs to the Jaxartes river.... Following the river upstream the road then passed through a fertile, oval-shaped valley.... The Ferghanan valley, source of the Jaxartes river, lies between two great mountain ranges, the Tianshan to the north-east and the Pamirs to the south-east. At the valley's head the road crossed the Terek Davan pass, where these two ranges join, and then dropped down into the city of Kashgar in the Taram basin. [Whitfield (1999) 22–23]

The optmal path did not summit the Tianshan and Pamir peaks, but took advantage of the Ferghanan Valley and Terek Davan pass between them. The spirit of inquiry that seeks the optimal path through a physical landscape—or an abstract problem—has been around for a long time.

Navigation and surveying have long depended on light rays traveling in straight lines. Through such practical experience, the optical laws of rectilinear propagation and reflection were known in antiquity. In what is perhaps the first unified theory in physics, Hero (or Heron) of Alexandria (c. 10–70 CE) set forth the principle that light rays follow the path of minimum distance. If that were so, then rectilinear propagation and the law of reflection would be explained. Two facts became instances of one principle [Hecht (2002); Neuenschwander (2014a)].

Refraction was also known and applied in antiquity. A partially immersed stick appearing to be sharply bent at the water's surface was mentioned in Plato's *Republic* (c. 360 BCE). "Burning glasses," lenses for starting fires by focusing sunlight, were part of ancient technology, as documented by artifacts such as a magnifier found in the ruins of the palace of Assyrian king Sennacherib (708–681 BCE) [Hecht (2002) ch. 1]. Refraction was made a quantitative science in the Middle Ages by Muslim scholars such as Ibn al-Haytham (c. 965–1040), known to us as Alhazen, who introduced the practice of measuring angles from the normal for reflected and refracted rays. Alhazen's contemporary Abu Sàd al-Alá ibn Sahl (c. 940–1000) in his book *On Burning Mirrors and Lenses*, written about 984, accurately describes refraction from air to glass in terms of two right triangles whose hypotenuse length ratio equals the index of refraction of glass [Rashed (1990)]. He successfully applied this quantitative description of refraction to lenses. Willebrord Snell (1580–1626) rediscovered the law of refraction in 1621, which René Descartes (1590–1650) rediscovered again and published in the well-known sine form in *Discourse on Method* in 1637.

Hero's principle does not explain refraction, but that crucial gap was filled by Pierre de Fermat (1601–1665) in 1657 through a broader unifying principle: of all possible paths connecting two fixed points, the path followed by a light ray minimizes the time for light to go between them. I suppose Fermat postulated the time to be a minimum, so that in a medium with a uniform refractive index, the principle would predict shortest-distance rectilinear propagation, containing Hero's principle as a special case. With the least-time principle, Fermat deduced the law of refraction in 1661.[19]

Turning to mechanics, paradigm-setting unifying principles came in the form of Isaac Newton's (1643–1727) laws of motion. He initially reasoned inductively from studies of specific systems, notably the lever (Archimedes, 287–212 BCE), freely falling bodies (Galileo Galilei, 1564–1642), the pendulum (Christiaan Huygens, 1629–1695), and planetary motion (Johannes Kepler (1571–1630) and Newton) [Millikan et al. (1937) 35–36]. Generalizing from these particular systems, studied with quantitative precision, Newton proposed a set of general principles, articulated in his celebrated axioms of motion that were published in the *Principia* of 1687. The concepts of kinetic and potential energies came along after the concept of force. In 1741 Emílie du Châtelet (1706–1749) extended Newtonian mechanics into the work-energy theorem, including *forces vives* (kinetic energy) and potential energy. Thomas Young (1773–1829) was among the first to use the term "energy" applied to mv^2 in 1801–1803.

Shortly after Newton published the *Principia*, the calculus of variations began with problems such as the brachistochrone[20] that seeks the shape of the frictionless path down which a particle slides in the least time. Such applications were extended by Johann (1667–1748) and Jacob Bernoulli (1654–1705) in 1686.

---

[19]Fermat's principle requires light to travel at finite speed. Astronomy offered the first meaningful estimate of this speed in 1676 when Olé Roemer (1644–1710) used as a clock the periodic emergence of Io from behind Jupiter's shadow (the moon has an orbital period of 42.5 hours). During the time of year when the Earth recedes from the Jupiter-Io system, after each orbit of Io around Jupiter the clock is seen from Earth to run slow. Roemer interpreted the delay as the extra time required for light to travel the additional distance between Earth and Io.

[20]From the Greek *brachistos*, short, and *chronos*, time.

As the trajectory of a light ray could be derived from Fermat's principle, could a similar principle be stated for a particle's trajectory in mechanics? Several proposals were forthcoming. These included Johann Bernoulli's 1717 principle of virtual work for statics, extended to dynamics by Jean le Rond d'Alembert (1717–1783) in 1743. In 1744 the French mathematican and astronomer Pierre Louis Moreau de Maupertius (1698–1759) suggested a "principle of least action" and applied it to a restatement of Fermat's principle, then to mechanics in 1747. Marpertius' notion of "action" was rather vague, and how it was to be minimized sounded rather teleological (through the "wisdom of God"). But his approach started a useful conversation. Action as defined since carries the units of (momentum)×(length), or equivalently the units of angular momentum, or (energy)×(time).

In the meantime, such an optimization approach was successfully demonstrated for central forces by Leonhard Euler (1707–1783) in 1744. Euler and Joseph Lagrange (1736–1813) developed the Euler-Lagrange equation in the 1750s, when Lagrange solved the tautochrone problem[21] and sent his solution to Euler. They applied their methods to mechanics, and Euler gave the discipline the name "calculus of variations."

Refinements to the calculus of variations were made by Adrien-Marie Legendre (1752–1833) with his Legendre transformation in 1786, and by Lagrange in 1788 with his Lagrange multipliers, which appeared in *Méchanique Analytique* (1788 and 1811), where he also generalized Maupertius's principle to all conservative forces and defined "action" as the line integral of momentum. Later minimization approaches included Karl Friedrich Gauss's (1777–1855) principle of least constraint introduced in 1828, refined into Heinrich Hertz's (1857–1894) principle of least curvature in 1894.

A version of Hamilton's equations (see chapter 9) were first obtained by Lagrange in 1809 and by Siméon-Denis Poisson (1781–1840) that same year. In 1831 Augustin Louis Cauchy (1789–1857) realized they were fundamental. Very quickly the action principle was generalized to all of mechanics in two papers published by William Rowan Hamilton (1805–1865) in 1834–1835. Hamilton's principle enlarges the notion of action, by postulating that of all the conceivable trajectories whereby a particle might travel between two fixed points, the trajectory actually followed minimizes the time-averaged difference between the particle's kinetic and potential energies (recall eq. [2.10]). Coming a century and a half after Newton's laws, Hamilton's principle was proposed with hindsight. But with that hindsight the principle offered enhanced elegance, unity, deeper understanding, and generalizability. It was eventually extended beyond mechanics to essentially

---

[21]The tautochrone problem seeks the curve such that the time required for a particle to slide down it and reach the bottom, without friction in a uniform gravitational field, will be independent of the starting point when released from rest.

all of physics. Hamilton-Jacobi theory (chapter 9), developed by Karl Gustav Jacob Jacobi (1804–1851) in 1837, builds on Hamilton's principle by making the covariance of Hamilton's equation in phase space a central tenet. Generalized coordinates, generalized velocities, and generalized momenta were introduced in 1867 by Sir William Thomson (Lord Kelvin, 1824–1907) and Peter G. Tait (1831–1901) in *Natural Philosophy*.[22]

The special theory of relativity, introduced by Albert Einstein (1879–1955) in 1905 in his famous paper "On the Electrodynamics of Moving Bodies," was translated into the language of four-dimensional spacetime by Hermann Minkowski (1864–1909) in 1908 [Einstein et al. (1952)]. With it James Clerk Maxwell's (1831–1879) equations of the electromagnetic field could be derived from a variational principle on a functional whose Lagrangian was made of tensors.

Einstein finished the general theory of relativity in 1915, a field theory of gravitation, and almost immediately David Hilbert showed how Einstein's equations for the gravitational field could be derived from a variational principle with a functional whose Lagrangian was made of tensors. In a general relativistic analog of Fermat's principle applied to a particle's trajectory, Hamilton's principle for classical mechanics emerges in the weak-field, low-velocity limit, an informal linkage of Fermat's principle, Hamilton's principle, and geodesics.

Clearly a deep and abiding interest has long existed in seeing physics in terms of optimization strategies. The generalizability of Hamilton's principle to the widest scope in physics, and its presentation of conservation laws for particles and fields, gives it depth and versatility. Noether's theorem reproduces not only the conservation laws picked up by Hamilton's principle, but finds some that Hamilton's principle overlooks.

3.5   Why Kinetic *Minus* Potential Energy?

Hamilton's principle (recall eq. [2.10]) forms a cornerstone postulate of physics. But what is this mechanical action, this quantity $\int_a^b (K - U)dt$, standing alone on its own identity, regardless of its consequences in the Euler-Lagrange equations and conservation laws? Why kinetic *minus* potential energy? One answer, necessary and practical, is that it works! But *why* does it work? One expects that such an abstract and nonintuitive statement—minimizing the time integral of kinetic minus potential energy—surely had some kind of motivation, or ties to something deeper. I can suggest two possible reasons that "explain" the relative minus sign between K and U in Hamilton's principle. The first takes the form of an equipartition argument.

---

[22]Tait was also a pioneer in knot theory, which helped lead to the development of topology.

The second sees Hamilton's principle of classical mechanics as the correspondence principle requirement that emerges from a relativistic "Fermat's principle."

(a) Equipartition

Hamilton's principle is equivalent to requiring the time averages of the kinetic and the potential energies to be as nearly equal as possible.[23] During the time interval from $t = a$ to $t = b$, the time average $\langle K \rangle$ of the instantaneous kinetic energy $K(t)$ is defined by

$$\langle K \rangle = \int_a^b K \frac{dt}{T} \tag{3.74}$$

where $T = b - a$. The time-averaged potential energy $\langle U \rangle$ is calculated similarly. The requirement that the time averages of the kinetic and potential energies distribute themselves as equally as possible (equipartition) gives $\langle K \rangle - \langle U \rangle = $ min., or

$$\int_a^b (K - U)dt = \text{min.} \tag{3.75}$$

A kind of symmetry between kinetic and potential energy has been hypothesized.

Such an approach offers good physics reasons for Hamilton's principle to require the functional to be minimized and not maximized.[24] Either a maximum or a minimum would yield the Euler-Lagrange equation. But to have the time average of $K - U$ be a maximum would mean that $\langle K \rangle = \langle U \rangle + $ max. That would be a problem, because it could require the average kinetic energy to be arbitrarily large in the simplest case of a free particle, for which $U = 0$. In the absence of other constraints, a principle that allows the kinetic energy to run off to infinity in the most elementary of cases makes no physical sense.[25]

(b) Relativity and the Correspondence Principle

In contrast to Hamilton's principle, relativistic mechanics (appendix B) postulates that, of all world lines through spacetime that a freely falling particle might follow from event a to event b, the world line actually followed is the one that maximizes the proper time:

$$\int_a^b d\tau = \text{max.} \tag{3.76}$$

---

[23][Neuenschwander et al. (2006)]

[24]Recall that the Euler-Lagrange equation emerges regardless of whether the functional is a maximum, minimum, or inflection point. In other words, to get the Euler-Lagrange equation the functional needs only to satisfy $[d\Gamma/d\varepsilon]_0 = 0$; see section 5.3.

[25]For another perspective on the question of "why minimum and not maximum," I refer you to the paper by C. G. Gray and Edwin F. Taylor [Gray & Taylor (2007)].

Although neither Fermat or Einstein ever said it, by analogy to Fermat's principle let us dignify eq. (3.76) with the name "Fermat's free-fall principle."[26] It has been enormously successful in every test of its application so far, making predictions consistent with observations on gravitational redshift, precession of orbit perihelion, deflection of starlight, radar echo delay, gravitational lensing, and the relativistic precession of gyroscopes. Recall our convention of summing over repeated indices from $\mu = 0$ (time) to 3 (three spatial coordinates). In any frame, the square of the proper time $d\tau$ between infinitesimally nearby events may be written in terms of the spacetime coordinates x^μ as

$$d\tau^2 = g_{\mu\nu}dx^\mu dx^\nu \tag{3.77}$$

where the $g_{\mu\nu}$ are the components of the metric tensor (see appendices A and B).

Consider a particle of mass m falling freely in a gravitational field. According to the principles of general relativity (see chapter 8), the particle falls along a geodesic in curved spacetime, such that the functional

$$\Delta\tau = \int_a^b \sqrt{g_{\mu\nu}u^\mu u^\nu}\, d\tau \tag{3.78}$$

is maximized, where $u^\mu \equiv dx^\mu/d\tau$ are velocities in spacetime. In the weak-field, low-velocity limit, where $d\tau \approx dt, u^0 \approx 1$, and $u^k \approx v^k$ for $k = 1, 2, 3$, it can be shown (as the reader will be invited to do in ex. 3.12) that, for a particle of mass m falling freely in a gravitational potential Φ,

$$m\sqrt{g_{\mu\nu}u^\mu u^\nu}\, d\tau \approx m\Phi - \tfrac{1}{2}mv^2 + \text{const.}$$
$$= U - K + \text{const.} \tag{3.79}$$

Therefore $m\Delta\tau = \text{max.}$ becomes, in the classical-mechanics limit, $\int_a^b (K - U)\, dt = \text{min.}$ The question "Why $K - U$?" becomes, "Why is $\Delta\tau$ a maximum?" But at least that question is the more straightforward one to answer (see ex. 3.12).

3.6 Extremals with External Constraints

In addition to the functional being made an extremal, sometimes supplementary constraints exist, as Queen Dido encountered when she wanted to enclose maximum area with a fixed perimeter. If there are N independent generalized coordinates, and A equations of constraint that relate them, then only $N - A$ coordinates are truly independent, resulting in $N - A$

---

[26]Taylor & Wheeler (2000) call this the "principle of maximum aging."

degrees of freedom. We can build the constraints into the Lagrangian, making a new Lagrangian that has $N - A$ degrees of freedom, but still treating all N coordinates as if they were independent. We implement this strategy by adding zero to the original Lagrangian as follows.

Let an equation of constraint be written in the form

$$h(t, x^\mu) = 0. \tag{3.80}$$

For example, when a tire of outer radius R rolls without slipping down an inclined plane, the location x of its center along the plane, and the angle θ through which the tire has turned while rolling, are related by $x = R\theta$, so that $h(x, \theta) = x - R\theta$. One introduces the constrained Lagrangian L_c by adding zero to the original Lagrangian, in the form of a term proportional to h. Thus we construct

$$L_c = L + \lambda h. \tag{3.81}$$

The coefficient λ is called a "Lagrange multiplier." The constraint h needs λ because L and h typically do not have the same dimensions. For instance, in a mechanics application, L is energy, and in the rolling tire illustration, h has dimensions of length, and thus λ must have the dimensions of force. With A equations of constraint, we add them all in,

$$L_c = L + \lambda_k h^k \tag{3.82}$$

with k summed from 1 to A. The N Euler-Lagrange equations of the constrained system are handled in the usual way: for the p-dot equation,

$$\frac{\partial L_c}{\partial q^\mu} = \dot{p}_\mu, \tag{3.83}$$

and for the H-dot equation,

$$\frac{\partial L_c}{\partial t} = -\dot{H}_c \tag{3.84}$$

where $H_c = p_\mu \dot{q}^\mu - L_c$.

Example: Let's pursue the case of the tire of mass m and outer radius R that rolls without slipping down a ramp inclined at an angle α above the horizontal, in the presence of a gravitational field g directed vertically downward. Place the zero of gravitational potential energy at the tire's center of mass when it stands at the bottom of the ramp. The Lagrangian of the tire that rolls and/or slides down the ramp without the constraint would be

$$L = \tfrac{1}{2}m\dot{x}^2 + \tfrac{1}{2}I\dot{\theta}^2 - mgx\sin\alpha, \tag{3.85}$$

where I denotes the tire's moment of inertia about the horizontal axis through its center. Without the constraint, the tire's translational velocity down the ramp and its angular velocity about its axis are uncoupled, and there would be two degrees of freedom.[27] With the constraint, the "constrained Lagrangian" may be written

$$L_c = \tfrac{1}{2}m\dot{x}^2 + \tfrac{1}{2}I\dot{\theta}^2 - mgx\sin\alpha + \lambda(x - R\theta). \qquad (3.86)$$

We still write separate Euler-Lagrange equations for x and for θ, but the constraint couples them. The x equation gives

$$-mg\sin\alpha + \lambda = m\ddot{x}, \qquad (3.87)$$

and for θ

$$-\lambda R = I\ddot{\theta}. \qquad (3.88)$$

Now λ will be recognized as the force of static friction exerted by the ramp on the tire. This is the force responsible for the "rolling without slipping" constraint.

Questions for Reflection and Discussion

Q3.a. Are principles such as Newton's laws of motion, Hamilton's principle, and Fermat's principle invented, or are they discovered?

Q3.b. In arriving at fundamental principles such as Newton's laws or Hamilton's principle or Fermat's principle, what are the roles of inductive and deductive reasoning? Does induction give knowledge? (The philosopher David Hume had much to say about the last question; see Godfrey-Smith (2003), 39–40.)

Q3.c. In the example of the tire rolling without slipping, how could the constraint be handled without invoking the constrained Lagrangian?

Q3.d. How would the Euler-Lagrange equation for a mechanical system be modified if forces not derivable from the gradient of a potential energy acted on the particle? For example, how would the Euler-Lagrange equation have to be modified to account for sliding friction?

Q3.e. Since Queen Dido founded Carthage many centuries before the invention of the calculus (Virgil writes about her in the *Aeneid*), how could she have determined the shape of an area with fixed perimeter that maximizes the area enclosed? How would you do it without calculus?

---

[27]Imagine a car coming down an icy hill, or trying to go up it, in the absence of friction. The spin of the wheels are not coupled to the car's forward progress.

Q3.f. Find out whatever you can about Heinrich Hertz's principle of least curvature, and how it is related to Gauss's principle of least constraint (see ex. 3.21).

Q3.g. How does a conventional usage of $\mathbf{F} = m\mathbf{a}$ deal with forces of constraint, and how does that approach compare to the handling of constraints with the constrained Lagrangian?

Q3.h. For central force motion, where $U = U(r)$ and $\mathbf{F} = \mathbf{F}(r)$, show that the angular momentum is constant, without invoking the Euler-Lagrange equations. Why can we say the particle moves in a plane? Does the choice of the plane matter?

Q3.i. Suppose the relation between a conservative force $\mathbf{F}$ and its potential energy U had been defined as $\mathbf{F} = \boldsymbol{\nabla}U$, lacking the usual minus sign.
a. How would this affect the work-energy theorem?
b. How would this affect the definition of mechanical energy, and how would its conservation be expressed?
c. How would Hamilton's principle be affected?
d. Why is the minus sign included in the usual definition, $\mathbf{F} = -\boldsymbol{\nabla}U$?

Q.3.j. Show that if Euler-Lagrange equation holds then the functional can be shown to be an extremal. In section 3.1 the ELE was shown to be necessary for Γ to be an extremal; here it is a question of whether the ELE being satisfied is sufficient for Γ to be an extremal.

Exercises

3.1. Return to the example of using Fermat's principle to find the equation of a light ray when the index of refraction, n, is spatially uniform. The Lagrangian is

$$L = L(y') = \frac{n}{c}\sqrt{1 + y'^2} \qquad (3.89)$$

where $y' = dy/dx$. We found that the p-dot form of the ELE, which here reads

$$\frac{\partial L}{\partial y} = \frac{d}{dx}\frac{\partial L}{\partial y'} \qquad (3.90)$$

leads to the equation of a straight line.
a. Find the Hamiltonian for this system.
b. Does the H-dot (or should I say H-prime) version of the ELE also lead to the equation of a straight line?

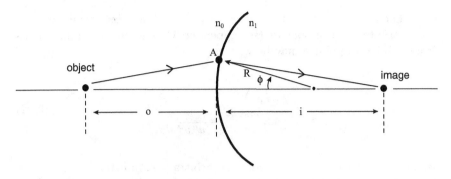

Figure 3.4: *Situation of ex. 3.2 for image formation by a spherical refractive surface.*

3.2. In this exercise you derive from Fermat's principle an important step toward the "lens maker's equation." To set this up, let two media, of uniform refractive indices n_o and n_i ("o is related to "object" and "i" to "image") lie to the left and right, respectively, of a spherical boundary of radius R. Let the object be located at the distance s_o and the image be located at the distance s_i from the boundary between the two media; both points lie along the line that passes through the center of curvature of the spherical surface. Let one light ray go from the object to the image point via some point A shown in Figure 3.4. As the location of point A changes, so does the angle ϕ. Starting from Fermat's principle, show that, in the small-angle limit,

$$\frac{n_o}{s_o} + \frac{n_i}{s_i} \approx \frac{n_o - n_i}{R}. \tag{3.91}$$

3.3. When sugar is mixed with water and the solution allowed to settle, a stratified index of refraction results. Consider an xy plane in a glass tank (like a fish aquarium), with x horizontal, y vertical, taking $y = 0$ at the bottom of the tank. Model the index of refraction as $n(y) = n_w + n_o(1 - y/h)$, where n_w denotes the refractive index of pure water, n_o is a constant, and h denotes the liquid's depth. A light ray is sent horizontally into the tank from $(x, y) = (0, \alpha h)$ where $0 < \alpha < 1$. Find the trajectory of the ray.

3.4. Consider the problem of measuring distance in an N-dimensional space that may or may not exhibit curvature. Denote the coordinates as x^μ, where μ ranges from 1 to N. With the convention of summing over repeated indices from 1 to N, the distance ds between infinitesimally nearby points, or rather its square, may be written as $ds^2 = g_{\mu\nu}dx^\mu dx^\nu$, where the $g_{\mu\nu}$ are metric tensor components (see appendix A). The distance functional s from point a to point b may be written

$$
\begin{aligned}
s &= \int_a^b \sqrt{g_{\mu\nu}dx^\mu dx^\nu} \\
&= \int_{\sigma(a)}^{\sigma(b)} \sqrt{g_{\mu\nu}u^\mu u^\nu}\, d\sigma,
\end{aligned}
\tag{3.92}
$$

where $u^\mu \equiv dx^\mu/d\sigma$ and σ denotes an arbitrary parameter.[28] The metric tensor components $g_{\mu\nu}$ may be functions of the coordinates, $g_{\mu\nu} = g_{\mu\nu}(x^\rho)$. Find the geodesic:

a. In three-dimensional Euclidean space described by rectangular coordinates;

b. On a cylinder of radius R in Euclidean space;

c. In the spacetime about a Schwarzschild star—a spherically symmetric, nonrotating, uncharged star of mass M, for which the metric, mapped with spherical spatial coordinates (r, θ, φ) centered on the star, is

$$
ds^2 \equiv c^2 d\tau^2 = A(r)c^2 dt^2 - \frac{dr^2}{A(r)} - r^2 d\theta^2 - r^2 \sin^2\theta\, d\varphi^2
\tag{3.93}
$$

where $d\tau$ denotes proper time and $A(r) = 1 - (GM/rc^2)$. The r-coordinate is defined, not by the distance from the origin to a spherical shell, but by the circumference of that spherical shell divided by 2π. The distance between two spherical shells is not $r_2 - r_1$, but $\int_{r_1}^{r_2} A^{-1/2}(r)dr$, a hallmark of curved space (see Taylor & Wheeler (2000)].

3.5. In a locally Euclidean coordinate system, with coordinates X^μ, metric tensor $\delta_{\mu\nu}$, and the $d\sigma$ of ex. 3.4, show that a transformation from the X^μ to a more general curvilinear global system of coordinates $x^\mu = x^\mu(X^\nu)$ turns

$$
\frac{d^2 X^\lambda}{d\sigma^2} = 0
\tag{3.94}
$$

into

$$
\frac{d^2 x^\rho}{d\sigma^2} + \left(\frac{\partial x^\rho}{\partial X^\lambda} \frac{\partial^2 X^\lambda}{\partial x^\mu \partial x^\nu} \right) u^\mu u^\nu = 0
\tag{3.95}
$$

where $u^\mu \equiv dx^\mu/d\sigma$. You may need to use

$$
\frac{\partial x^\rho}{\partial X^\lambda} \frac{\partial X^\lambda}{\partial x^\mu} = \delta^\rho{}_\mu.
\tag{3.96}
$$

---

[28]Typically σ is the distance along a curve through space, or the proper time along a world line in spacetime.

Compare this result to ex. 3.20—same mathematics, different context. The term in parentheses, called the "affine connection" and usually denoted $\Gamma^\rho_{\mu\nu}$, will become an important player for us later.

3.6. Newton's second law applied to the damped oscillator of mass m, assuming linear restoring and damping forces, says

$$-kx - b\dot{x} = m\ddot{x} \tag{3.97}$$

where k and b are constants.
a. Find a Lagrangian whose Euler-Lagrange equation gives this equation of motion. (Recall that by "Lagrangian" some authors mean strictly $L = K - U$, whereas we take "the Lagrangian" to mean more generically the integrand of the functional.)
b. Identify the canonical momentum and the Hamiltonian.
c. Determine what, if anything, is conserved.

3.7. Two identical pendulums, each of moment of inertia I about one end, are suspended in tandem, with the upper pendulum suspended from a hook in the ceiling. The compound pendulum is set swinging in a plane, in a uniform gravitational field $\mathbf{g}$. Using a set of generalized coordinates, write the Lagrangian for the two-pendulum system, and from it construct the equations of motion.

3.8. A point mass m is attached to one end of a spring having negligible mass and spring constant k. The other end of the spring is suspended from a hook in the ceiling in the presence of a uniform gravitational field $\mathbf{g}$, to make a pendulum of variable length.
a. Assuming the pendulum swings in a plane, write the Lagrangian in terms of appropriate generalized coordinates and their Euler-Lagrange equations.
b. Generalize the problem to where the pendulum need not swing in a plane.

3.9. Solve Dido's problem using the calculus of variations: Given a perimeter of fixed length w, find the shape that encloses the maximum area in the plane. What is the meaning of the Lagrange multiplier λ?

3.10. In the xy plane find the shape of a flexible hanging chain of uniform linear mass density μ and length l, suspended between arbitrary but fixed endpoints in a uniform gravitational field of magnitude g. Compare your answer for this shape (a *catenary*) to the shape of the Gateway Arch in St. Louis, Missouri, and comment on the insight of the arch's engineers.

3.11. Solve the brachistochrone problem: Find the shape of a hill down which you could ski such that, without friction, you would make the descent between two fixed endpoints in the least time, pulled along only by a vertical uniform gravitational field **g**. You will need to assume conservation of energy.

3.12. a. For a particle in free fall, show why $\int_a^b d\tau$ of special and general relativity must be a maximum, where τ denotes proper time. Absorbing the speed of light c into the times, $d\tau$ is given by

$$d\tau^2 = dt^2 - dx^2 - dy^2 - dz^2 \tag{3.98}$$

in special relativity, and, more generally,

$$d\tau^2 = g_{\mu\nu}dx^\mu dx^\nu \tag{3.99}$$

in general relativity. Hint: Consider a billiard ball sitting on a pool table, at rest in an inertial frame. Watch the billiard ball for an hour (or better, imagine watching it) as it moves through time, but not through space, in this reference frame.

b. Fill in the steps that were described at the end of section 3.5 for a particle of mass m falling with low speed in a weak gravitational field. Assume that $d\tau \approx dt, dx^k/d\tau \approx v^k$ for $k = 1, 2, 3$, and (anticipating the result of ex. 8.5) that $g_{00} \approx 1 + 2\Phi$ with Φ the gravitational potential. Under these conditions, show that

$$m \int_{\tau_1}^{\tau_2} \sqrt{g_{\mu\nu}\frac{dx^\mu}{d\tau}\frac{dx^\nu}{d\tau}}\ d\tau \approx m\Phi - \frac{1}{2}mv^2. \tag{3.100}$$

3.13. For a particle of mass m moving through a potential energy function $U(x)$, the time-independent Schrödinger equation is

$$-\frac{\hbar^2}{2m}\frac{d^2\psi}{dx^2} + U\psi = E\psi, \tag{3.101}$$

where $\psi = \psi(x)$ denotes a stationary state wave function having energy eigenvalue E. Show this to be the Euler-Lagrange equation for the Lagrangian

$$L(x, \psi, \psi^*, \psi_x, \psi_x^*) = \frac{\hbar^2}{2m}(\psi_x^*\psi_x) - \psi^*(U - E)\psi, \tag{3.102}$$

where $\psi_x \equiv d\psi/dx$ and * denotes complex conjugate. Notice that ψ and ψ^* are to be treated as distinct generalized coordinates, dependent on x.

3.14. The special relativity extension of Newton's second law reads

$$f^\mu = \frac{dp^\mu}{d\tau}, \tag{3.103}$$

where $d\tau$ denotes the invariant proper time, the relativistic momentum is $p^\mu = m dx^\mu/d\tau \equiv m u^\mu$, and f^μ denotes a component of the force four-vector. Consider a free particle of mass m, for which $f^\mu = 0$. The relativistic Newton's second law becomes

$$0 = \frac{du^\mu}{d\tau}. \tag{3.104}$$

a. Verify that this result follows from the Lagrangian (with $c = 1$)

$$L = -m\sqrt{1 - v^2} \equiv -m\gamma^{-1}. \tag{3.105}$$

b. Show that the canonical momenta are $p^\mu = m v^\mu \gamma$ where $v^\mu = dx^\mu/dt$.
c. Show that the Hamiltonian is $H = m\gamma$ (or $H = mc^2\gamma$ in conventional units) which numerically equals the particle's energy E.
d. Show that the functional using the Lagrangian just cited may be written as the integral

$$\Gamma = \int_a^b d\tau. \tag{3.106}$$

The proper time interval a to b is the time between two events as recorded by a clock carried along with the freely moving particle. Therefore, the requirement that a freely moving particle suffers no acceleration is the same as requiring $\int_a^b d\tau$ to be an extremum.

3.15. Consider a particle of mass m and electric charge q moving in an electric potential V and vector potential $\mathbf{A}$. The electric field $\mathbf{E}$ and magnetic field $\mathbf{B}$ are derived from the potentials according to $\mathbf{E} = -\nabla V - \partial \mathbf{A}/\partial t$ and $\mathbf{B} = \nabla \times \mathbf{A}$. The Lagrangian (with $c = 1$) is[29]

$$L = -m\gamma^{-1} + q\mathbf{A} \cdot \mathbf{v} - qV, \tag{3.107}$$

where $\gamma \equiv (1 - v^2)^{-1/2}$ and v denotes the particle's speed relative to an inertial frame.
a. Show that the canonical momentum components p_k for $k = 1, 2, 3$ are

$$p_k = m v_k \gamma + q A_k. \tag{3.108}$$

---

[29]Notice this is only the Lagrangian of a particle in a background field. This Lagrangian does not include the effects of the particle on the field or the field's own energy. See chapters 7 and 8.

Notice that the canonical momentum is not the "mv" momentum.

b. Show that the Hamiltonian is[30]

$$H = \left(\frac{(\mathbf{p} - q\mathbf{A})^2}{m} + m \right) \gamma^{-1} + qV, \qquad (3.109)$$

and that H numerically equals $m\gamma + qV$.

c. With $H = E$ numerically, show that $E^2 - p^2 \neq m^2$. In special relativity, $E^2 - p^2 = m^2$ only for noninteracting particles, where in this context $p = mv\gamma$.

3.16. Show that if the Lagrangian depends on the second derivative of the dependent variable, and if we require

$$\delta \int_a^b L(t, q^\mu, \dot{q}^\mu, \ddot{q}^\mu) dt = 0, \qquad (3.110)$$

then the Euler-Lagrange equation becomes

$$\frac{\partial L}{\partial q^\mu} = \frac{d}{dt} \frac{\partial L}{\partial \dot{q}^\mu} - \frac{d^2}{dt^2} \frac{\partial L}{\partial \ddot{q}^\mu}. \qquad (3.111)$$

What must be assumed about the q^μ and ζ^μ and their derivatives?

b. Suggest a generalization of the Euler-Lagrange equation when the Lagrangian depends on the nth derivative of q^μ with respect to t.

3.17. Consider the case where the endpoints of the definite integral functional includes variable endpoint T,

$$\Gamma(T) = \int_a^T L(t, q^\mu, \dot{q}^\mu) dt. \qquad (3.112)$$

What result does $\delta\Gamma = 0$ give now?

3.18. a. Show that the functional of Fermat's principle, eq. (3.2), may be written in parametric form as

$$\int_a^b n(\mathbf{r}) \left| \frac{d\mathbf{r}}{ds} \right| ds, \qquad (3.113)$$

where $n(\mathbf{r})$ is a position-dependent index of refraction with $\mathbf{r}$ a position vector, and s an arbitrary parameter. In what follows denote $|d\mathbf{r}/ds| \equiv r'$.

b. Show that the momentum canonically conjugate to the coordinate x_k is $p_k = nx_k'/r'$.

----

[30]In units where $c = 1$.

c. Compute the Hamiltonian. Does a distinction need to be made between H as a function and the numerical value of H?

d. If you *impose* $H = K + U$ and define an "optical kinetic energy" as $K = \frac{1}{2}r'^2$, show the optical potential energy to be $U(\mathbf{r}) = -\frac{1}{2}n^2$. Construct the optical Lagrangian $L = K - U$ and show that the Euler-Lagrange equation results in a "Newton's second law for geometrical optics" [see Turner (1991) and references therein],

$$-\nabla U = \frac{d^2\mathbf{r}}{ds^2}. \tag{3.114}$$

e. Use the result of part (d) to derive Snell's law of refraction.

f. Derive the trajectory of a ray where, for an xy coordinate system, $n(y) = n_0 + \alpha y$, where n_0 and α are constants. This offers a reasonable model of the variable index above a heated roadway that produces road mirage.

3.19. Consider a Lagrangian

$$L = \tfrac{1}{2}mg_{jk}\dot{q}^j\dot{q}^k - U(q^i) \tag{3.115}$$

where the g_{jk} are the component of the metric tensor in three-dimensional space mapped by an arbitrary coordinate system, and the Latin indices range over 1, 2, 3. Show that the Euler-Lagrange equation gives

$$-\partial^n U = m[\ddot{x}^n + \Gamma^n{}_{pq}\dot{x}^p\dot{x}^q] \tag{3.116}$$

where

$$\Gamma^n{}_{pq} = \tfrac{1}{2}g^{nk}[\partial_p g_{qk} + \partial_q g_{pk} - \partial_k g_{pq}] \tag{3.117}$$

and $g^{ab}g_{bc} = \delta^a{}_c$ (see appendix F for more on the $\Gamma^\lambda{}_{\mu\nu}$).

3.20. Start from the postulate

$$\Delta\tau = \int_{\tau(a)}^{\tau(b)} \sqrt{g_{\mu\nu}u^\mu u^\nu}\, d\tau = \text{max.}, \tag{3.118}$$

where a and b label events in spacetime, the metric tensor components $g_{\mu\nu}$ are functions of the coordinates x^μ, and $u^\mu \equiv dx^\mu/d\tau$. Derive via the Euler-Lagrange equation the general relativistic equation for a particle in free fall,

$$\frac{du^\mu}{d\tau} + \Gamma^\mu{}_{\rho\sigma}u^\rho u^\sigma = 0, \tag{3.119}$$

where (see appendix F)

$$\Gamma^\mu{}_{\rho\sigma} = \frac{1}{2}g^{\mu\lambda}[\partial_\rho g_{\lambda\sigma} + \partial_\sigma g_{\lambda\rho} - \partial_\lambda g_{\rho\sigma}]. \tag{3.120}$$

Compare this result to exercise 3.5—same mathematics, different context.

3.21. Although they are now mostly of historical interest as stepping-stones to Hamilton's principle, this problem deals with Bernoulli's principle of virtual work, its extension by d'Alembert to dynamics, and Gauss's principle of least constraint.

a. Johann Bernoulli introduced the concept of "virtual work" into statics in 1717, by imagining that the nth particle (among N particles) undergoes an arbitrary imagined displacement $d\mathbf{r}_n$, where these displacements are consistent with forces of constraints, such as imposed by rigid rods, normal forces, and so forth. For example, all virtual displacements may add up to zero,

$$\sum_{n=1}^{N} C_n d\mathbf{r}_n = 0, \tag{3.121}$$

as in an Atwood machine (weights on frictionless pulleys), where the constraint of fixed rope length means that displacements of the two weights are equal and opposite; as one drops the other rises the same distance. The principle of virtual work says the system of particles is in mechanical equilibrium if and only if the virtual work done on all the particles vanishes,

$$\sum_{n=1}^{N} \mathbf{F}_n \cdot d\mathbf{r}_n = 0, \tag{3.122}$$

where only external forces such as gravity are included in the sum over $\mathbf{F}_n$ and the forces of constraint are handled by eq. (3.121); the $d\mathbf{r}$'s in eqs. (3.121) and (3.122) are the same.[31] Consider an Atwood machine consisting of two particles, of mass m_1 and m_2, connected by a nonstretchable cord of negligible mass, drooped over a frictionless peg. Apply the principle of virtual work,

$$m_1 g \, dy_1 + m_2 g \, dy_2 = 0, \tag{3.123}$$

and the equation of constraints to derive the condition for static equilibrium, $m_1 g = m_2 g$.

b. If the forces in the virtual work principle are derivable from potential energy functions, show that eq. (3.122) becomes

$$\sum_n U_n = \text{max. or min.}, \tag{3.124}$$

a precursor of Hamilton's principle.

c. Jean le Rond d'Alembert extended the principle of virtual work to

---

[31]In contrast, in the familiar Newtonian paradigm, forces of constraint are included explicitly in $\mathbf{F} = m\mathbf{a}$, and their magnitudes and directions inferred by invoking the third law.

dynamics in 1743 [Goldstein (1965) 14–16]. Split the net force on a particle into the net "impressed" force $\mathbf{F}^{(i)}$ derivable from potential energy functions, and "constraint" forces $\mathbf{F}^{(c)}$ such as contact forces. With this distinction, Newton's second law says, for the nth particle,

$$\mathbf{F}_n^{(i)} + \mathbf{F}_n^{(c)} = m_n \mathbf{a}_n. \tag{3.125}$$

Transpose and multiply by Bernoulli's virtual displacements, for which eq. (3.121) holds, to obtain

$$(\mathbf{F}_n^{(i)} - m_n \mathbf{a}_n) \cdot d\mathbf{r}_n = -\mathbf{F}_n^{(c)} \cdot d\mathbf{r}_n. \tag{3.126}$$

By summing over all the system's particles, d'Alembert's principle effectively reduces dynamics to statics, with $-m_n \mathbf{a}_n$ treated as a "force" opposing $\mathbf{F}_n^{(i)}$. Because forces of contraint perform no work, we obtain the generalization of the principle of virtual work:

$$\sum_n (\mathbf{F}_n^{(i)} - m_n \mathbf{a}_n) \cdot d\mathbf{r}_n = 0. \tag{3.127}$$

Applied to the two-particle Atwood machine, show the system accelerates with acceleration

$$a = g \left(\frac{m_1 - m_2}{m_1 + m_2} \right), \tag{3.128}$$

the same as found by using good old $\mathbf{F} = m\mathbf{a}$ in the usual way. Show that d'Alembert's principle may also be stated

$$\sum_n (K + U)_n = \text{max. or min.} \tag{3.129}$$

d. Suppose the nth particle in a system of particles would undergo the displacement $\Delta \mathbf{s}_n$ in time Δt if it were subject to no forces of constraint. With constraints, let the actual displacement be $\Delta \mathbf{r}_n$. Let $\Delta \mathbf{u}_n = \Delta \mathbf{s}_n - \Delta \mathbf{r}_n$. Gauss's principle of least constraint (1828)[32] asserts that

$$G \equiv \sum_n m_n (\Delta \mathbf{u}_n)^2 = \text{min.} \tag{3.130}$$

Applied to the two-particle Atwood machine, removing the constraints by cutting the rope means both particles fall freely. In that case, each particle's displacement after time t would be $\frac{1}{2}gt^2$. But with the rope intact, their

---

[32]One cannot help but be struck by the similarity between Gauss's principle of least constraint for mechanics and the least squares method of regression analysis used in data fitting, to which Gauss greatly contributed.

actual displacements in this time are $\frac{1}{2}at^2$ for particle 1 and $-\frac{1}{2}at^2$ for particle 2. Determine a by invoking Gauss's principle of least constraint.

3.22. a. Find the Euler-Lagrange equation for a Lagrangian of the form $L = L(x, \dot{x}, \ddot{x}, \dddot{x})$ where $\dot{x} \equiv dx/dt$ (recall ex. 3.16).
b. The y-deflection of a uniform cantilevered beam of length ℓ, oriented along a horizontal x-axis and subjected to a vertical load $W(x)$, is well modeled by the solution of the fourth-order differential equation [Rainville & Bedient (1974) 220]

$$YI\frac{d^4y}{dx^4} = W(x) \tag{3.131}$$

for $0 \leq x \leq \ell$, where Y is the beam's modulus of elasticity and I its moment of inertia about the supporting point. Construct a Lagrangian for which eq. (3.131) is the ELE.

Part II

WHEN FUNCTIONALS ARE INVARIANT

Chapter 4

Invariance

We know that if the Lagrangian L does not depend explicitly on t *... then [the Hamiltonian] is a constant. To say that L does not depend explicitly on* t *could be reinterpreted as to say that the functional J is invariant under the transformation which takes* t *to* t + ε. *The Noether theorem... can to a certain degree be considered a generalization of the above fact. It states that there is, in general, a connection between the existence of the first integrals of the Euler-Lagrange equation and the invariance of the fundamental integral.* —John D. Logan, *Invariant Variational Principles*, 1977, 27–28

4.1   Formal Definition of Invariance

Theories of relativity are built on quantities that are postulated to be invariant among members of a well-defined set of coordinate systems. Newtonian relativity assumes length and time intervals to be separately invariant among all inertial reference frames; consequently, in the Newtonian world, the speed of light is frame dependent. In contrast, the special theory of relativity postulates the speed of light to be invariant among all inertial frames; therefore, length and time intervals are frame dependent. "Invariance" in this context means the same numerical value *exactly* in all members of the set of coordinate systems.

Consider transformations from one coordinate system to another, described by a parameter ε that can be varied continuously from zero. The identity transformation ε = 0 makes no change whatsoever. As ε increases from zero, the difference between the original and the new coordinate system becomes larger. Noether's theorem deals with continuous transformations.

71

Such mappings imply the existence of functions T and Q^μ that take the original independent variable t and the original dependent variables $q^\mu(t)$, and produce new independent and dependent variables:

$$t \to t' = T(t, q^\mu, \varepsilon), \tag{4.1}$$

$$q^\mu \to q'^\mu = Q^\mu(t, q^\nu, \varepsilon). \tag{4.2}$$

Examples include rotations of orthogonal axes, for instance through the angle ε about the z-axis,

$$\begin{aligned} x' &= x \cos\varepsilon + y \sin\varepsilon, \\ y' &= -x \sin\varepsilon + y \cos\varepsilon, \\ z' &= z. \end{aligned} \tag{4.3}$$

Another example is found in the simplest Lorentz transformation, where ε is the relative velocity v between two inertial frames, and in units where $c = 1$,

$$\begin{aligned} t' &= \gamma(t - vx), \\ x' &= \gamma(x - vt), \\ y' &= y, \\ z' &= z, \end{aligned} \tag{4.4}$$

with $\gamma \equiv (1 - v^2)^{-1/2}$.

Example: Consider the distance functional in the xy plane

$$s = \int_a^b \sqrt{1 + \left(\frac{dy}{dx}\right)^2}\, dx. \tag{4.5}$$

Under a rotation about the z-axis, this becomes

$$s' = \int_{a'}^{b'} \sqrt{1 + \left(\frac{dy'}{dx'}\right)^2}\, dx'. \tag{4.6}$$

Notice the transformation does not change the form of the Lagrangian, but merely the independent and dependent variables stuffed into it.

In general, under the transformation the old functional, Γ, becomes the new functional, Γ', where Γ' means

$$\Gamma' = \int_{a'}^{b'} L\left(t', q'^\mu(t'), \frac{dq'^\mu(t')}{dt'}\right) dt' \tag{4.7}$$

which is abbreviated as $\Gamma' = \int_{a'}^{b'} L' dt'$. How shall we define invariance of the functional?

We could require $\Gamma' = \Gamma$ exactly as our definition of invariance, similar to the meaning of the invariance of the speed of light among inertial reference frames in special relativity. Such a definition requires Γ' to be the same as Γ to an infinite number of decimal places. Such a strict requirement is appropriate for the founding assumption of the special theory of relativity ("$c' = c$ whatever v happens to be"), because it postulates a fundamental principle about light. The requirement that $\Gamma' = \Gamma$ was Emmy Noether's definition of invariance in 1918 (she used $I' = I$).

But in matters of measurements in the physical world, there is room for more tolerance, in the sense of the laboratory and the machine shop. Two experimental observations are accepted as the same if they agree to within experimental uncertainty. Two machine parts are considered identical if their micrometer measurements are the same within the tolerance of the instrument's precision. If the two measurements agree to, say, four significant figures, that may be close enough to be considered identical, even though we have no idea what the numbers are in the ninth or fifteenth decimal places. Unlike the special relativity postulate about the speed of light being exactly invariant as a matter of principle, in the machine shop the two parts will work equally well since any difference between them is too small to matter. We distinguish matters of *principle* from what is *detectable*.

Recall our discussion in chapter 1, where we pondered the requirement for the potential energy to be unchanged under a translation. We wrote the derivative of the potential energy as

$$\frac{dU}{dx} \approx \frac{\Delta U}{\Delta x} = \frac{U(x + \varepsilon) - U(x)}{\varepsilon}. \tag{4.8}$$

For this to be vanishingly small as $\varepsilon \to 0$, the numerator does not have to be exactly zero, but only proportional to ε^s, where $s > 1$. In that case $dU/dx \sim \varepsilon^{s-1}$, which still goes to zero as $\varepsilon \to 0$. Likewise, instead of the strict definition $\Gamma' = \Gamma$, let us take as our definition of invariance the notion that, for ε sufficiently small, it is adequate to have the difference between the new and old functional lie below the threshold of detectability, even if that difference is not zero exactly.[1] Of course, any transformation that stipulates the strict definition $\Gamma' - \Gamma = 0$ exactly would be included in the less strict definition for which the difference between Γ' and Γ is merely too small to detect.

---

[1]Although invariance is defined in terms of infinitesimal transformations, a finite transformation can be build up by a succession of infinitesimal ones; see ex. 4.7.

To express an infinitesimal transformation explicitly, expand $T(t, q^\mu, \varepsilon)$ and $Q^\mu(t, q^\nu, \varepsilon)$ in Taylor series about $\varepsilon = 0$:

$$t' = t + \varepsilon \left(\frac{dT}{d\varepsilon} \right)_0 + O(\varepsilon^2) + \cdots, \tag{4.9}$$

$$q'^\mu = q^\mu + \varepsilon \left(\frac{dQ^\mu}{d\varepsilon} \right)_0 + O(\varepsilon^2) + \cdots. \tag{4.10}$$

Clearly $\varepsilon = 0$ corresponds to the identity transformation, where $t' = t$ and $q'^\mu = q^\mu$. The coefficients of ε to the first power are called the "generators" of the transformation. Let's denote them τ and ζ^μ:

$$\tau \equiv \left(\frac{dT}{d\varepsilon} \right)_0 = \tau(t, q^\mu), \tag{4.11}$$

and

$$\zeta^\mu \equiv \left(\frac{dQ^\mu}{d\varepsilon} \right)_0 = \zeta^\mu(t, q^\nu). \tag{4.12}$$

Unlike the situation in deriving the Euler-Lagrange equation, here ζ is not required to vanish at the endpoints. That is the difference between a "variation" and a "transformation."

Now we can write the transformations succinctly:

$$t' = t + \varepsilon\tau + O(\varepsilon^2) + \ldots, \tag{4.13}$$

and

$$q'^\mu = q^\mu + \varepsilon\zeta^\mu + O(\varepsilon^2) \ldots. \tag{4.14}$$

As implied in eqs. (4.11) and (4.12), the generators may be functions of the independent and the dependent variables. For example, in the rotation of orthogonal axes about the z-axis, upon expanding in powers of ε we obtain

$$\begin{aligned} x' &= x + y\varepsilon + \ldots \\ y' &= y - x\varepsilon + \ldots, \end{aligned} \tag{4.15}$$

so that $\tau = y$ and $\zeta = -x$. For the simple Lorentz transformation, we find to first order in v

$$\begin{aligned} t' &= t - vx + \ldots, \\ x' &= x - vt + \ldots, \end{aligned} \tag{4.16}$$

and thus $\tau = -x$ and $\zeta^x = -t$.

A "group" is a set of elements on which a binary operation is defined with closure and associativity, an identity element exists, and each element has an inverse. A set of transformations like those we are considering has an identity element, each transformation has an inverse that reverses it, and two successive transformations equal a single transformation. Therefore the transformations typically form a group. Such continuously parameterized transformations are "Lie groups,"[2] named after the Norwegian mathematician Sophus Lie (1842–1899) who pioneered their study. Emmy Noether was an expert in Lie groups; hence, Noether's theorem was, to her, an application of Lie group theory.[3]

Now the formal definition of the invariance of a functional can be stated.

Definition of Invariance: The functional

$$\Gamma = \int_a^b L(t, q^\mu, \dot{q}^\mu)dt, \qquad (4.17)$$

where $q^\mu = q^\mu(t)$ and $\dot{q}^\mu \equiv dq^\mu/dt$, is said to be invariant under the infinitesimal transformation

$$t' = t + \varepsilon\tau + \dots, \qquad (4.18)$$

$$q'^\mu = q^\mu + \varepsilon\zeta^\mu \dots \qquad (4.19)$$

if and only if

$$\Gamma' - \Gamma \sim \varepsilon^s \qquad (4.20)$$

where $s > 1$.

It is possible to recast this definition in terms of the Lagrangian since T and Q^μ, and thus the generators τ and ζ^μ, are, in general, functions of t. Thus $\Gamma' - \Gamma$ can be written as one integral over t, from $t = a$ to $t = b$, by a change of variable in Γ':

$$\Gamma' - \Gamma = \int_{a'}^{b'} L' \, dt' - \int_a^b L \, dt$$
$$= \int_a^b \left[L'\frac{dt'}{dt} - L \right] dt. \qquad (4.21)$$

Because a and b are arbitrary, the definition of the functional being invariant can be restated in terms of the integrand.

Definition: The functional

$$\Gamma = \int_a^b L(t, q^\mu, \dot{q}^\mu)dt, \qquad (4.22)$$

---

[2]Pronounced "Lee."

[3]Although the study of invariance is usefully framed in terms of group theory, for our purposes the transformations are not *required* to be a group.

where $q^\mu = q^\mu(t)$ and $\dot{q}^\mu \equiv dq^\mu/dt$, is said to be invariant under the infinitesimal transformation

$$t' = t + \varepsilon\tau + \ldots,$$
$$q'^\mu = q^\mu + \varepsilon\zeta^\mu \ldots$$

if and only if

$$L'\frac{dt'}{dt} - L \sim \varepsilon^s, \tag{4.23}$$

where $s > 1$.

This condition for invariance is both necessary and sufficient. If the invariance condition holds, the functional is invariant; if the functional is invariant, eq. (4.23) is satisfied.

Example (Distance Functional): The Lagrangian of the distance functional in the xy plane is $L = [1 + (dy/dx)^2]^{1/2}$. Working directly from the definition of invariance, we will evaluate $L'(dx'/dx) - L$,

$$L'\frac{dx'}{dx} - L = \sqrt{1 + \left(\frac{dy'}{dx'}\right)^2}\frac{dx'}{dx} - \sqrt{1 + \left(\frac{dy}{dx}\right)^2} \tag{4.24}$$

under the rotation of axes through angle ε about the z-axis, and see whether it meets meets the definition of invariance. To first order in ε, $x' = x + \varepsilon y$ and $y' = y - \varepsilon x$. Therefore,

$$\frac{dx'}{dx} = 1 + \varepsilon\frac{dy}{dx} \tag{4.25}$$

and

$$\frac{dy'}{dx'} = \frac{dy'}{dx}\frac{dx}{dx'} = \frac{\left(\dfrac{dy'}{dx}\right)}{\left(\dfrac{dx'}{dx}\right)} = \frac{\dfrac{dy}{dx} - \varepsilon}{1 + \varepsilon\dfrac{dy}{dx}}. \tag{4.26}$$

For notational brevity let us temporarily put $dy/dx \equiv \sigma$. With the help of binomial expansions, our job is to see the leading power of ε that emerges from the definition of invariance. We have

$$\frac{dy'}{dx'} = \frac{\sigma - \varepsilon}{1 + \varepsilon\sigma} \approx \sigma - \varepsilon(1 + \sigma^2) = \sigma - \varepsilon L^2, \tag{4.27}$$

so that

$$(dy'/dx')^2 \approx \sigma^2 - 2\varepsilon\sigma L^2. \tag{4.28}$$

Therefore

$$L'\frac{dx'}{dx} - L = L[1 - 2\varepsilon\sigma]^{1/2}(1 + \varepsilon\sigma) - L \approx -L\sigma^2\varepsilon^2 \qquad (4.29)$$

plus other terms of order ε^3 and higher. Therefore $L'(dx'/dx) - L \sim \varepsilon^2$, which satisfies the definition of invariance.

This example illustrates how the formal definition of invariance, while precise, can be cumbersome to implement. A more efficient criterion can be built from this definition, an identity necessary and sufficient for invariance.

4.2 The Invariance Identity

Can we tell merely from the Lagrangian, and from the transformation generators, whether or not the functional is invariant? In this section we answer in the affirmative, through the invariance identity.[4] We state this important result as a theorem.

Theorem: The functional

$$\Gamma = \int_a^b L(t, q^\mu, \dot{q}^\mu)dt \qquad (4.30)$$

is invariant under the infinitesimal transformation

$$t' = t + \varepsilon\tau + \dots$$
$$q'^\mu = q^\mu + \varepsilon\zeta^\mu + \dots,$$

if and only if the following identity holds:

$$\frac{\partial L}{\partial q^\mu}\zeta^\mu + p_\mu\dot{\zeta}^\mu + \frac{\partial L}{\partial t}\tau - H\dot{\tau} = 0 \qquad (4.31)$$

where $\dot{\tau} \equiv d\tau/dt$ and $\dot{\zeta}^\mu \equiv d\zeta^\mu/dt$, $p_\mu \equiv \partial L/\partial\dot{q}^\mu$ is a canonical momentum, and $H \equiv p_\mu\dot{q}^\mu - L$ is the Hamiltonian.

Proof: First, let us prove that invariance implies the identity. By definition, if the functional is invariant, then eq. (4.23) holds, $L'(dt'/dt) - L \sim \varepsilon^s$, where $s > 1$. Differentiate this definition with respect to ε, then set $\varepsilon = 0$, which presents us with

$$L\frac{d}{d\varepsilon}\left[\frac{dt'}{dt}\right]_0 + \left[\frac{dL'}{d\varepsilon}\right]_0 = 0. \qquad (4.32)$$

---

[4]Published in its original form by Noether in her 1918 paper [Noether (1918)], the form presented here is a more streamlined version, also called the Rund-Trautman identity [Logan (1977)], because Rund and Trautman rederived it in the 1960s and 1970s [Rund (1972); Trautman (1967)].

In the first term, from $t' = t + \varepsilon\tau$, we have $dt'/dt = 1 + \varepsilon\dot{\tau}$, and with it

$$\frac{d}{d\varepsilon}\left(\frac{dt'}{dt}\right) = \dot{\tau}. \tag{4.33}$$

Turning to the second term in eq. (4.32), we expand $dL'/d\varepsilon$ with the chain rule (remembering that q'^μ is a function of t'):

$$\frac{dL'}{d\varepsilon} = \frac{\partial L'}{\partial t'}\frac{dt'}{d\varepsilon} + \frac{\partial L'}{\partial q'^\mu}\frac{dq'^\mu}{d\varepsilon} + \frac{\partial L'}{\partial \dot{q}'^\mu}\frac{d\dot{q}'^\mu}{d\varepsilon}$$

$$= \frac{\partial L'}{\partial t'}\tau + \frac{\partial L'}{\partial q'^\mu}\zeta^\mu + \frac{\partial L'}{\partial \dot{q}'^\mu}\frac{d\dot{q}'^\mu}{d\varepsilon}.$$

Insert this and eq. (4.33) into eq. (4.32), then set $\varepsilon = 0$, and we have

$$L\dot{\tau} + \frac{\partial L}{\partial t}\tau + \frac{\partial L}{\partial q^\mu}\zeta^\mu + \frac{\partial L}{\partial \dot{q}^\mu}\left[\frac{d\dot{q}'^\mu}{d\varepsilon}\right]_0 = 0. \tag{4.34}$$

The last term in eq. (4.34) requires more work. First, write out $\dot{q}'^\mu$ explicitly:

$$\dot{q}'^\mu = \frac{dq'^\mu}{dt'} = \frac{dq^\mu + \varepsilon d\zeta^\mu}{dt + \varepsilon d\tau}$$

$$= \frac{\dot{q}^\mu + \varepsilon\dot{\zeta}^\mu}{1 + \varepsilon\dot{\tau}},$$

so that

$$\frac{d}{d\varepsilon}\left[\frac{d\dot{q}'^\mu}{dt'}\right]_0 = \frac{d}{d\varepsilon}\left[\frac{\dot{q}^\mu + \varepsilon\dot{\zeta}^\mu}{1 + \varepsilon\dot{\tau}}\right]_0 = \dot{\zeta}^\mu - \dot{q}^\mu\dot{\tau}. \tag{4.35}$$

Now the differentiated definition of invariance, eq. (4.32), has become

$$L\dot{\tau} + \frac{\partial L}{\partial t}\tau + \frac{\partial L}{\partial q^\mu}\zeta^\mu + \frac{\partial L}{\partial \dot{q}^\mu}(\dot{\zeta}^\mu - \dot{q}^\mu\dot{\tau}) = 0. \tag{4.36}$$

Factor out the $\dot{\tau}$ and identify the definitions of canonical momenta and the Hamiltonian, and the result claimed in eq. (4.31) follows. Invariance implies the identity.

To show the identity implies invariance, assume the identity holds and demonstrate, by reversing the above steps, that the identity leads back to the definition of invariance (see ex. 4.5). Therefore, if the functional is invariant, the identity holds; conversely, if the identity holds, the functional is invariant. The invariance identity is both necessary and sufficient for the functional to be invariant.

Example (Distance Functional): In the distance functional example above, the independent variable is x, and $y(x)$ carries the role of the dependent variable. For the arc length Lagrangian, $\partial L/\partial x = 0$ and $\partial L/\partial y = 0$.

The generators under rotation are $\tau = y$ and $\zeta = -x$. Therefore, the quantity on the left-hand side of the invariance identity gives $-p - H(dy/dx)$. By computing p and H it follows at once that this quantity vanishes, satisfying the invariance identity.

The invariance identity may be presented in two forms. The first one we have just seen:

$$\frac{\partial L}{\partial q^\mu}\zeta^\mu + p_\mu \dot{\zeta}^\mu + \frac{\partial L}{\partial t}\tau - H\dot{\tau} = 0. \tag{4.37}$$

Another way to say the same thing follows from going back to eq. (4.36) and using the product rule for derivatives, giving an alternative version:[5]

$$-(\zeta^\mu - \dot{q}^\mu \tau)\left[\frac{\partial L}{\partial q^\mu} - \frac{d}{dt}\frac{\partial L}{\partial \dot{q}^\mu}\right] = \frac{d}{dt}[p_\mu \zeta^\mu - H\tau]. \tag{4.38}$$

Caution! Do not be hasty about invoking the Euler-Lagrange equation in the invariance identity. Whether the functional is extremal, and whether it is invariant, are two separate issues. Functionals can be extremal but not invariant, and they can be invariant but not extremal.[6]

4.3   A More Liberal Definition of Invariance

The functional

$$\Gamma = \int_a^b L \, dt \tag{4.39}$$

is unchanged by adding to L a term dS/dt for which $S(b) = S(a)$. Let

$$\begin{aligned}\Gamma' &= \int_a^b \left(L + \frac{dS}{dt}\right) dt \\ &= \Gamma + [S(b) - S(a)] \\ &= \Gamma.\end{aligned} \tag{4.40}$$

This means the definition of invariance can be liberalized to allow the integrand of Γ' and Γ to differ by a term linear in ε, provided the coefficient

---

[5] Notice the Euler-Lagrange operator, $\psi_\mu \equiv \frac{\partial L}{\partial q^\mu} - \frac{d}{dt}\left(\frac{\partial L}{\partial \dot{q}^\mu}\right)$. At the equivalent point in her 1918 paper, this operator appears in Noether's equation for invariance. The invariance identity does not require ψ_n to vanish; it vanishes when and only when the functional is an extremal, a separate consideration from invariance. In chapter 8, where Noether's second theorem is presented, this distinction between invariance and extremal is paramount.

[6] See section 5.3 for a discussion of a functional being "extremal," "invariant," and "stationary."

of ε is a derivative with respect to the independent variable. In other words, if we write $S = -\varepsilon F$, this "extra" term in the functional shows up in the invariance identity. Accordingly, a functional is said to be "divergence invariant" if and only if, for some function $F(t)$,

$$L'\frac{dt'}{dt} - L = \varepsilon\frac{dF}{dt} + O(\varepsilon^s), \quad s > 1. \tag{4.41}$$

When the invariance identity is derived again, this inhomogeneous term survives:

$$\frac{\partial L}{\partial q^\mu}\zeta^\mu + p_\mu\dot{\zeta}^\mu + \frac{\partial L}{\partial t}\tau - H\dot{\tau} - \frac{dF}{dt} = 0. \tag{4.42}$$

Example: Consider projectile motion without air resistance. In the xy plane usually used to describe this system,

$$L(y, \dot{x}, \dot{y}) = \frac{1}{2}m(\dot{x}^2 + \dot{y}^2) - mgy, \tag{4.43}$$

where g denotes the magnitude of a uniform gravitational field. Newtonian relativity allows Galilean transformations between inertial reference frames. When a second inertial reference frame with coordinates (t', x', y') moves horizontally with uniform velocity v_o relative to the first, then

$$t' = t,$$
$$x' = x - v_o t, \tag{4.44}$$
$$y' = y.$$

The relative velocity v_o assumes the role of ε, and the generators are $\tau = 0$, $\zeta^x = -t$, and $\zeta^y = 0$. Noting that $dt'/dt = 1$ and $dx'/dt' = \dot{x} - v_o$, we find $L'(dt/dt) - L = -m\dot{x}v_o$ to first order in v_o, which satisfies the liberalized definition of invariance with $dF/dt = -m\dot{x}$.

Questions for Reflection and Discussion

Q4.a. We described a transformation that takes us from (t, x) to (t', x') with generators τ and ζ. How would one write the reverse transformation from (t', x') to (t, x) in terms of the original τ and ζ? If the functional is invariant under the "forward" transformation, is it also invariant under the "reverse" transformation?

Q4.b. Is it necessary in general for the transformations to form a group?

Q4.c. Why do we write the inhomogeneous term in the liberal definition of invariance as dF/dt instead of merely F? In a similar move in chapter 6, the reason for the name "divergence invariance" should become clear (recall Gauss's divergence theorem).

Q4.d. You may have noticed that if $\tau = 0$, then in deriving the invariance identity, we would come across

$$\frac{\partial L}{\partial q^\mu}\zeta^\mu + \frac{\partial L}{\partial \dot{q}^\mu}\dot{\zeta}^\mu = 0, \tag{4.45}$$

which is strikingly similar to an intermediate step in the derivation of the Euler-Lagrange equation, where we encountered

$$\left[\frac{d\Gamma}{d\varepsilon}\right]_0 = \int_a^b \left[\frac{\partial L}{\partial q^\mu}\zeta^\mu + \frac{\partial L}{\partial \dot{q}^\mu}\dot{\zeta}^\mu\right] = 0. \tag{4.46}$$

What is different about ζ in the derivations of the Euler-Lagrange equation and the invariance identity?

Exercises

4.1. A particle of mass m and electric charge q moves through an electromagnetic field described by the potentials $V(t, \mathbf{r})$ and $\mathbf{A}(t, \mathbf{r})$. When the particle moves with low speeds compared to the speed of light, its Lagrangian is

$$L = \frac{1}{2}m\dot{\mathbf{r}}^2 + q\mathbf{v} \cdot \mathbf{A} - qV. \tag{4.47}$$

Determine whether the functional based on the time integral of this Lagrangian is invariant (original or liberal definition) under
a. the simplest Galilean transformation

$$t' = t,$$
$$x' = x - vt$$

and
b. The simplest Lorentz transformation,

$$t' = \gamma(t - vx),$$
$$x' = \gamma(x - vt),$$

with $y' = y$ and $z' = z$ in both cases.

4.2. Consider the damped harmonic oscillator, a particle of mass m acted on by a linear restoring force with a time-dependent spring constant

k (e.g., metal fatigue), and a damping force proportional to the velocity, having damping coefficient b. The Lagrangian is

$$L(t, x, \dot{x}) = \left[\frac{1}{2} m \dot{x}^2 - \frac{1}{2} k(t) x^2 \right] e^{bt/m}. \tag{4.48}$$

Determine whether the functional based on the time integral of this Lagrangian could be invariant under the rescaling $t' = t(1 + \varepsilon)$ and $x' = x(1 + \kappa \varepsilon)$, where $\kappa = \text{const}$.

4.3. Show that the two versions of the invariance identity, eqs. (4.37) and (4.38), are identical.

4.4. Fill in the steps in the derivation of the invariance identity for the inhomogeneous (liberal) definition of invariance, eq. (4.42).

4.5. We showed that if the functional is invariant, then the invariance identity holds. Show that if the identity holds, the functional is invariant.

4.6. Show from the invariance identity that central force motion is invariant under a change of latitude and longitude coordinates.

4.7 Show how a finite continuous transformation can be built incrementally from a sequence of infinitesimal ones. To first order in ε, consider

$$t' = t + \varepsilon \tau,$$
$$q'^\mu = q^\mu + \varepsilon \zeta^\mu.$$

For convenience, arrange these variables and generators into column vectors

$$|X\rangle \equiv \begin{bmatrix} t \\ q^1 \\ \cdot \\ \cdot \\ \cdot \\ q^N \end{bmatrix}, \qquad |K\rangle \equiv \begin{bmatrix} \tau \\ \zeta^1 \\ \cdot \\ \cdot \\ \cdot \\ \zeta^N \end{bmatrix}. \tag{4.49}$$

The vector $|K\rangle$ in some contexts is called a "Killing vector."[7]
a. Let $|K\rangle \equiv M|X\rangle$ define a matrix M. Show that $|X'\rangle = (1 + \varepsilon M)|X\rangle$, where in this context "1" denotes the $(N+1) \times (N+1)$ unit matrix.
b. Consider another transformation, $|X'\rangle \to |X''\rangle = (1 + \varepsilon M)|X'\rangle$. Iterate this process to the nth transfomation, and show that

$$|X^{(n)}\rangle = (1 + \varepsilon M)^n |X\rangle. \tag{4.50}$$

---

[7]Named after Wilhelm Killing (1847–1923), the Killing vector components are generators of injective mappings between metric spaces that preserve distance.

c. Consider a finite transformation made of n successive ones, resulting in $\theta \equiv n\varepsilon$. As $n \to \infty$ (the number of steps goes to infinity, while the size of each step becomes infinitesimal), show that[8] $|X^{(\infty)}\rangle = e^{\theta M} |X\rangle$.

d. Discuss how any finite continuous transformation is equivalent to a sequence of infinitesimal ones, and thus why we lose nothing by considering only infinitesimal transformations in the definition of invariance.

4.8. Recall the Lagrangian for a Newtonian particle moving in one spatial dimension x in a time-dependent potential:

$$L(t, x, \dot{x}) = \frac{1}{2} m \dot{x}^2 - U(t, x). \qquad (4.51)$$

Consider the rescalings $t' = t + \varepsilon t$ and $x' = x + \frac{1}{2}\varepsilon x$, so that $\tau = t$ and $\zeta = \frac{1}{2}x$. What further stipulations on the potential energy are necessary to make the functional invariant?

---

[8] Placing an operator or matrix in the argument of a transcendental function makes sense in terms of a Taylor series.

Chapter 5

Emmy Noether's Elegant (First) Theorem

Where ψ represents the Lagrangian expressions, *that is to say, the left-hand side of the Lagrangian equations of the associated variational problem $\delta I = 0$.... If we pass from these identity relations to the associated variational problem, that is to say, if we set $\psi = 0$, then Theorem I states... the existence of ρ first integrals.... [that] have often been referred to as "conservation laws."* —Emmy Noether, "Invariant Variational Problems" (1918), translated by M. A. Tavel, 1971

5.1 Invariance + Extremal = Noether's Theorem

We have seen that if a functional Γ is invariant under the transformation having generators τ and ζ^μ, then the invariance identity holds:

$$\frac{\partial L}{\partial q^\mu} \zeta^\mu + p_\mu \dot{\zeta}^\mu + \frac{\partial L}{\partial t} \tau - H\dot{\tau} - \frac{dF}{dt} = 0 \tag{5.1}$$

where p_μ is the momentum conjugate to q^μ,

$$p_\mu \equiv \frac{\partial L}{\partial \dot{q}^\mu}, \tag{5.2}$$

H denotes the Hamiltonian,

$$H \equiv p_\mu \dot{q}^\mu - L, \tag{5.3}$$

84

and F comes from the liberalized definition of invariance, the divergence invariance. We have also seen, as a separate consideration, that if Γ is an extremal, then the Euler-Lagrange equation holds, and comes in two forms: the p-dot version

$$\frac{\partial L}{\partial q^\mu} = \dot{p}_\mu, \tag{5.4}$$

and the H-dot form,

$$\frac{\partial L}{\partial t} = -\dot{H}. \tag{5.5}$$

Now comes the punch line: If Γ is *both* invariant and extremal, then the Euler-Lagrange equations, when inserted into the invariance identity, turn the latter into a broad conservation law:[1]

$$\frac{d}{dt}[p_\mu \zeta^\mu - H\tau - F] = 0 \tag{5.6}$$

so that

$$p_\mu \zeta^\mu - H\tau - F = \text{const.} \tag{5.7}$$

We have just proved Noether's theorem, now stated formally:

Noether's Theorem: If under the infinitesimal transformation

$$t' = t + \varepsilon\tau + \dots$$
$$q'^\mu = q^\mu + \varepsilon\zeta^\mu + \dots$$

the functional

$$\Gamma = \int_a^b L(t, q^\mu, \dot{q}^\mu)\, dt \tag{5.8}$$

is both invariant and extremal, then the following conservation law holds:[2]

---

[1] For a list of numerous sources on Noether's theorem (first and second), including biographical and historical notes, see Neuenschwander (2014*b*).

[2] For comparing this discussion with that of Noether's "second theorem" in chapter 8, Noether's "first theorem" before us here may be written

$$-(\zeta^\mu - \dot{q}^\mu\tau)\psi_\mu = \frac{d}{dt}(p_\mu\zeta^\mu - H\tau) \tag{5.9}$$

where

$$\psi_\mu \equiv \frac{\partial L}{\partial x} - \dot{p}_\mu \tag{5.10}$$

denotes the Euler-Lagrange operator, or what Noether called the "Lagrangian expression." Here one sees how Γ being extremal (which means $\psi_\mu = 0$) and Γ being invariant are separate considerations. In chapter 8, conservation laws are derived even if $\psi_\mu \neq 0$.

$$p_\mu \zeta^\mu - H\tau - F = \text{const.} \qquad (5.11)$$

Example 1: Let a particle of mass m move through a space mapped by cylindrical coordinates, with the Lagangian of mechanical motion including a potential energy that may depend on r and z,

$$L = \frac{1}{2}m[\dot{r}^2 + r^2\dot{\theta}^2 + \dot{z}^2] - U(r, z). \qquad (5.12)$$

Consider a time translation, for which $t' = t + \varepsilon$ and all spatial coordinates are unchanged. Therefore $\tau = 1$ and all $\zeta^\mu = 0$. The reader may confirm the invariance identity is satisfied, and on an extremal path Noether's theorem says the conserved quantity is the Hamiltonian, $H = p_r^2/2m + p_z^2/2m + p_\theta^2/2mr^2 + U$, the mechanical energy. Invariance of this functional under a time translation yields conservation of energy.

Example 2: For the same Lagrangian as example 1, consider a rotation about the z-axis, $\theta \to \theta + \Delta\theta$. In this case $\tau = 0, \zeta^r = 0, \zeta^z = 0$, and $\zeta^\theta = 1$. According to Noether's theorem, the conserved quantity is the momentum conjugate to $\theta, p_\theta = mr^2\dot{\theta}$. Invariance under a spatial rotation implies the isotropy of space, and with it the conservation of the component of angular momentum about the axis of rotation.

The conservation laws in examples 1 and 2 also follow from their respective Euler-Lagrange equations. The next example picks up a conservation law that both versions of the Euler-Lagrange equation, by themselves, overlook.

Example 3: Consider the damped oscillator with the Lagrangian

$$L = \left[\frac{1}{2}m\dot{x}^2 - \frac{1}{2}kx^2\right]e^{bt/m}. \qquad (5.13)$$

After confirming the system to be invariant under the infinitesimal transformation $t' = t + \varepsilon\tau$ and $x' = x + \varepsilon\zeta$, where $\tau = 1$ and $\zeta = -bx/2m$, a Noether conservation law is revealed:

$$\left[\frac{1}{2}bx\dot{x} + \frac{1}{2}m\dot{x}^2 + \frac{1}{2}kx^2\right]e^{bt/m} = \text{const.} \qquad (5.14)$$

An interesting feature of this system is that neither the energy nor the momentum are separately conserved, yet a conservation law exists that includes both of them, despite the damping! As time advances, the energy and momentum damp out together according to

$$\frac{b}{2m}x(mv)p + E = (\text{const.})e^{-bt/m}. \qquad (5.15)$$

Notice in this example the Hamiltonian is not merely the mechanical energy $K + U$, but it is still the energy of *something*, $H = (K + U)e^{bt/m} = Ee^{bt/m}$ (see q. 5.h and ex. 5.13).

Example 4: Next consider a particle of mass m and electric charge e moving in an electromagnetic field described by the potential V and the vector potential $\mathbf{A}$. For nonrelativistic motion the Lagrangian for the particle's motion is

$$L = \frac{1}{2}m\dot{\mathbf{r}}^2 + e\dot{\mathbf{r}} \cdot \mathbf{A} - eV. \tag{5.16}$$

The canonical momentum is not the $m\mathbf{v}$ momentum, but is rather

$$\mathbf{p} = m\dot{\mathbf{r}} + e\mathbf{A}, \tag{5.17}$$

with Hamiltonian

$$H = \frac{(\mathbf{p} - e\mathbf{A})^2}{2m} + eV, \tag{5.18}$$

which numerically equals $\frac{1}{2}m\mathbf{v} \cdot \mathbf{v} + eV = K + U$, the particle's total mechanical energy. You can verify the invariance identity to be satisfied under the transformation

$$t' = t(1 + \varepsilon),$$

$$x'^\mu = x^\mu(1 + \frac{1}{2}\varepsilon),$$

which produces the Noether conservation law

$$\frac{1}{2}\mathbf{p} \cdot \mathbf{r} - Ht = const. \tag{5.19}$$

Example 5: Recall the projectile example near the end of chapter 4. It featured the Lagrangian

$$L(y, \dot{x}, \dot{y}) = \frac{1}{2}m(\dot{x}^2 + \dot{y}^2) - mgy. \tag{5.20}$$

Under the Galilean transformation

$$t' = t,$$

$$x' = x - v_o t,$$

$$y' = y,$$

with generators $\tau = 0, \zeta^x = -t$, and $\zeta^y = 0$, the functional is invariant under the liberal definition of invariance, with $dF/dt = -m\dot{x}$, so that $F = -mx + const.$ Ex. 5.10 invites you to relate this invariance to the conservation of linear momentum.

5.2   Executive Summary
of Noether's Theorem

"Extremal" and "invariance" have been treated as separate considerations, and so they are. But the alert reader will have noticed that the steps in making the functional an extremal,

$$\left[\frac{d\Gamma'}{d\varepsilon}\right]_0 = \frac{d}{d\varepsilon}\left[\int_a^b L'dt\right]_0 = 0, \tag{5.21}$$

and the steps in working out the invariance identity,

$$\left[\frac{d(\Gamma' - \Gamma)}{d\varepsilon}\right]_0 = \frac{d}{d\varepsilon}\left[\int_a^b \left(L'\frac{dt'}{dt} - L\right)dt\right]_0 = 0, \tag{5.22}$$

are almost but not quite identical. Let's take another look.

Strip away all indices! We can always put them back later, where repeated indices are summed. Right now we are getting at the essence of the ideas. Let

$$\Gamma = \int_a^b L(t, q, \dot{q})\,dt \tag{5.23}$$

and consider the transformation

$$t' = t + \varepsilon\tau$$
$$q' = q + \varepsilon\zeta.$$

Recall what we did to investigate the functional being an extremal: we had $\zeta \neq 0$ but $\tau = 0$. Then the functional became a function of ε, in other words $\Gamma' = \Gamma'(\varepsilon)$, and $\delta\Gamma = 0$ meant

$$\left[\frac{d\Gamma'}{d\varepsilon}\right]_0 = \int_a^b \left(\frac{\partial L}{\partial q}\zeta + \frac{\partial L}{\partial \dot{q}}\dot{\zeta}\right)dt = 0. \tag{5.24}$$

Defining $p \equiv \partial L/\partial \dot{q}$, an integration by parts turned this into

$$\int_a^b \left(\frac{\partial L}{\partial q} - \dot{p}\right)\zeta\,dt + (\zeta p)|_a^b = 0. \tag{5.25}$$

In making Γ an extremal, normally we set $\zeta(b) = \zeta(a) = 0$, which kills off the boundary terms and leaves the integral equaling zero, resulting in the Euler-Lagrange equation. But suppose we leave $(\zeta p)|_a^b$ in the game. Let's set eq. (5.25) aside momentarily; we'll return to it shortly.

Next, consider what we did to derive the invariance identity. Here we allowed τ to be nonzero. The definition of invariance,

$$\left[\frac{d(\Gamma' - \Gamma)}{d\varepsilon}\right]_0 = 0, \tag{5.26}$$

in other words, $\delta\Gamma' = \delta\Gamma$, gave us

$$\int_a^b \left(\frac{\partial L}{\partial q}\zeta + \frac{\partial L}{\partial \dot{q}}\frac{d}{d\varepsilon}\left[\frac{dq'}{dt'}\right]_0 + \frac{\partial L}{\partial t}\tau + L\dot{\tau}\right) dt = 0. \tag{5.27}$$

Now τ and $\dot{\tau}$ are in the game. Identifying the canonical momentum and working out the transformed velocity gave, to first order in ε, $dq'/dt' = \dot{q} + \varepsilon(\dot{\zeta} - \dot{q}\dot{\tau})$, and the invariance identity becomes

$$\int_a^b \left(\frac{\partial L}{\partial q}\zeta + p(\dot{\zeta} - \dot{q}\dot{\tau}) + \frac{\partial L}{\partial t}\tau + L\dot{\tau}\right) dt = 0. \tag{5.28}$$

Identifying the Hamiltonian $H = p\dot{q} - L$, this may be written

$$\int_a^b \left(\frac{\partial L}{\partial q}\zeta + p\dot{\zeta} + \frac{\partial L}{\partial t}\tau - H\dot{\tau}\right) dt = 0. \tag{5.29}$$

With integration by parts applied to the $p\zeta$ and $H\tau$ terms, this becomes

$$\int_a^b \left[\left(\frac{\partial L}{\partial q} - \dot{p}\right)\zeta + \left(\frac{\partial L}{\partial t} + \dot{H}\right)\tau\right] dt + (p\zeta - H\tau)|_a^b = 0. \tag{5.30}$$

This is the invariance identity in integrated form. When the Euler-Lagrange equations hold because the functional is also an extremal, we may invoke the p-dot version $\partial L/\partial q = \dot{p}$ and the H-dot version $\partial L/\partial t = -\dot{H}$, which kills off the integral and leaves the boundary terms,

$$(p\zeta - H\tau)|_a^b = 0, \tag{5.31}$$

or

$$[p\zeta - H\tau]|_b = [p\zeta - H\tau]|_a. \tag{5.32}$$

But a and b are arbitrary. Therefore,

$$p\zeta - H\tau = \text{const.} \tag{5.33}$$

This is Noether's (first) theorem.

Returning to eq. (5.25), we see why the derivation of the Euler-Lagrange equation imposed the boundary condition $\zeta(b) = \zeta(a) = 0$, for then the Euler-Lagrange equation made the functional extremal. Had $\zeta(b)$ and $\zeta(a)$

not been required to vanish at the boundary of the region of integration, and
the Euler-Lagrange equation still been imposed, this would have resulted in
$p(b)\zeta(b) = p(a)\zeta(a)$, in other words $p\zeta$ = const. If that result were carried
over into the invariance identity, it would follow that $H\tau$ and $p\zeta$ would
only be *separately* constant. That would have overlooked the distinguishing
feature of Noether's theorem, that $p\zeta - H\tau$ = const. even if $p\zeta$ and $H\tau$ are
not separately constant.

It is crucial to notice that the ζ used in the derivation of the Euler-
Lagrange equation is not necessarily the same ζ used in the derivation of
the invariance identity. Let's temporarily rename the Euler-Lagrange ζ as
something else, say η. To derive the Euler-Lagrange equation, we require
$[d\Gamma'/d\varepsilon]_0 = 0$ when $t' = t, q' = q + \varepsilon\eta$ and impose $\eta(b) = \eta(a) = 0$. To
derive the invariance identity we require $[d(\Gamma'-\Gamma)/d\varepsilon]_0 = 0$ when $t' = t+\varepsilon\tau$
and $q' = q + \varepsilon\zeta$, but no requirements are made that ζ or τ vanish at the
integral's boundary. Thus η and ζ do not have to be the same. That is
why the functional being extremal and the functional being invariant are
distinct issues.

To sum up, "extremal" and "invariance" are similar, but a fine distinction exists between them. One of them does not transform the independent
variable, the other does. One of them requires the dependent variable generators to vanish at the boundary of integration, the other does not. When
the functional is both extremal and invariant, Emmy Noether's elegant first
theorem results.

5.3   "Extremal" or "Stationary"?

We have been describing the procedure and payoff of making a functional an
"extremal." Some authors prefer to speak of the functional as being "stationary." Sometimes the term "invariance" gets thrown into the mix too, without
the invariance identity being mentioned explicitly. Terminology is important,
and when real or perceived ambiguity prevails, even a small term can become
a sticking point to new students of a subject, just as the small chock blocks
wedged against the tires of a parked airliner impede the powerful machine
from getting underway.

In his epic quantum mechanics book, Paul Dirac writes

> The principle of least action in classical mechanics says that the
> action function $[\int_a^b Ldt]$ remains stationary for small variations
> of the trajectory of the system which do not alter the endpoints.
> [Dirac (1947) 128]

In stating Hamilton's principle for mechanics, Dirac says "least" action,
but describes the action functional as "stationary." In a recent textbook

on general relativity, the authors also use the word "stationary" [Hobson et al. (2006) 79] in the context of deriving the Euler-Lagrange equations:

> By demanding that the action
>
> $$S = \int_{t_i}^{t_f} L \, dt \qquad (5.34)$$
>
> is stationary with respect to small variations in the functions $x^a(t)$, the equations of motion of the system are then found as the Euler-Lagrange equations
>
> $$\frac{d}{dt}\left(\frac{\partial L}{\partial \dot{x}^a}\right) - \frac{\partial L}{\partial x^a} = 0. \qquad (5.35)$$

In the venerable classical mechanics textbook by Herbert Goldstein we find these words:

> We can now state the integral Hamilton's Principle for conservative systems. ... The motion of the system from time t_1 to time t_2 is such that the line integral
>
> $$I = \int_{t_1}^{t_2} L \, dt \qquad (5.36)$$
>
> where $L = T - V$,[3] is an extremum for the path of motion.

> That is, out of all possible paths by which the system point could travel from its position at time t_1 to its position at time t_2, it will actually travel along that path for which the integral [just cited] is an extremum, whether a minimum or a maximum. [Goldstein (1965) 30]

In contrast, in multiple editions of the classical mechanics book originally written by Jerry Marion, Hamilton's principle is described with minimization made explicit:

> In two papers published in 1834 and 1835, Hamilton announced the dynamical principle on which it is possible to base all of mechanics and, indeed, most of classical physics. Hamilton's Principle may be stated as follows:

> "Of all the possible paths along which a dynamical system may move from one point to another within a specified time interval (consistent with any constraints), the actual path followed is that which minimizes the time integral of the difference between the kinetic and potential energies."[4]

---

[3]Goldstein uses T for kinetic energy and V for potential energy.
[4][Marion (1970) 198; repeated in Marion & Thornton (2004) 230].

But Marion goes on to explain that "the variational statement of the principle requires only that the integral of $T - U$ be an *extremum*, not necessarily a *minimum*. But in almost all applications of importance in dynamics the minimum condition obtains" [Marion (1970) 198].[5]

The classic text on optics by Francis Jenkins and Harvey White[6] introduces Fermat's principle with more than the usual caveats:

> A correct and complete statement of this principle is seldom found in textbooks, because the tendency is to cite it in Fermat's original form, which was incomplete. Using the concept of optical path,[7] the principle should read

> "The path taken by a light ray in going from one point to another through any set of media is such as to render its optical path equal, in the first approximation, to other paths closely adjacent to the actual one."

Jenkins and White explain:

> The "other paths" must be possible ones in the sense that they may only undergo deviations where there are reflecting or refracting surfaces. Now Fermat's principle will hold for a ray whose optical path is a *minimum* with respect to adjacent hypothetical paths. Fermat himself stated that the time required by the light to traverse the path is a minimum, and the optical path is a measure of this time. But there are plenty of cases in which the optical path is a *maximum*, or else neither a maximum nor a minimum but merely *stationary* (at a point of inflection) at the position of the true ray. ... The essential condition involved in Fermat's principle is that any slight variation of the actual path ... must at most cause only a second-order variation in the optical path. The term "stationary" expresses this condition, and also includes the possibilities of having the optical path a maximum or a minimum.

Another interesting comment that throws "invariance" into the mix comes from the general relativity book of Hobson et al. mentioned above. In the context of the calculus of variations applied to fields, the authors write,

---

[5]The fifth edition of Marion & Thornton (2004) p. 321 gives the identical statement of Hamilton's principle, but slightly rewords the last sentence to read "But in almost all important applications in dynamics, the minimum condition occurs."

[6][Jenkins & White (1950) 6–7].

[7]As usually stated, Fermat's principle is cited as $\int_{t_a}^{t_b} dt = \min.$ Since the distance ds traveled is related to the speed v, refractive index n and time dt according to $ds/dt = v = c/n$, the principle can also be stated $\int_a^b n\,ds = \min.$, where $n\,ds$ is the "optical path."

We now derive the form of the field equations for (some subset of) the fields Φ^a by demanding that the action is *stationary*, or invariant, under small variations in (the same subset of) the fields of the form[8]

$$\Phi^a(x) \to \Phi'^a(x) = \Phi^a(x) + \delta\Phi^a(x). \tag{5.37}$$

"Stationary" is usually understood to mean that a first-order variation in the path makes less than a first-order change to the functional. This sounds like our definition of invariance, where under transformations $t' = t+\varepsilon\tau+\ldots$ and $x' = x+\varepsilon\zeta+\ldots$, invariance obtains whenever $\Gamma'-\Gamma \sim \varepsilon^s$ with $s > 1$. Since $\varepsilon << 1$, any ε^s with $s > 1$ is smaller than ε itself. For instance, Jenkins and White say, in effect, that being stationary means $\Gamma' - \Gamma \sim \varepsilon^2$.

It appears that a functional being "stationary" can be interpreted "extremal" or "invariant." This is why I have refrained from using the word "stationary" in describing the conditions that lead to the Euler-Lagrange equations. However, it may be safe to say that, in deriving the Euler-Lagrange equation, "stationary" means

$$\Gamma' - \Gamma \sim \varepsilon^s, \quad s > 1, \tag{5.38}$$

which can be written more fully as

$$\int_a^b L\left(t, q'(t), \frac{dq'(t)}{dt}\right) dt - \int_a^b L\left(t, q(t), \frac{dq(t)}{dt}\right) dt \sim \varepsilon^s, \tag{5.39}$$
$$s > 1.$$

In this context the dependent functions are varied but the independent variables are not changed in any way, so that $q' = q+\varepsilon\zeta+\ldots$ but $t' = t$. To make the functional an extremal is to carry this definition of "stationary" into the derivative,

$$\left[\frac{d\Gamma}{d\varepsilon}\right]_0 = 0. \tag{5.40}$$

The derivative of $d\Gamma(\varepsilon)/d\varepsilon$ could not be zero at $\varepsilon = 0$ unless the difference between the varied and original Γ were proportional to a power of ε greater than 1. Otherwise, in the limit as $\varepsilon \to 0$, the derivative would blow up. In other words, the functional has to be stationary in order to be an extremal.

Setting a derivative equal to zero does not, by itself, tell us if the function is a maximum, minimum, or an inflection point, and the Euler-Lagrange equation works out to be the same each case. When a functional

---

[8]The authors continue, "It is important to note that we are *not* performing any coordinate transformation here." In other words, they are considering a gauge transformation.

is said to be a minimum and not a maximum, or vice versa, it is for physical reasons, not mathematical ones.

Furthermore, the definition of "invariance" borrows the definition of "stationary," but allows the independent variable to also be transformed according to $t' = t + \varepsilon\tau + \ldots$, so that

$$\int_{a'}^{b'} L\left(t', q'(t'), \frac{dq'(t')}{dt'}\right) dt' - \int_{a}^{b} L\left(t, q(t), \frac{dq(t)}{dt}\right) dt \sim \varepsilon^{s}, \quad s > 1.$$

$$(5.41)$$

"Stationary" is a vernable word, widely used in the literature. It serves as a root concept from which grow the distinct meanings of "extremal" and "invariant" as we use them in this book.

5.4 An Inverse Problem: Finding Invariances

In section 5.1 we inquired whether a given functional was invariant under a given transformation, and if so the Noether conservation law followed. These relationships can be inverted in at least two ways: (1) given a transformation, seek a Lagrangian whose functional is invariant; or (2) given a Lagrangian, seek transformations that lead to invariance. To proceed strategically,[9] one imposes the invariance identity and solves for L in case (1), or the generators in case (2). In this section we pursue case (2).

The invariance identity depends on the coordinate velocities, but in the context of finding generators of transformations that lead to invariances, the identity must hold whatever the $\dot{q}^{\mu}$ happen to be. Because it forms a polynomial in the velocity components, the identity is enforced by setting to zero each coefficient of the various powers of velocity.

In so doing, hidden velocities must be made to show themselves explicitly. The Lagrangian typically depends on $\dot{q}^{\mu}$ explicitly; furthermore, the momentum

$$p_{\mu} = \frac{\partial L}{\partial \dot{q}^{\mu}} \tag{5.42}$$

contains $\dot{q}^{\mu}$, as does the Hamiltonian when written in terms of the $\dot{q}^{\mu}$ instead of the p_{μ},

$$H = \dot{q}^{\mu} \frac{\partial L}{\partial \dot{q}^{\mu}} - L(t, q^{\mu}, \dot{q}^{\mu}). \tag{5.43}$$

---

[9]Of course, one can always guess the solution. A theorem in differential equations says a solution that satisfies the differential and initial (or boundary) conditions can be expected to be unique; see for example, Ritger & Rose (1968), 17.

Furthermore, τ and ζ may be functions of the independent and dependent variables,

$$\tau = \tau(t, q^{\mu}),$$
$$\zeta^{\mu} = \zeta^{\mu}(t, q^{\nu}). \tag{5.44}$$

Therefore, the derivatives of the generators that appear in the invariance identity produce velocities because of the chain rule:

$$\dot{\tau} = \frac{\partial \tau}{\partial t} + \frac{\partial \tau}{\partial q^{\mu}} \dot{q}^{\mu} \tag{5.45}$$

and

$$\dot{\zeta}^{\mu} = \frac{\partial \zeta^{\mu}}{\partial t} + \frac{\partial \zeta^{\mu}}{\partial q^{\nu}} \dot{q}^{\nu}. \tag{5.46}$$

Once all the terms in the invariance identity are written in terms of velocities, their various powers gathered up to form a polynomial, and their coefficients set separately equal to zero, the set of equations that results, the so-called Killing equations, may be solved for the generators.

Example: Allow me illustrate how this works with a Newtonian mechanics example, for a particle moving in one spatial dimension, whose doings are described by the Lagrangian $L = \frac{1}{2}m\dot{x}^2 - U(t, x)$. The canonical momentum is $p = m\dot{x}$, so that H, technically $p^2/2m + U$ in this case, numerically equals $\frac{1}{2}m\dot{x}^2 + U$. Since we require invariance, the invariance identity, eq. (4.42), is invoked (with $F = 0$), which becomes

$$-\frac{\partial U}{\partial x}\zeta + m\dot{x}\left[\frac{\partial \zeta}{\partial t} + \frac{\partial \zeta}{\partial x}\dot{x}\right]$$
$$-\frac{\partial U}{\partial t}\tau - \left(\frac{1}{2}m\dot{x}^2 + U\right)\left[\frac{\partial \tau}{\partial t} + \frac{\partial \tau}{\partial x}\dot{x}\right] = 0. \tag{5.47}$$

Regroup the terms as a polynomial in $\dot{x}$:

$$\left(-\frac{1}{2}m\frac{\partial \tau}{\partial x}\right)\dot{x}^3 + \left(-\frac{1}{2}m\frac{\partial \tau}{\partial t} + m\frac{\partial \zeta}{\partial x}\right)\dot{x}^2 + \left(-U\frac{\partial \tau}{\partial x} + m\frac{\partial \zeta}{\partial t}\right)\dot{x}^1$$
$$+ \left(-\tau\frac{\partial U}{\partial t} - U\frac{\partial \tau}{\partial t} - \zeta\frac{\partial U}{\partial x}\right)\dot{x}^0 = 0. \tag{5.48}$$

This identity must hold whatever $\dot{x}$ happens to be. That will be guaranteed if the coefficients of each power of $\dot{x}$ separately vanish. This procedure gives the Killing equations that I will denote Kn for each coefficient of $\dot{x}^n$:

$$\frac{\partial \tau}{\partial x} = 0 \quad (K3)$$

$$-\frac{1}{2}\frac{\partial \tau}{\partial t} + \frac{\partial \zeta}{\partial x} = 0 \quad (K2)$$

$$-U\frac{\partial \tau}{\partial x} + m\frac{\partial \zeta}{\partial t} = 0 \quad (K1)$$

$$\frac{\partial (U\tau)}{\partial t} + \zeta\frac{\partial U}{\partial x} = 0 \quad (K0).$$

(5.49)

If we can solve this system of equations for τ and ζ, we will have generators of transformations that leave the functional invariant.[10]

Equation $(K3)$ says that τ does not depend explicitly on x, so that $\tau = \tau(t)$. Using this result in $(K1)$ implies that $\zeta = \zeta(x)$. Therefore, in $(K2)$ the partial derivatives may be replaced with total derivatives:

$$\frac{d\tau}{dt} = 2\frac{d\zeta}{dx} = C, \tag{5.50}$$

where C is a separation constant. These integrate to

$$\tau = Ct + t_0 \tag{5.51}$$

and

$$\zeta = \frac{1}{2}Cx + x_0 \tag{5.52}$$

where t_0 and x_0 are integration constants.

But we are not finished. To ensure invariance, $(K0)$ must also be satisfied. It puts a constraint on the potential energy that must hold for invariance and the Noether conservation law to go through. With this constraint in mind, and the generators in hand, we now let the functional be an extremal, and appeal to Noether's theorem, which yields the conservation law

$$p\left(\frac{1}{2}Cx + x_0\right) - H(Ct + t_0) = \text{const.} \tag{5.53}$$

This is perhaps the world's most generic conservation law for systems described by the Lagrangian $L = \frac{1}{2}m\dot{x}^2 - U(t,x)$. It features a super-position of p and H. In contrast, they appear *independently* as possibly

---

[10]Since a functional may be invariant under more than one transformation, the solutions of the Killing equations are not unique, as we shall see.

conserved quantities in the p-dot and H-dot versions of the Euler-Lagrange equations. When the *Lagrangian* is invariant under a transformation, the Euler-Lagrange equation gives the conservation of p or H separately. When the *functional* is invariant under a transformation, the conservation law generalizes those coming from the Euler-Lagrange equations by themselves. The invariance of the functional will pick up conservation laws, thanks to Noether's theorem, that the invariance of the Lagrangian overlooks. In the example before us, let's consider special cases:

1. Suppose $C = 0$ and $x_0 = 0$ but $t_0 \neq 0$, a time translation. Then $\tau = t_0$ and $\zeta = 0$, and for $(K0)$ to be satisfied requires $\partial U/\partial t = 0$; therefore, invariance obtains only if the potential energy contains no explicit time dependence, leaving the functional invariant under a time translation. In that case the conserved quantity is the Hamiltonian, $H = $ const., which with the given Lagrangian is the particle's mechanical energy, $H = p^2/2m + U = E$. We saw this result also emerge in the context of the Euler-Lagrange equation $\partial L/\partial t = -\dot{H}$ when the Lagrangian was invariant under a time translation.

2. If $C = 0, t_0 = 0$, but $x_0 \neq 0$, then the transformation is a spatial translation, so that $\tau = 0, \zeta = x_0$. The constraint $K(0)$ requires $\partial U/\partial x = 0$ (no force), which if satisfied yields the conservation of momentum, $p = $ const., signifying the homogeniety of space, as also seen in $\partial L/\partial x = \dot{p}$.

3. Suppose $t_0 = 0, x_0 = 0$, but $C \neq 0$. This describes a rescaling of the time and position variables,[11]

$$t' = t(1 + \varepsilon C)$$
$$x' = x(1 + \frac{1}{2}\varepsilon C), \tag{5.54}$$

so that $\tau = Ct$ and $\zeta = \frac{1}{2}Cx$. To have invariance, the potential energy function must satisfy $(K0)$, which says

$$\frac{\partial(tU)}{\partial t} + \frac{1}{2}x\frac{\partial U}{\partial x} = 0. \tag{5.55}$$

Should this be satisfied (or imposed), the Noether conservation law becomes

$$\frac{1}{2}px - Ht = \text{const.} \tag{5.56}$$

This conservation law may be unfamiliar, for it does not readily emerge from either of the two versions of the Euler-Lagrange equations. It requires the functional to be invariant, not just the Lagrangian.

Eq. (5.56) has an application in the context of "adiabatic invariance," previously known through approaches that predate Noether's theorem. We

---

[11]Also called a conformal transformation; see Logan (1974).

now show how adiabatic invariance follows from earlier methods, and then from Noether's theorem, illustrating again the power of Emmy Noether's elegant theorem as a unifying principle in physics.

5.5 Adiabatic Invariance in Noether's Theorem

We therefore conclude that in an adiabatic change of the Hamiltonian, the action J remains constant. It is, in fact, a well-known theorem of classical mechanics that J does indeed remain constant in an adiabatic process.
— David Bohm,*Quantum Theory*, 1951, 500

"Adiabatic invariance" refers to situations where the system parameters are slowly changed, such that the product of two quantities is approximately conserved, even though the two quantities themselves are not separately conserved. Perhaps the most cited example arises in the simple pendulum whose length slowly changes. Imagine a simple pendulum consisting of a point mass m suspended from a string of length ℓ, swinging to and fro with period T, the restoring force provided by a gravitational field g. As the pendulum swings, suppose the string gradually stretches, or someone slowly winches the string up, changing its length. The pendulum's energy E does not stay constant, because of the work performed on it. The period T does not stay constant either, because Newton's second law predicts that the period depends on the length according to

$$T = 2\pi\sqrt{\frac{\ell}{g}}. \tag{5.57}$$

For a pendulum of varying length, neither E nor T are separately conserved. However, if the length changes very slowly, so that $|\Delta\ell|/\ell << 1$ in each cycle, then over a few cycles the product ET remains approximately constant. Other simple examples featuring adiabatic invariance for ET are demonstrated in a mass on a spring whose spring constant slowly varies, and a particle making elastic collisions with the slowly moving walls of a confining box.

To illustrate the robustness of adiabatic invariance, let us first see, without Noether's theorem, how ET is approximately constant for the simple pendulum under the conditions described above. Denote the pendulum's angular frequency as $\omega = 2\pi/T$ and let its amplitude be A. It energy is $E = \frac{1}{2}m\omega^2 A^2$ so that

$$ET = \frac{2\pi^2 mA^2}{T}. \tag{5.58}$$

Take the derivative of ET with respect to time:

$$\frac{d(ET)}{dt} = 2\pi^2 m \left(\frac{2A\dot{A}}{T} - \frac{A^2 \dot{T}}{T^2} \right)$$

$$= ET \left(\frac{2\dot{A}}{A} - \frac{\dot{\ell}}{2\ell} \right).$$

Approximate the differentials with small but finite differences,

$$\frac{\Delta(ET)}{ET} \approx \frac{2\Delta A}{A} - \frac{\Delta \ell}{2\ell}. \tag{5.59}$$

If $|\Delta A|/A << 1$ and $|\Delta \ell|/\ell << 1$ over one cycle, then the fractional change in ET is also very small over a few cycles and therefore $ET \approx$ const., even though E and T are not separately constant.

The concept of adiabatic invariance carries a rich historical legacy. Around 1870, Ludwig Boltzmann and Rudolf Clausius appealed to it in their valiant but unsuccessful attempts to reduce the second law of thermodynamics to Newtonian mechanics. Boltzmann showed that, for a periodic system, $2\langle K \rangle T$ is adiabatically invariant, where $\langle K \rangle$ denotes the time average of the system's kinetic energy. In the early days of atomic physics Paul Ehrenfest showed that $2\langle K \rangle T$ may be written as the volume of phase space (see chapter 9) enclosed by one circuit of a periodic motion, the abbreviated action integrated over one period. In particular, for a system mapped with generalized coordinates q^μ and their canonical momenta p_μ,

$$2\langle K \rangle T = \oint p_\mu dq^\mu. \tag{5.60}$$

The adiabatic invariance of the abbreviated action formed the foundation of what is now called the "old quantum theory" (c. 1913–1926), which preceded the wave mechanics of Louis de Broglie, Erwin Schrödinger, Wolfgang Heisenberg, and their colleagues. In old quantum theory, periodic mechanical systems are quantized by invoking the Bohr-Sommerfeld-Wilson quantization postulate, illustrated here for one-dimensional motion,

$$\oint p\,dx = nh, \tag{5.61}$$

where n denotes an integer and h Planck's constant. It was thought necessary to quantize an adiabatic invariant so small changes in a system's parameters would not result too readily in quantum jumps [Bohm (1951) 500]. The quantization suggested by eq. (5.61) is implemented by writing p in terms of the energy E,

$$p = \pm\sqrt{2m}\sqrt{E - U(x)}, \tag{5.62}$$

evaluating the integral over one period, then solving for E to obtain the energy spectrum in terms of the quanta nh.

Before bringing adiabatic invariance into the orbit of Noether's theorem, let's see how the generators derived above from the Killing equations, $\tau = t$ and $\zeta = \frac{1}{2}x$,[12] lead to $ET \approx$ const. Noting that the period is a time, and the amplitude is the maximum value of x, we may write from the transformation equations $T' = T(1 + \varepsilon)$ and $A' = A(1 + \frac{1}{2}\varepsilon)$, so that, from eq. (5.58),

$$(ET)' = \frac{2\pi^2 m A^2}{T} \frac{(1 + \frac{1}{2}\varepsilon)^2}{1 + \varepsilon}. \tag{5.63}$$

To lowest order in ε this gives $(ET)' \approx ET(1 - \varepsilon^2)$, which satisfies the formal definition of invariance. We now investigate how Noether's theorem predicts adiabatic invariance [Starkey & Neuenschwander (1993)].

We have already shown that invariances possible with the Lagrangian $\frac{1}{2}m\dot{x}^2 - U$ include a scale invariance that manifests itself in the Noether conservation law of eq. (5.56),

$$\frac{1}{2}px - Ht = \text{const.} \tag{5.64}$$

Consider such a system undergoing periodic motion. Average eq. (5.64) over one cycle:

$$\frac{1}{2}\oint px\frac{dt}{T} - \oint Ht\frac{dt}{T} = \text{const.} \tag{5.65}$$

Notice that, for this Lagrangian, $p = mdx/dt$, so that the first integral in eq. (5.65) may be written

$$\frac{m}{2T}\oint \frac{dx}{dt}x\,dt = \frac{m}{2T}\oint xdx, \tag{5.66}$$

which integrates to $\frac{1}{4}mx^2/T$, evaluated over one period. If the system's parameters change very little during one cycle, then $x(t + T) \approx x(t)$ and this integral essentially vanishes.

Now turn to the second integral in eq. (5.65). If the system changes slowly over one cycle, then during that cycle H may be approximated by its average value $\langle H \rangle$, so that

$$-\oint Ht\frac{dt}{T} \approx -\frac{\langle H \rangle}{T}\oint tdt$$
$$= -\tfrac{1}{2}\langle H \rangle T. \tag{5.67}$$

---

[12]Notice the rescaling constant C cancels out of the Noether conservation law.

Noether's theorem has given the conservation law

$$\langle H \rangle T \approx \text{const.} \tag{5.68}$$

Let us push this example a little farther. Recall the Legendre transformation that relates the Hamiltonian to the Lagrangian:

$$H = p\dot{x} - L. \tag{5.69}$$

Multiply this by dt and integrate over one cycle of periodic motion:

$$\oint H \, dt = \oint p\dot{x}dt - \oint L dt$$
$$= \oint p dx - \oint L dt. \tag{5.70}$$

By the definition of time average, the integral on the left side equals $\langle H \rangle T$, and we have just seen that $\langle H \rangle T$ is an adiabatic invariant. Since we began with the assumption that the functional $\int_0^T L dt$ is already invariant, we see that $\oint p dx$ is an adiabatic invariant also.

Questions for Reflection and Discussion

Q5.a. Notice the similarity between the phase of a harmonic wave, $kx - \omega t$, and the Noether conserved quantity, $p\zeta - H\tau$. Is this similarity merely a coincidence, or might it suggest something deeper?

Q5.b. Are all conservation laws derivable from Noether's theorem? If you think so, try it with the Runge-Lenz vector $\mathbf{R}$, where in spherical coordinates $\mathbf{R} = [(\mathbf{r} \times \mathbf{p}) \times \mathbf{p}] + k\hat{\mathbf{r}}$, which is conserved in orbits controlled by the potential energy $U = -k/r$, and $\mathbf{p} = m\mathbf{v}$.

Q5.c. Does *every* symmetry correspond to a conservation law?

Q5.d. Some authors call conserved quantities "integrals of motion" or "first integrals." How does finding a conservation law amount to integrating a differential equation?

Q5.e. Speculate on where the term "adiabatic" as used in "adiabatic invariance" may have come from.

Q5.f. If nonconservation means "broken symmetry," what symmetry is broken in the second law of thermodynamics that leads to nonconservation of entropy? Given that thermodynamic adiabatic processes conserve entropy, why might Clausius and Boltzmann have looked to adiabatic invariance in trying to reduce the second law of thermodynamics to mechanics?

Q5.g. To illustrate another version of adiabatic invariance, if you're up to a little side trip into quantum mechanics let's consider the so-called Berry phase. Consider a system described by a Hamiltonian H. The time-independent Schrödinger equation reads $H\psi_n = E_n\psi_n$, with E_n the eigenvalue of H corresponding to the eigenstate ψ_n. Through some process, interaction, or transformation, let the Hamiltonian change slowly from H to H' (note that the system's energy is not conserved). We want to know, for the same state n before and after the change (e.g., old ground state and new ground state), how the new ψ'_n compares to the old ψ_n. The adiabatic theorem of quantum mechanics [Griffiths (2005) ch. 10] shows that

$$\psi'_n = e^{i(\gamma_n - \theta_n)}\psi_n, \tag{5.71}$$

where

$$\theta_n(t) = -\frac{i}{\hbar} \int_0^t E_n(t') \, dt' \tag{5.72}$$

denotes the so-called dynamic phase,[13] and

$$\gamma_n(t) = i \oint \langle \psi_n | \dot{\psi}_n \rangle \, dt' \tag{5.73}$$

denotes the geometric phase, with $\dot{\psi}_n \equiv d\psi_n/dt'$.

a. Comment on the similarities and differences between the product of the dynamic and geometric phases, $e^{i(\gamma_n - \theta_n)}$ on the one hand, and, on the other hand, the general phase that appears in the wave function corresponding to the motion of a free particle, $\psi(t, x) = e^{i(px - Et)/\hbar}\psi(0,0)$.
b. When the wave function depends on N parameters $\{R_1, R_2, \ldots, R_N\} \equiv \mathbf{R}$, the quantum textbooks show how the geometric phase shift may be written in terms of the "Berry phase" [M. V. Berry (1984); see Griffiths (2005) 376–391],

$$\gamma_n(t) = i \oint \mathbf{A}_n \cdot d\mathbf{R}, \tag{5.74}$$

where $\mathbf{A}_n(t) \equiv \langle \psi_n | \nabla_\mathbf{R} \psi_n \rangle$. Show that the geometric phase may also be written

$$\gamma_n(t) = \int (\mathbf{B}_n \cdot \hat{\mathbf{n}}) \, da, \tag{5.75}$$

where $\hat{\mathbf{n}}$ denotes a normal vector. Identify $\mathbf{B}_n$ in terms of $\mathbf{A}_n$. The Berry phase equals the flux of $\mathbf{B}_n$ through an area in the N-dimensional parameter space.

---

[13]Here the t' is a dummy variable of integration, not a transformed time coordinate.

c. Now do some library or internet research (if necessary) to make yourself familiar with the Aharonov-Bohm effect. It describes something weird that happens when a beam of charged particles passes by a current-carrying solenoid. Even though $\mathbf{B} \approx \mathbf{0}$ outside the solenoid, the particles moving by it somehow "know" about the magnetic field inside the solenoid.

d. Provide an argument that shows how the Aharanov-Bohm effect is a Berry phase phenomenon, for the motion of a charged particle moving in the vicinity of a current-carrying solenoid.

Q.5.h. In example 3, about the damped oscillator with the Lagrangian

$$L = \left[\frac{1}{2}m\dot{x}^2 - \frac{1}{2}kx^2\right]e^{bt/m}, \tag{5.76}$$

it was mentioned that "in this example the Hamiltonian is not merely the mechanical energy $K + U$, but it is still the energy of *something*, $H = (K + U)e^{bt/m} = Ee^{bt/m}$." What might that "something" be for which this is the energy?

Look at it this way: Suppose the particle is released from $x = A$ at $t = 0$. The energy starts out at $E(0) = \frac{1}{2}kA^2 \equiv U_0$. When the particle completes one oscillation at time $T > 0$, its energy is $E(T) < U_0$. Where did the lost energy go? Here the particle interacts with the rest of the world; with dissipative interactions the particle's initial energy does not remain the property of the particle. We encounter analogous situations in the context of the interactions of matter with the electromagnetic and graviational fields in chapters 7 and 8 (see also ex. 5.13).

Exercises

5.1. Consider our old friend, the damped oscillator, with the Lagrangian of eq. (5.76). Requring invariance under the infinitesimal transformation $t' = t + \varepsilon\tau$ and $x' = x + \varepsilon\zeta$, set up the Killing equations, and derive the generators $\tau = 1$ and $\zeta = -bx/2m$ (they were given in example 3 of this chapter; here you derive them).

5.2. Consider the functional with Lagrangian $L = L(t, \dot{x}) = t \cot \dot{x}^2$.
a. Find the canonical momentum and the Hamiltonian.
b. Impose the invariance identity and from the Killing equations derive $\zeta = -\frac{1}{2}Cx + x_0$ (where C and x_0 are constants) and $d\tau/dt = \tau/t - C$.
c. Show that $\tau = At$ and $\zeta = x_0$ if $C = 0$, where $A = $ const.
d. Confirm the result of part (c) by substituting it back into the invariance identity.
e. Confirm the result of part (c) from the definition of invariance, which considers $L'(dt'/dt) - L$. Must divergence-invariance be invoked?

f. When the functional is an extremal, what Noether conservation law results with the generators of part (c)?

5.3. Emden's equation (Robert Emden, 1907) is encountered in modeling the gravitational collapse of a dust of particles, when thermodynamics is taken into account[Logan (1977)]. After some dimensional constants are absorbed to make dimensionless parameters, one version of Emden's equation takes the form

$$\ddot{x} + \frac{2\dot{x}}{t} + x^5 = 0. \tag{5.77}$$

a. Find a Lagrangian that gives Emden's equation as an Euler-Lagrange equation.
b. Demand invariance under the transformation $t' = t + \varepsilon\tau$, $x' = x + \varepsilon\zeta$, set up the Killing equations, and find the generators. Construct the Noether conservation law.

5.4. Newton's second law, when transformed from an inertial reference frame to one that rotates with constant angular velocity $\boldsymbol{\omega}$ relative to the inertial frame, becomes

$$-\boldsymbol{\nabla}U - 2m(\boldsymbol{\omega} \times \mathbf{v}) - m\boldsymbol{\omega} \times (\boldsymbol{\omega} \times \mathbf{r}) = m\frac{d^2\mathbf{r}}{dt^2}, \tag{5.78}$$

where U is the potential energy function ($-\boldsymbol{\nabla}U$ is the force "felt" by the particle in the inertial frame), $\mathbf{r}$ the location and $\mathbf{v}$ the particle's velocity relative to the rotating frame's coordinate system.
a. Show that the Lagrangian

$$L = \frac{1}{2}m[\mathbf{v} + (\boldsymbol{\omega} \times \mathbf{r})]^2 - U(t, \mathbf{r}) \tag{5.79}$$

gives the above equation of motion.
b. Impose the invariance identity, write the Killing equations, and solve them for a set of generators that yield invariance. If you cannot find the most general set of generators, make assumptions to find some set of generators. For example, does a solution result if you assume that ζ does not depend on time?[14]
c. From the generators found in part (b), find the Noether conservation law.

5.5. Consider a generic Lagrangian written in generalized coordinates, for a Newtonian particle:

$$L(t, q^\mu, \dot{q}^\mu) = \frac{1}{2}mg_{\mu\nu}\dot{q}^\mu\dot{q}^\nu - U(t, q^\nu) \tag{5.80}$$

---

[14]See Dallen and Neuenschwander (2011) for further suggestions.

where the $g_{\mu\nu}$ are the components of the metric tensor (see appendix A).
a. Show that the Killing equations are

$$\frac{\partial \tau}{\partial q^\mu} = 0,$$

$$-\frac{1}{2}g_{\mu\nu}\frac{\partial \tau}{\partial t} + g_{\mu\rho}\frac{\partial \zeta^\rho}{\partial q^\nu} + \frac{1}{2}\frac{\partial g_{\mu\nu}}{\partial x^\rho}\zeta^\rho = 0,$$

$$U\frac{\partial \tau}{\partial q^\mu} + mg_{\mu\nu}\frac{\partial \zeta^\nu}{\partial t} = 0,$$

$$\frac{\partial (U\tau)}{\partial t} + \frac{\partial U}{\partial q^\mu}\zeta^\mu = 0.$$

(5.81)

b. Apply these to a particle moving in a central potential mapped by spherical coordinates (r, θ, φ).
c. Show how the conservation of energy and angular momentum emerge as special cases.

5.6. Show how the conservation of angular momentum follows from rotational invariance for a Newtonian particle, using rectangular coordinates and a rigid body rotating about the z-axis.

5.7. Show that if one differentiates with respect to time the Noether conservation law $p_\mu\zeta^\mu - H\tau = $ const., in order to recover

$$\frac{d}{dt}[p_\mu\zeta^\mu - H\tau] = 0,$$

(5.82)

eq. ($K0$) must be invoked.

5.8. Show that $\oint p\,dx = 2\langle K\rangle T$ for a particle of mass m moving in a one-dimensional box of width w, reflecting elastically off the opposite walls. Demonstrate the adiabatic invariance of these quantities when the walls move slowly apart.

5.9. Derive the quantized energies of the one-dimensional simple harmonic oscillator, for which $U = \frac{1}{2}m\omega^2 x^2$, using old quantum theory, the Bohr-Sommerfield-Wilson quantization rule. Compare your result to that predicted by the Schrödinger equation applied to this system, $E_n = (n + \frac{1}{2})\hbar\omega$, where $n = 0, 1, 2, 3, \ldots$

5.10. Recall example 5 of section 5.1, an illustration of divergence invariance with projectile motion, where $\tau = 0, \zeta^x = -t, \zeta^y = 0,$ and

$F = -mx + \text{const.}$, that featured in the conservation law

$$p_\mu \zeta^\mu - H\tau - F = \text{const.} \tag{5.83}$$

Show the equivalence of this result to the conservation of the x-component of linear momentum.

5.11. Solve eq. $(K0)$ for U (within integration constants) in these cases:
a. $U = U(x)$.
b. $U = U(t)$.
c. $U = U(t, x) = g(t) + f(x)$.
d. $U = U(t, x) = g(t)f(x)$.
e. For the simple harmonic oscillator with $U = \frac{1}{2}kx^2$, what condition on k as a function of t must hold in order for eq. $(K0)$ to be satisfied?

5.12 Consider the Lagrangian of a particle of mass m and electric charge q moving through a region with electromagnetic potentials V and $\mathbf{A}$:

$$L = -mc^2\sqrt{1 - \frac{v^2}{c^2}} + q\mathbf{v} \cdot \mathbf{A} - qV. \tag{5.84}$$

a. Find the canonical momenta and the Hamiltonian.
b. Under the infinitesimal transformation $t' = t + \varepsilon\tau + \cdots$ and $x'^\mu = x^\mu + \varepsilon\zeta^\mu + \cdots$, impose the invariance identity, write the Killing equations, and show that a solution to them is

$$\tau = Ct$$
$$\zeta^\mu = \frac{1}{2}Cx^\mu \tag{5.85}$$

where C is a separation constant. When the functional is also extremal, show that this transformation produces the Noether conservation law

$$\frac{1}{2}\mathbf{p} \cdot \mathbf{r} - Ht = \text{const.} \tag{5.86}$$

c. Consider a spatially uniform magnetic field $\mathbf{B} = B(t)\hat{\mathbf{z}}$ in a cylindrical coordinate system (ρ, θ, z). Suppose $B(t)$ varies slowly compared to the particle's period T for one revolution around the helical path, for which $2\pi/T = eB/m$. Show the following three quantities to be adiabatic invariants:[15]
d. The time-averaged rotational kinetic energy divided by the magnetic field,

$$\frac{\langle \frac{1}{2}m\rho^2\dot{\theta}^2 \rangle}{B} = \text{const.} \tag{5.87}$$

---

[15]These are the adiabatic invariants of plasma physics. See Jackson (1975) 588–593; Taylor & Neuenschwander (1996).

where the brackets denote time averaging;

e. The magnetic flux through one turn of the helix,

$$\pi \rho^2 B = \text{const.}, \tag{5.88}$$

and

f. The magnetic dipole moment μ of the current loop,

$$\mu = \frac{q}{T} \pi \rho^2 = \text{const.} \tag{5.89}$$

g. Show that for the Noether conservation law to hold, in other words for

$$\frac{d}{dt} \left[\frac{1}{2} \mathbf{p} \cdot \mathbf{r} - Ht \right] = 0, \tag{5.90}$$

the Killing equation ($K0$) must be invoked [Taylor et al. (1998)].

5.13. Return to the damped oscillator, whose Lagrangian is given by eq. (5.76). Write down the Euler-Lagrange equation, which is Newton's second law for this system. Multiply the result by dx and integrate to derive the work-energy theorem. Although the energy of the oscillator is not conserved, how can the conservation of energy be enlarged, beyond mechanical energy $K + U$, to include other forms of energy?

Part III

THE INVARIANCE
OF FIELDS

Chapter 6

Noether's Theorem and Fields

It was the year [1918] of the Noether theorem.
In November 1915, neither Hilbert nor Einstein was aware of this royal road to the conservation laws—Abraham Pais, *Subtle Is the Lord*, 1982, 274

6.1 Multiple-Integral Functionals

Electrodynamics, gravitation, and fluid mechanics are among the physics topics discussed in the language of *fields*—functions of space and time. Newtonian gravitational fields include the vector field **g** and its potential Φ, which are static functions of position. Fluid mechanics of laminar flow features pressure $\mathcal{P}$ and the velocity field **v** as functions of position and time. Electrodynamics has the electric field **E** and the magnetic field **B**, along with their potentials V and **A**. In general relativity, the metric tensor components $g_{\mu\nu}$ depend on spacetime coordinates. In quantum mechanics, the de Broglie wave Ψ is a function of position and time.

For the purposes of mathematical physics these fields are well-behaved functions of time and space coordinates (x^0, x^1, x^2, x^3) such as (t, x, y, z) or (t, r, θ, φ) or simply x^μ for short. Most fields change with space and time according to first- or second-order partial differential equations. For example, Maxwell's four first-order partial differential equations of electromagnetism can be combined into two second-order wave equations. Newtonian gravity has Poisson's equation; the Navier-Stokes equation is Newton's second law applied to fluids. In quantum theory we find the Schrödinger equation for the nonrelativistic de Broglie wave function, or its relativistic

111

extensions in the equations of Dirac (first order) and Klein-Gordon (second order). General relativity features Einstein's field equations.

These fields respect various conservation laws, including energy and momentum. Newtonian fluid mechanics also postulates local mass conservation; quantum mechanics depends crucially on the conservation of probability. Conservation laws for fields find expression in "equations of continuity," so we must become familiar with them.

The conservation laws for these fields can be derived by manipulating the equations of motion. But the equations of motion themselves, and the conservation laws, follow elegantly from field theory versions of functionals, through their versions of Euler-Lagrange equations, invariance identities, and Noether's theorem. Because fields depend on spacetime coordinates, the functionals are multiple integrals, and Noether's conservation laws are equations of continuity. In this chapter we motivate and develop these ideas.

The field theories considered here are so-called classical fields, which in this context means we will not delve into quantum field theory operators that create and annihilate quanta through so-called second quantization.[1] Quantum field theories with second quantization build on the standard Lagrangian formalism (see, e.g., Sakurai [1967], Roman [1968], Srednicki [2007], and Weinberg [1995]).

In this chapter we consider only scalar fields. Their quanta carry spin-0. Of course, the lessons of Noether's theorem apply to vector fields and other tensor fields of higher order, and to spinor fields. Same principles, different Lagrangians.

A prototypical example of a field theory that emerges within a familiar Newtonian context may be found in a guitar string. Let the undisturbed string define the x-axis, and let a transverse wave travel down the string, with the vibration plane-polarized in the xy plane (Figure 6.1).

The disturbance at location x on the string, at time t, will be described by a wave function $y(t, x)$. Suppose the string carries linear density μ and has been tightened to tension T. Apply the y-component of $\mathbf{F} = m\mathbf{a}$ to a small segment of the string of length Δs. If we assume that T and μ remain constant (which means we neglect any stretching of the string) and neglect

---

[1] "First quantization" is the replacement of the momentum vector $\mathbf{p}$ with the operator $(\hbar/i)\boldsymbol{\nabla}$ and the Hamiltonian H with $i\hbar(\partial/\partial t)$. Second quantization introduces operators that create and destroy elementary particles, the field quanta. The photon, the quantum of the electromagnetic field, offers the conceptual prototype. If $|0\rangle$ denotes the no-photon ground state, or "vacuum," then a "creation operator" or "raising operator" $a^{\dagger}$ is introduced (the second quantization), when operating on the vacuum state produces the one-photon state $|1\rangle$, $a^{\dagger}|0\rangle \sim |1\rangle$. Conversely, the "annihilation" or lowering operator a, lowers the number of quanta by one, $a|n\rangle \sim |n-1\rangle$ where, by definition, $a|0\rangle = 0$. For quanta that are bosons, the raising and lowering operators satisfy commutation relation; fermion operators respect anticommutation relations.

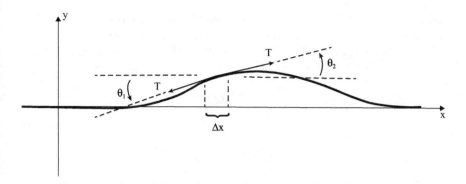

Figure 6.1: *The geometry of the disturbance on a guitar string.*

the string's weight, the y-component of Newton's second law applied to our little string segment gives

$$T(\sin\theta_2 - \sin\theta_1) = \mu\Delta s\frac{\partial^2 y}{\partial t^2}. \tag{6.1}$$

For small-amplitude oscillations, $\Delta s \approx \Delta x$ and $\sin\theta \approx \tan\theta = \partial y/\partial x$, and eq. (6.1) becomes

$$\frac{\Delta(\partial y/\partial x)}{\Delta x} = \frac{1}{v^2}\frac{\partial^2 y}{\partial t^2} \tag{6.2}$$

where

$$v^2 = \frac{T}{\mu}. \tag{6.3}$$

In the limit as $\Delta x \to 0$, the equation of motion becomes the *linear homogeneous wave equation*

$$\frac{\partial^2 y}{\partial x^2} = \frac{\partial^2 y}{\partial(vt)^2}. \tag{6.4}$$

Dimensional analysis shows v to be a speed—the wave's speed as it runs down the x-axis. Any function that solves eq. (6.4) can be put into a form where the x and t arguments fit together as $y(t,x) = y(x \pm vt)$.[2]

Our task is to find the Lagrangian whose Euler-Lagrange equation produces this wave equation. Imagine tracking the motion of an infinitesimal

---

[2]To see this formally, let $s = x \pm vt$ and take derivatives of y with respect to x and t, using the chain rule with s. Informally, notice that the denominators of the wave equation say that $\Delta x = \pm v\Delta t$ and $x \pm vt = \text{const}$.

bit of string, a task made easier to visualize by mentally placing a little dot of paint on the string, and follow the motion of that dot. As the wave travels down the x-axis, our paint dot oscillates vertically along the y-direction. Its contribution to the string's kinetic energy is straightforward:

$$\Delta K = \frac{1}{2}\Delta m \left(\frac{\partial y}{\partial t}\right)^2$$

$$= \frac{1}{2}\mu(\Delta x)v^2 \left(\frac{\partial y}{\partial(vt)}\right)^2. \tag{6.5}$$

To find the potential energy, think of our little increment of string as a small mass oscillating on a spring, where the y-component of the tension provides the restoring force. The potential energy of the mass Δm on this "spring" is

$$\Delta U = \frac{1}{2}k(\Delta y)^2. \tag{6.6}$$

By definition of a spring constant, $k = T_y/\Delta y = T\sin\theta/\Delta y$. Again we approximate $\sin\theta \approx \tan\theta = \partial y/\partial x$ for small displacements, so that

$$\Delta U = \frac{1}{2}T\frac{\partial y}{\partial x}(\Delta y). \tag{6.7}$$

Finally, since $\Delta y = (\partial y/\partial x)\Delta x$ and $T = \mu v^2$, the potential energy of our string increment becomes

$$\Delta U = \frac{1}{2}\mu v^2 \left(\frac{\partial y}{\partial x}\right)^2 \Delta x. \tag{6.8}$$

The Lagrangian ΔL for our little increment of string may now be written

$$\Delta L = \Delta K - \Delta U$$

$$= \frac{1}{2}\mu v^2 \left[\left(\frac{\partial y}{\partial(vt)}\right)^2 - \left(\frac{\partial y}{\partial x}\right)^2\right]\Delta x. \tag{6.9}$$

With the velocity absorbed into a redefinition of the t coordinate ($vt \to t$ defines a new t, with time measured in meters), ΔL may be rewritten

$$\Delta L = \mu v^2 \left[\frac{1}{2}\left(\frac{\partial y}{\partial t}\right)^2 - \frac{1}{2}\left(\frac{\partial y}{\partial x}\right)^2\right]\Delta x. \tag{6.10}$$

The Lagrangian for the entire string of length ℓ will be obtained by summing the Lagrangians for all the little string segments, because energy is additive and these Lagrangian terms are energies. Letting $\Delta x \to dx$, our sum becomes an integral over dx from $x = 0$ to $x = \ell$,

$$L = \mu v^2 \int_0^\ell \left[\frac{1}{2} \left(\frac{\partial y}{\partial t} \right)^2 - \frac{1}{2} \left(\frac{\partial y}{\partial t} \right)^2 \right] dx. \tag{6.11}$$

The Lagranian is an integral over x; therefore the functional, as the integral of the Lagrangian over t, becomes a double integral:

$$\Gamma = \int_a^b L \, dt$$
$$= \mu v^2 \int_a^b \int_0^\ell \left[\frac{1}{2} \left(\frac{\partial y}{\partial t} \right)^2 - \frac{1}{2} \left(\frac{\partial y}{\partial x} \right)^2 \right] dx \, dt. \tag{6.12}$$

This functional boasts two independent variables, t and x. Absorbing the μv^2 into a redefiniton of Γ, we have[3]

$$\Gamma = \int_a^b \int_0^\ell \mathcal{L} \, dx \, dt, \tag{6.13}$$

where the *Lagrangian denisty* $\mathcal{L}$ is given by

$$\mathcal{L} \equiv \frac{1}{2} \left(\frac{\partial y}{\partial t} \right)^2 - \frac{1}{2} \left(\frac{\partial y}{\partial x} \right)^2. \tag{6.14}$$

Our functional is a double integral of $\mathcal{L}$, integrated over domains of x and t. What Euler-Lagrange equation results when there are two (or more) independent variables? To that question we now turn.

6.2 Euler-Lagrange Equations for Fields

Consider a wave field φ that is a function of two independent variables t and x, $\varphi = \varphi(t, x)$. The functional Γ will be a double integral,

$$\Gamma = \int_{\mathcal{R}} \mathcal{L} \, dt \, dx \tag{6.15}$$

where the multiple integral is implicit, and the region $\mathcal{R}$ describes some domain $a \leq t \leq b$ and $A \leq x \leq B$. Let us develop the Euler-Lagrange equation that emerges when this functional is made an extremal. When summing over all coordinates, it's convenient to label the time and space coordinates as

---

[3]The factor of $\frac{1}{2}$ could also be absorbed into Γ, but it's traditional to leave it in; more about this later.

$t = x^0$ and $x = x^1$, a notation that easily generalizes to higher-dimensional spaces. Since the wave equation is second order in the partial derivatives, its Lagrangian density must depend on first partial derivatives of the field. Partial derivatives of φ with respect to time or space are denoted with a variety of notations:

$$\frac{\partial \varphi}{\partial t} = \frac{\partial \varphi}{\partial x^0} \equiv \partial_0 \varphi \equiv \varphi_0 \tag{6.16}$$

$$\frac{\partial \varphi}{\partial x} = \frac{\partial \varphi}{\partial x^1} \equiv \partial_1 \varphi \equiv \varphi_1 \tag{6.17}$$

and in the spacetime of special relativity, where $x_0 = x^0$ but $x_1 = -x^1$ (see appendix B), we also need

$$\frac{\partial \varphi}{\partial x_0} \equiv \partial^0 \varphi \equiv \varphi^0 \equiv \frac{\partial \varphi}{\partial t}, \tag{6.18}$$

$$\frac{\partial \varphi}{\partial x_1} \equiv \partial^1 \varphi \equiv \varphi^1 \equiv -\frac{\partial \varphi}{\partial x}. \tag{6.19}$$

In chapter 7, where vector field components carry subscripts or superscripts even without differentiation, the subscript/superscript notation for derivatives will have to be reconsidered.

For now let us take φ to be a real field (we'll consider complex-valued fields later in this chapter). We can picture the field $\varphi(t, x)$ as a surface over the tx plane, as in Figure 6.2. Our task is to find the surface that makes Γ an extremum.

As in the approach to the single-integral problem, imagine that varied surfaces can be obtained through a continuous deformation that, away from the boundary of $\mathcal{R}$, departs from the extremal surface $\varphi(t, x)$ according to the parameterization

$$\varphi'(t, x) \equiv \varphi(t, x) + \varepsilon \zeta(t, x). \tag{6.20}$$

On the boundary of $\mathcal{R}$ we require all varied surfaces to coincide with the extremal surface, so that $\zeta(a, x) = \zeta(b, x) = 0$ and $\zeta(t, A) = \zeta(t, B) = 0$. The varied functional is now a function of ε:

$$\Gamma(\varepsilon) = \int_{\mathcal{R}} \mathcal{L}(t, x, \varphi', \varphi'_0, \varphi'_1) \, dt \, dx. \tag{6.21}$$

It is not necessary to include both φ^μ and φ_μ when explicitly listing the arguments of the Lagrangian density, because φ^μ, for example, is not independent of φ_μ; the former can be written in terms of the latter with the

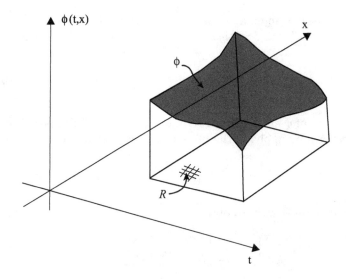

Figure 6.2: *The φ surface above the domain $\mathcal{R}$ in the t-x plane.*

aid of the metric tensor: $\varphi^\mu = g^{\mu\nu}\varphi_\nu$, and, conversely, $\varphi_\mu = g_{\mu\nu}\varphi^\nu$.[4] For example, the following Lagrangian density can be written in multiple ways:

$$\begin{aligned}
\mathcal{L} &= \frac{1}{2}\left(\frac{\partial\varphi}{\partial t}\right)^2 - \frac{1}{2}\left(\frac{\partial\varphi}{\partial x}\right)^2 \\
&= \tfrac{1}{2}\varphi^\mu\varphi_\mu \\
&= \tfrac{1}{2}g^{\mu\nu}\varphi_\mu\varphi_\nu \\
&= \tfrac{1}{2}g_{\mu\nu}\varphi^\mu\varphi^\nu \\
&= \tfrac{1}{2}(\partial_\mu\varphi)(\partial^\mu\varphi).
\end{aligned} \tag{6.22}$$

Returning to the task at hand, for our scalar field functional we require

$$\left[\frac{d\Gamma(\varepsilon)}{d\varepsilon}\right]_{\varepsilon=0} = 0. \tag{6.23}$$

Now we do what eq. (6.23) says: first, differentiate eq. (6.21) to obtain

$$\frac{d\Gamma(\varepsilon)}{d\varepsilon} = \int_\mathcal{R}\left[\frac{\partial\mathcal{L}}{\partial\varphi'}\frac{d\varphi'}{d\varepsilon} + \frac{\partial\mathcal{L}}{\partial\varphi'_\mu}\frac{d\varphi'_\mu}{d\varepsilon}\right] dt\,dx, \tag{6.24}$$

---

[4]In general, the metric tensor components may be functions of the x^μ, this feature alone can make $\mathcal{L}$ an explicit function of the coordinates.

where the repeated indices are summed. From eq. (6.20), $d\varphi'/d\varepsilon = \zeta$ and $d\varphi'_\mu/d\varepsilon = \partial\zeta/\partial x^\mu \equiv \zeta_\mu$. Setting $\varepsilon = 0$ after evaluating the derivatives, we find

$$\left[\frac{d\Gamma(\varepsilon)}{d\varepsilon}\right]_0 = \int_\mathcal{R} \left[\frac{\partial\mathcal{L}}{\partial\varphi}\zeta + \frac{\partial\mathcal{L}}{\partial\varphi_\mu}\zeta_\mu\right] dt\, dx = 0. \qquad (6.25)$$

The derivative of ζ in the second term of the integrand can be shifted to a boundary term by applying the product rule for derivatives to

$$\partial_\mu\left(\zeta\frac{\partial\mathcal{L}}{\partial\varphi_\mu}\right), \qquad (6.26)$$

where ∂_μ denotes $\partial/\partial x^\mu$, to obtain

$$\left[\frac{d\Gamma(\varepsilon)}{d\varepsilon}\right]_0 = \int_\mathcal{R} \left[\frac{\partial\mathcal{L}}{\partial\varphi} - \partial_\mu\left(\frac{\partial\mathcal{L}}{\partial\varphi_\mu}\right)\right]\zeta\, dt\, dx$$
$$+ \int_\mathcal{R} \partial_\mu\left[\zeta\frac{\partial\mathcal{L}}{\partial\varphi_\mu}\right] dt\, dx = 0. \qquad (6.27)$$

By Gauss's divergence theorem, the second integral gives $\zeta(\partial\mathcal{L}/\partial\varphi_\mu)$ evaluated on the boundary of $\mathcal{R}$, where ζ vanishes. Because $[d\Gamma(\varepsilon)/d\varepsilon]_0 = 0$ for any ζ that satisfies the boundary conditions, the fundamental lemma of the calculus of variations demands

$$\frac{\partial\mathcal{L}}{\partial\varphi} = \partial_\mu\left(\frac{\partial\mathcal{L}}{\partial\varphi_\mu}\right), \qquad (6.28)$$

the Euler-Lagrange equation for fields.[5] Writing out the implied sum explicitly for t and x in our simple example of the scalar field Lagrangian density, we obtain the Euler-Lagrange equation for $\varphi(t, x)$:

$$\frac{\partial\mathcal{L}}{\partial\varphi} = \partial_0\left(\frac{\partial\mathcal{L}}{\partial\varphi_0}\right) + \partial_1\left(\frac{\partial\mathcal{L}}{\partial\varphi_1}\right). \qquad (6.29)$$

A function $\varphi(x^\mu)$ that satisfies this Euler-Lagrangian equation makes the functional an extremal. Note the appearance of partial derivatives with respect to the independent variables. You can verify that, for the wave on the string, where φ is y, the Lagrangian density

$$\mathcal{L} = \tfrac{1}{2}(\partial_0 y)^2 - \tfrac{1}{2}(\partial_1 y)^2 = \tfrac{1}{2}\varphi^\mu\varphi_\mu \qquad (6.30)$$

produces the linear homogeneous wave equation, eq. (6.4).

---

[5]To write out eq. (6.28) when $\mathcal{L}$ contains both φ_μ and φ^μ, write φ^μ as $g^{\mu\nu}\varphi_\nu$.

In the derivation of the Euler-Lagrange equation of eq. (6.28), the specific form of the Lagrangian never had to be considered, other than that it depended on no derivatives of higher order than the first.[6] Thus, the field φ may be a scalar, a component of a vector field, or some other kind of field (tensor, spinor). We consider some of these options for φ in later chapters. In addition, the summation convention used in eq. (6.28) is not limited to two independent variables, but can be extended to sums over N independent coordinates. Therefore for any field φ that depends on N spacetime coordinates $x^1, x^2, \ldots, x^N$, the multiple-integral functional looks like this:

$$\Gamma = \int_{\mathcal{R}} \mathcal{L}(x^\mu, \varphi, \varphi_\mu) \, dx^1 dx^2 \cdots dx^N, \quad \mu = 1, 2, \ldots, N. \tag{6.31}$$

When this is made an extremal, the Euler-Lagrange equation of eq. (6.28) results. Applications to relativity usually feature four spacetime coordinates labeled x^μ with $\mu = 0, 1, 2, 3$, where $\mu = 0$ denotes the time coordinate.

6.3 Canonical Momentum and the Hamiltonian Tensor for Fields

The canonical momentum, or rather the canonical momentum density for a field, is defined the usual way in terms of the Lagrangian density,[7]

$$p^\mu = \frac{\partial \mathcal{L}}{\partial \varphi_\mu} \tag{6.32}$$

where $\varphi_\mu \equiv \partial_\mu \varphi$.

In special relativity, with Cartesian spatial coordinates, the metric tensor is a simple one, $g_{\mu\nu} = \eta_{\mu\nu} \equiv diag(1, -1, -1, -1)$. Therefore, for the Lagrangian density of eq. (6.30),

$$
\begin{aligned}
p^\mu &= \frac{1}{2}\eta^{\rho\sigma} \frac{\partial}{\partial \varphi_\mu} [\varphi_\rho \varphi_\sigma] \\
&= \frac{1}{2}\eta^{\rho\sigma} [\delta^\mu{}_\rho \varphi_\sigma + \varphi_\rho \delta^\mu{}_\sigma] \\
&= \varphi^\mu.
\end{aligned}
\tag{6.33}
$$

Specifically, for each field derivative, the corresponding momenta are

$$
\begin{aligned}
p_0 &= \varphi_0 = \tfrac{\partial \varphi}{\partial t}, & p^0 &= \varphi^0 = \tfrac{\partial \varphi}{\partial t} \\
p_1 &= \varphi_1 = \tfrac{\partial \varphi}{\partial x}, & p^1 &= \varphi^1 = -\tfrac{\partial \varphi}{\partial x}
\end{aligned}
\tag{6.34}
$$

---

[6]Extensions can be made, of course, to Lagrangian densities with higher-order derivatives.

[7]Some authors denote the canonical momentum density as π^μ.

In terms of the momentum density, the Euler-Lagrange equation reads

$$\frac{\partial \mathcal{L}}{\partial \varphi} = \partial_\mu p^\mu. \tag{6.35}$$

Based on our experience with single-integral functionals, we might assume the *Hamiltonian density*, in terms of the Lagrangian density, would be

$$\mathcal{H} \equiv \varphi_\mu p^\mu - \mathcal{L}. \tag{6.36}$$

However, this causes a problem; namely, this expression results in $\mathcal{H} = \mathcal{L}$. In classical mechanics cases where $H = K + U$ and $L = K - U$, the condition $H = L$ occurs if and only if there are no interactions, $U = 0$. Eq. (6.36) also makes this occur for the real scalar field, where $\mathcal{L} = \frac{1}{2}\varphi_\mu \varphi^\mu$ and $p_\mu = \varphi_\mu$, because

$$p_\mu \varphi^\mu - \mathcal{L} = \varphi_\mu \varphi^\mu - \tfrac{1}{2}\varphi_\mu \varphi^\mu$$

$$= \frac{1}{2}\varphi_\mu \varphi^\mu \tag{6.37}$$

$$= \mathcal{L}.$$

The way to define a Hamiltonian is to use different indices, thereby defining the *Hamiltonian tensor*, with components $\mathcal{H}_\mu{}^\nu$:

$$\mathcal{H}_\mu{}^\nu \equiv \varphi_\mu p^\nu - \delta_\mu{}^\nu \mathcal{L} \tag{6.38}$$

where $\delta_\mu{}^\nu$ is the Kronecker delta.

Example: In the case where φ represents the scalar field, or a wave on a string, the Hamiltonian tensor components form the array

$$\mathcal{H}_\mu{}^\nu = \begin{pmatrix} \mathcal{H}_0{}^0 & \mathcal{H}_0{}^1 \\ \mathcal{H}_1{}^0 & \mathcal{H}_1{}^1 \end{pmatrix}$$

$$= \begin{pmatrix} \eta & p_0 p^1 \\ -p_0 p^1 & -\eta \end{pmatrix}$$

where

$$\eta \equiv \tfrac{1}{2}[(p^0)^2 + (p^1)^2] \tag{6.39}$$

denotes the energy density of the field (or the wave on the string).

We saw that with the single-integral functional, two forms of the Euler-Lagrange equation, the p-dot and the H-dot versions, could be obtained. In the field version of the Euler-Lagrange equation, so far we have derived the analog of the p-dot equation. Let's develop the field analog of the H-dot equation, by taking the divergence of the Hamiltonian tensor

$$\partial_\nu \mathcal{H}_\mu{}^\nu = \partial_\nu \left(\varphi_\mu \frac{\partial \mathcal{L}}{\partial \varphi_\nu} - \delta_\mu{}^\nu \mathcal{L} \right)$$
$$= (\partial_\nu \varphi_\mu) p^\nu + \varphi_\mu (\partial_\nu p^\nu) - \partial_\mu \mathcal{L} \qquad (6.40)$$

where we have identified the momentum density, $p^\nu = \partial \mathcal{L}/\partial \varphi_\nu$. To swap out the first term that contains the second derivative of φ, let's take a step analogous to one taken in deriving the H-dot version of the Euler-Lagrange equations: expand $\partial_\mu \mathcal{L}$ with the chain rule, recalling that $\mathcal{L} = \mathcal{L}(x^\nu, \varphi, \varphi_\nu)$:

$$\partial_\mu \mathcal{L} = \left(\frac{\partial \mathcal{L}}{\partial x^\nu} \right) \frac{\partial x^\mu}{\partial x^\nu} + \left(\frac{\partial \mathcal{L}}{\partial \varphi} \right) \varphi_\mu + \left(\frac{\partial \mathcal{L}}{\partial \varphi_\nu} \right) (\partial_\mu \varphi_\nu). \qquad (6.41)$$

Recognizing $\partial x^\mu / \partial x^\nu$ as $\delta_\nu{}^\mu$, we can cancel the $\partial_\mu \mathcal{L}$ terms in eq. (6.41), which leaves

$$(\partial_\mu \varphi_\nu) p^\nu = -\varphi_\mu \left(\frac{\partial \mathcal{L}}{\partial \varphi} \right). \qquad (6.42)$$

Carry this back to the divergence of the Hamiltonian tensor, and it becomes

$$\partial_\nu \mathcal{H}_\mu{}^\nu = \varphi_\mu \left(\partial_\nu p^\nu - \frac{\partial \mathcal{L}}{\partial \varphi} \right) - \partial_\mu \mathcal{L}. \qquad (6.43)$$

By virtue of the Euler-Lagrange equation for fields, eq. (6.35), the term in the parentheses vanishes, leaving the field theory analog of the H-dot equation,

$$\partial_\nu \mathcal{H}_\mu{}^\nu = -\partial_\mu \mathcal{L}. \qquad (6.44)$$

This, along with the analog of the p-dot version of the Euler-Lagrange equation for fields,

$$\frac{\partial \mathcal{L}}{\partial \varphi} = \partial_\nu p^\nu, \qquad (6.45)$$

give $N + 1$ equations in N-dimensional space or spacetime. Two conservation laws immediately present themselves: If and only if $\mathcal{L}$ does not depend on φ explicitly, then a conservation law results, in the form of an "equation of continuity" (see section 6.4) for the canonical momentum density:

$$\partial_\nu p^\nu = 0 \quad \Longleftrightarrow \quad \frac{\partial \mathcal{L}}{\partial \varphi} = 0. \qquad (6.46)$$

Similarly, the components $\mathcal{H}_\mu{}^\nu$ of the Hamiltonian tensor enter an equation of continuity if and only if the Lagrangian does not depend explicitly on x^μ, which means the Lagrangian is invariant under a transformation of this coordinate:

$$\partial_\nu \mathcal{H}_\mu{}^\nu = 0 \quad \Longleftrightarrow \quad \partial_\mu \mathcal{L} = 0. \tag{6.47}$$

These equations of continuity for the canonical momenta and the Hamiltonian tensor components arise separately from their respective versions of the Euler-Lagrange equations. More generally, Noether's theorem for fields produces an equation of continuity that features a combination of canonical momenta and the Hamiltonian tensor. Let us review equations of continuity.

6.4   Equations of Continuity

A *global* conservation law would allow for some amount of, say, electric charge Q to disappear from Earth right now, and the same amount Q to appear on Jupiter, without charge ever flowing through any surface that encloses Earth. But with *local* conservation, whenever an amount of charge within a volume V decreases by dQ, the same amount of dQ must flow outward through the closed surface S that forms the boundary of V. An equation of continuity puts into mathematics this simple idea.

Consider some quantity Q, say an electric charge, or a mass of fluid, or a quantity of energy, initially enclosed within a volume V bounded by the closed surface S. Let ρ denote the density of Q, and let $\mathbf{j}$ denote its current density.[8] Local conservation of Q means that the rate at which Q decreases inside V equals the flux of Q-current through S:

$$-\frac{d}{dt}\int_V \rho\, dV = \oint_S \mathbf{j} \cdot \hat{\mathbf{n}}\, da, \tag{6.48}$$

where $\hat{\mathbf{n}}$ denotes the unit normal vector pointing outward from the surface element of area da on S (Figure 6.3).

In eq. (6.48), use Gauss's divergence theorem on the flux integral, and pull the time derivative under the volume integral, to obtain

$$-\int_V \frac{\partial \rho}{\partial t} dV = \int_V (\boldsymbol{\nabla} \cdot \mathbf{j}) dV. \tag{6.49}$$

Local conservation in general means this relation must hold whatever the volume V; therefore the integrands must be equal. Local conservation

---

[8]The current density may be written $\mathbf{j} = \rho \mathbf{v}$ where $\mathbf{v}$ is the velocity of the charge, mass, or energy; ρ and $\mathbf{j}$ are, in general, functions of position and time.

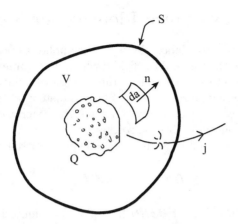

Figure 6.3: *The volume V and its surface S described in the discussion of equations of continuity.*

thereby finds differential expression as the *equation of continuity*:

$$\nabla \cdot \mathbf{j} + \frac{\partial \rho}{\partial t} = 0. \tag{6.50}$$

In spacetime, define the current density four-vector (see ex. 6.11)

$$\{j^{\nu}\} \equiv (\rho, \mathbf{j}) \tag{6.51}$$

to express the equation of continuity in covariant form,

$$\partial_{\nu} j^{\nu} = 0. \tag{6.52}$$

The steps leading from eq. (6.48) to (6.50) are reversible. Local conservation implies an equation of continuity, and conversely, an equation of continuity implies local conservation. Therefore, "equation of continuity" and "local conservation" are equivalent.

Notice in eq. (6.48) that if the volume V is all space then the surface S resides at infinity. Since physically realistic current densities are finite in extent and therefore vanish at infinity,[9] the surviving contribution in the conservation law says that the total charge, or total mass, or total energy, and so forth, remains constant:

$$\int_{V} \rho dV = \text{const.} \tag{6.53}$$

---

[9]For example, in Euclidean spaces the surface area goes at r^2. For the integral to vanish at infinity the integrand must drop off faster than $1/r^2$.

6.5   The Invariance Identity for Fields

We now turn to the invariance identity as it applies to field theory, where the spacetime coordinates and the field itself undergo infinitesimal transformations. For now let's consider one-parameter transformations. The discussion can easily be generalized to a set of M parameters $\{\varepsilon^k, k = 1, 2, \ldots, M\}$ with another set of repeated indices on ε and the generators, and imposing the summation convention over k, so that $\varepsilon\tau \to \varepsilon^k\tau_k$ and $\varepsilon\zeta^\mu \to \varepsilon^k\zeta^\mu{}_k$.[10]

Consider a functional in N-dimensional space (or spacetime):[11]

$$\Gamma = \int_{\mathcal{R}} \mathcal{L}(x^\mu, \varphi, \varphi_\mu) \, dx^0 dx^1 \cdots dx^{N-1}, \tag{6.54}$$

where $\varphi = \varphi(x^\mu)$ and $\varphi_\mu \equiv \partial\varphi/\partial x^\mu$. Deploy an infinitesimal transformation of the coordinates and the field, recalling that the generators may be functions of both:

$$x'^\mu = x^\mu + \varepsilon\tau^\mu(x^\rho, \varphi), \tag{6.55}$$

$$\varphi' = \varphi + \varepsilon\zeta(x^\rho, \varphi). \tag{6.56}$$

Frequently occuring examples are found in Lorentz transformations (see appendix B). Consider for definiteness a two-dimensional spacetime $(t, x) = (x^0, x^1)$; one can always extend to more dimensions by summing over repeated spacetime indices from 0 to $N - 1$. We take as the definition of invariance

$$\Gamma' - \Gamma \sim \varepsilon^s, \quad s > 1, \tag{6.57}$$

or more explicitly,

$$\int_{\mathcal{R}'} \mathcal{L}' \, dt' \, dx' - \int_{\mathcal{R}} \mathcal{L} \, dt \, dx \sim \varepsilon^s, \quad s > 1, \tag{6.58}$$

---

[10]In her 1918 paper, Emmy Noether wrote of this step, "The group will be called a finite continuous group $\underline{G}_\rho$ if its transformations are contained in a most general (transformation) depending analytically on ρ essential parameters ε"(Tavel translation [Noether & Tavel (1971)]). The ε parameters are continuous, but the set of transformations, each labeled by ρ (our k) is finite, k (or ρ) = 1 to M.

[11]In relativity, Γ is to be a scalar under transformations among reference frames. The volume element $d^N x \equiv dx^0 dx^1 \cdots dx^{N-1}$ is not invariant, but $\sqrt{||g||}d^N x$ is invariant, where $||g||$ denotes the absolute value of the determinant $|g|$ of the metric tensor (the square root is often written in textbooks as $\sqrt{-|g|}$ because the metric tensor of flat spacetime, $diag(1, -1, -1, -1)$ has determinant $|g| = -1$; another common notation is merely g for the determinant of the metric tensor). Some authors [Hobson et al. (2006) 528] define the "Lagrangian L" through $\mathcal{L} \equiv L\sqrt{-|g|}$ so that $L\sqrt{-|g|}d^4 x$ is a scalar field, which means that the Lagrangian density $\mathcal{L}$ is not a scalar. It is, rather, a *scalar density* of weight +1. To say a quantity Q is a "tensor density of weight W" means that under the transformation, $Q' = J^W Q$ where J is the Jacobian. The usual definition of the Lagrangian L is the integral of the Lagrangian density $\mathcal{L}$ over three-dimensional space. We will not focus on these technicalities here, but want to be aware of them.

and $\mathcal{L}'$ means the same Lagrangian density, with the new independent and dependent variables going into it:

$$\mathcal{L}' \equiv \mathcal{L}\left(x'^{\mu}, \varphi'(x'^{\mu}), \frac{\partial \varphi'(x'^{\mu})}{\partial x'^{\nu}}\right). \tag{6.59}$$

As we saw with single-integral functionals, both the new and original functionals can be combined into one integral over the original domain $\mathcal{R}$. The integration measure of the primed integral may be written

$$dt' \, dx' = J \, dt \, dx \tag{6.60}$$

where J denotes the Jacobian of the transformation (see appendix E), the determinant of the matrix of coordinate transformation coefficients:

$$\begin{aligned}
J &\equiv \left|\frac{\partial x'^{\mu}}{\partial x^{\nu}}\right| \\
&\equiv \begin{vmatrix} \frac{\partial x'^0}{\partial x^0} & \frac{\partial x'^0}{\partial x^1} \\ \frac{\partial x'^1}{\partial x^0} & \frac{\partial x'^1}{\partial x^1} \end{vmatrix} \\
&= \begin{vmatrix} 1 + \varepsilon \partial_0 \tau^0 & \varepsilon \partial_1 \tau^0 \\ \varepsilon \partial_0 \tau^1 & 1 + \varepsilon \partial_1 \tau^1 \end{vmatrix} + O(\varepsilon^2).
\end{aligned} \tag{6.61}$$

Thus the Jacobian becomes unity plus a divergence of the spacetime generators:

$$J = 1 + \varepsilon(\partial_{\mu} \tau^{\mu}) + \cdots . \tag{6.62}$$

Our deliberations so far allow us to express the definition of invariance as

$$\int_{\mathcal{R}} (\mathcal{L}' J - \mathcal{L}) \, dt \, dx \sim \varepsilon^s, \quad s > 1. \tag{6.63}$$

Because the definition of invariance must hold whatever the region $\mathcal{R}$, the functional will be invariant if and only if

$$\mathcal{L}' J - \mathcal{L} \sim \varepsilon^s, \quad s > 1. \tag{6.64}$$

From this we derive the invariance identity for fields.

Invariance identity for fields: A multiple-integral functional in the form of eq. (6.54) is invariant under the infinitesimal transformation $x'^{\mu} = x^{\mu} + \varepsilon \tau^{\mu}$ and $\varphi' = \varphi + \varepsilon \zeta$ if and only if the following identity holds:

$$\left(\frac{\partial \mathcal{L}}{\partial \varphi}\right) \zeta + p^{\mu}(\partial_{\mu} \zeta) + (\partial_{\mu} \mathcal{L}) \tau^{\mu} - \mathcal{H}_{\mu}{}^{\nu}(\partial_{\nu} \tau^{\mu}) = 0 \tag{6.65}$$

where

$$p^\mu = \frac{\partial \mathcal{L}}{\partial \varphi_\mu} \tag{6.66}$$

and

$$\mathcal{H}_\mu{}^\nu = p_\mu \varphi^\nu - \delta_\mu{}^\nu \mathcal{L}. \tag{6.67}$$

The invariance identity follows from differentiating the definition of invariance with respect to ε, then setting $\varepsilon = 0$:

$$\frac{d}{d\varepsilon} \left[\mathcal{L}' J - \mathcal{L} \right]|_0 = 0. \tag{6.68}$$

The only terms that depend on ε are $\mathcal{L}'$ and the Jacobian, so the differentiation presents us with

$$\left(\frac{d\mathcal{L}'}{d\varepsilon} \right)_0 + \mathcal{L} \left(\frac{dJ}{d\varepsilon} \right)_0 = 0. \tag{6.69}$$

The contribution from the Jacobian is straightforward. From eq. (6.62) it follows at once:

$$\left(\frac{dJ}{d\varepsilon} \right)_0 = \partial_\mu \tau^\mu. \tag{6.70}$$

The calculation of $(\partial \mathcal{L}'/\partial \varepsilon)_0$ requires more work. Let's begin by evaluating $d\mathcal{L}'/d\varepsilon$ using the chain rule, recalling that $\mathcal{L}'$ depends on the x'^μ, φ', and $\partial \varphi'/\partial x'^\mu$:

$$\begin{aligned}
\left[\frac{d\mathcal{L}'}{d\varepsilon} \right]_0 &= \left[\frac{\partial \mathcal{L}'}{\partial x'^\mu} \frac{dx'^\mu}{d\varepsilon} + \frac{\partial \mathcal{L}'}{\partial \varphi'} \frac{d\varphi'}{d\varepsilon} + \frac{\partial \mathcal{L}'}{\partial \varphi'_\mu} \frac{d}{d\varepsilon} \left(\frac{\partial \varphi'}{\partial x'^\mu} \right) \right]_0 \\
&= \frac{\partial \mathcal{L}}{\partial x^\mu} \tau^\mu + \frac{\partial \mathcal{L}}{\partial \varphi} \zeta + \frac{\partial \mathcal{L}}{\partial \varphi_\mu} \left[\frac{d}{d\varepsilon} \frac{\partial \varphi'}{\partial x'^\mu} \right]_0.
\end{aligned} \tag{6.71}$$

Let's take the last term apart in steps. First, display the ε's explicitly in $\partial \varphi'/\partial x'^\mu$. Think of x'^μ as a function of all the x^ν and invoke the chain rule:

$$\frac{\partial \varphi'}{\partial x'^\mu} = \frac{\partial \varphi'}{\partial x^\nu} \frac{\partial x^\nu}{\partial x'^\mu}. \tag{6.72}$$

Use the reverse coordinate transformation to write x^ν in terms of x'^ν:

$$\begin{aligned}
\frac{\partial \varphi'}{\partial x'^\mu} &= \frac{\partial \varphi'}{\partial x^\nu} \frac{\partial (x'^\nu - \varepsilon \tau^\nu)}{\partial x'^\mu} \\
&= \frac{\partial \varphi'}{\partial x^\nu} \left[\delta^\nu{}_\mu - \varepsilon \frac{\partial \tau^\nu}{\partial x'^\mu} \right].
\end{aligned} \tag{6.73}$$

Similarly, use the transformation of the field to write φ' in terms of φ:

$$\frac{\partial \varphi'}{\partial x'^\mu} = [\varphi_\nu + \varepsilon(\partial_\nu \zeta)] \left[\delta^\nu{}_\mu - \varepsilon \frac{\partial \tau^\nu}{\partial x'^\mu} \right]. \tag{6.74}$$

To first order in ε we now have

$$\frac{\partial \varphi'}{\partial x'^\mu} = \varphi_\mu - \varepsilon \varphi_\nu \left(\frac{\partial \tau^\nu}{\partial x'^\mu} \right) + \varepsilon \partial_\mu \zeta. \tag{6.75}$$

At last we have reduced the complicated last term in $[\partial \mathcal{L}/\partial \varepsilon]_0$ to something managable:

$$\left[\frac{d}{d\varepsilon} \frac{\partial \varphi'}{\partial x'^\mu} \right]_0 = -\varphi_\nu (\partial_\mu \tau^\nu) + \partial_\mu \zeta. \tag{6.76}$$

Now put eqs. (6.76) and (6.70) back into eq. (6.69). We thereby assemble the invariance identity for a field, eq. (6.65).

By evaluating derivatives of products, you can show that the field invariance identity may also be written in the alternative form

$$-(\zeta - \varphi_\nu \tau^\nu) \left[\frac{\partial \mathcal{L}}{\partial \varphi} - \partial_\mu p^\mu \right] = \partial_\mu \left[p^\mu \zeta - \mathcal{H}_\nu{}^\mu \tau^\nu \right]. \tag{6.77}$$

A more liberal definition of invariance, called *divergence invariance*, includes the possibility of an inhomogeneity, linear in ε, expressed as a divergence:

$$\mathcal{L}' J - \mathcal{L} = \varepsilon(\partial_\mu F^\mu) + O(\varepsilon^s), \quad s > 1. \tag{6.78}$$

This comes from adding to the integrand in the functional a divergence,

$$\int_\mathcal{R} [\mathcal{L} J - \mathcal{L} - \varepsilon \partial_\mu F^\mu] \, d^N x \sim \varepsilon^s \tag{6.79}$$

which can be shifted to the boundary of $\mathcal{R}$ because of the divergence theorem,

$$\int_\mathcal{R} \partial_\mu F^\mu \, d^N x = \oint_S F^\mu n_\mu \, da \tag{6.80}$$

where n_μ is the component of the outward-pointing unit vector normal to the patch of area da on the surface S that encloses $\mathcal{R}$. If $\mathcal{R}$ is all space (or spacetime) then S is at infinity, where the fields vanish.

Including $\partial_\mu F^\mu$, and repeating the steps in the derivation of the invariance identity, leads to

$$\frac{\partial \mathcal{L}}{\partial \varphi} \zeta + p^\mu (\partial_\mu \zeta) + (\partial_\mu \mathcal{L}) \tau^\mu - \mathcal{H}_\mu{}^\nu (\partial_\nu \tau^\mu) - \partial_\mu F^\mu = 0, \tag{6.81}$$

or alternatively

$$-(\zeta - \varphi_\nu \tau^\nu) \left[\frac{\partial \mathcal{L}}{\partial \varphi} - \partial_\mu p^\mu \right] = \partial_\mu \left[p^\mu \zeta - \mathcal{H}_\nu{}^\mu \tau^\nu - F^\mu \right]. \tag{6.82}$$

6.6   Noether's Theorem for Fields

The invariance identity holds whenever the functional Γ is invariant under a transformation. If the functional is an extremal, the Euler-Lagrange equation holds. When the functional is both invariant and extremal, then the invariance identity, together with the Euler-Lagrange equation, gives at once the Noether conservation law, which in field theory is an equation of continuity:

$$\partial_\nu j^\nu = 0, \tag{6.83}$$

where the current j^ν stands for

$$j^\nu \equiv p^\nu \zeta - \mathcal{H}_\mu{}^\nu \tau^\mu - F^\nu. \tag{6.84}$$

Henceforth I will take $F^\nu = 0$ unless explicitly stated otherwise.

Example: In the traveling wave example with one spatial coordinate, the equation of continuity may be written

$$\frac{\partial j^0}{\partial t} + \frac{\partial j^1}{\partial x} = 0. \tag{6.85}$$

The global version of this conservation law follows by integrating eq. (6.85) over the entire x-axis:

$$\frac{d}{dt} \int_{-\infty}^{+\infty} j^0 dx + [j^1(+\infty) - j^1(-\infty)] = 0. \tag{6.86}$$

If the current density vanishes at spatial infinity (as it would in a physically realistic situation), then we are left with the conservation law

$$\int_{-\infty}^{+\infty} j^0 dx = \text{const.} \tag{6.87}$$

For time translations, $\tau^0 = 1, \tau^1 = 0$, and $\zeta = 0$, and invariance of the functional is guaranteed when the Lagrangain does not depend explicitly on t. Then eq. (6.87), with the j^0 of eq. (6.84) (with $F^\nu = 0$), gives the Noether conservation law

$$\int_{-\infty}^{+\infty} \mathcal{H}_\mu{}^0 \tau^\mu \, dx = \text{const.} \tag{6.88}$$

or

$$\int_{-\infty}^{+\infty} \eta \, dx = \text{const.}, \tag{6.89}$$

where $\eta \equiv \frac{1}{2}[(p^0)^2 + (p^1)^2]$, the energy density. As with particle mechanics, time translation invariance for fields implies conservation of energy.

6.7 Complex Fields

In the previous discussion, we assumed the transverse wave on the string to be plane polarized: as the wave runs down the x-axis, the bits of string vibrate to and fro parallel to the y-axis only. Alternatively, we could imagine choosing the disturbances to go only parallel to the z-axis. And, of course, the wave could be in a superposition of y-displacements and z-displacements. We already have the wave equation for the y-displacement, eq. (6.4),[12]

$$\frac{\partial^2 y}{\partial x^2} = \frac{\partial^2 y}{\partial t^2} \tag{6.90}$$

Similarly, for the oscillations parallel to the z-axis we would write

$$\frac{\partial^2 z}{\partial x^2} = \frac{\partial^2 z}{\partial t^2}. \tag{6.91}$$

Putting these two orthogonal oscillations together as the components of a vector gives us

$$\mathbf{A} = y\hat{\mathbf{j}} + z\hat{\mathbf{k}}. \tag{6.92}$$

We may write a wave equation for $\mathbf{A}$:

$$\frac{\partial^2 \mathbf{A}}{\partial x^2} = \frac{\partial^2 \mathbf{A}}{\partial t^2} \tag{6.93}$$

of which eqs. (6.90) and (6.91) are the components. Together they give a wave equation for the vector field $\mathbf{A}$, eq. (6.93). These two components of eq. (6.93) are Euler-Lagrange equations for some Lagrangian density. Since energies are additive, that Lagrangian density would be

$$\mathcal{L} = \frac{1}{2}\left[\left(\frac{\partial y}{\partial t}\right)^2 + \left(\frac{\partial z}{\partial t}\right)^2\right] - \frac{1}{2}\left[\left(\frac{\partial y}{\partial x}\right)^2 + \left(\frac{\partial z}{\partial x}\right)^2\right]. \tag{6.94}$$

Another way to account for the polarization information would be to use complex variables. Write

$$\chi = y + iz, \tag{6.95}$$

where $i^2 = -1$. Now eqs. (6.90) and (6.91) are the real and imaginary parts of the complex equation

$$\frac{\partial^2 \chi}{\partial x^2} = \frac{\partial^2 \chi}{\partial t^2}. \tag{6.96}$$

---

[12]Recall that t denotes time measured in meters, as the wave speed v as been absorbed into the t-coordinate.

Its real part gives the y wave, and its imaginary part gives the z wave. What happens to the Lagrangian in this way of doing it? First make note of the complex conjugate of χ,

$$\chi^* = y - iz. \tag{6.97}$$

Therefore,

$$y = \tfrac{1}{2}(\chi + \chi^*) \tag{6.98}$$

and

$$z = \tfrac{1}{2i}(\chi - \chi^*), \tag{6.99}$$

where

$$\chi^*\chi = y^2 + z^2 = A^2. \tag{6.100}$$

Trading y and z for χ and χ^* in the Lagrangian density of eq. (6.94), we obtain

$$\mathcal{L} = \tfrac{1}{2}(\partial_\mu \chi^*)(\partial^\mu \chi). \tag{6.101}$$

So far we have merely found a second way to describe a wave with two possible polarizations. We can write it either as a vector $\mathbf{A}$ or as a complex number χ. No fundamental difference distinguishes a two-dimensional vector $\mathbf{A}$ from the complex variable χ, because any two-dimensional vector space can be mapped into the complex plane. It's a matter of taste and convenience, because $y\hat{\mathbf{j}} + z\hat{\mathbf{k}}$ and $y + iz$, each with their respective algebra rules, carry the same information.

Be that as it may, one may wonder, other than "because we can," why consider a complex field χ? Thanks to Euler's formula,

$$e^{i\theta} = \cos\theta + i\sin\theta, \tag{6.102}$$

any discipline that works with periodic signals, from acoustics to antenna theory, from seismology and hydraulics to electrical circuits and optics, routinely uses complex-valued fields to model signals. This works because periodic signals can be represented as a superposition of harmonics (Fourier's theorem), provided the underlying wave equation is linear.[13] Each harmonic with its one frequency and one wavelength has an amplitude and a phase, and can therefore be expressed as a complex number. Of course, one could eschew complex numbers and stick with trigonometry functions. But sines and cosines can be awkward to manipulate, compared to their equivalents expressed as complex numbers. Complex numbers offer *convenience* in applications to subjects such as acoustics, optics, and electrical engineering.

---

[13]In our example of the wave on the string, the wave equation derived from $F = ma$ would have been nonlinear had the small-angle approximation not been made [Holmes (1994)]. Mechanical wave equations are approximately linear if the amplitudes of the oscillations are small.

In contrast, quantum mechanics finds complex numbers *compelling*. The existence of the tunneling effect illustrates why. If a beam of mono-energetic electrons approaches a repulsive energy barrier that is 1 eV greater than each electron's kinetic energy, and if the barrier is 0.1 nm wide, then about half of the electrons get past the barrier, even though, according to Newtonian physics, none of them have enough kinetic energy to get over it. It's as if they tunnel through the barrier. The quantum mechanical wave function offers a sensible interpretation of this bizarre effect when expressed in complex numbers. For total energy E, kinetic energy K, and barrier height U_0, if $E < U_0$ then inside the barrier, $K = E - U_0 < 0$. But $K = p^2/2m$, and since there is no interpretation of negative mass, $K < 0$ requires $p = i|p|$, which makes no Newtonian sense. That is why Newtonian physics says "impossible!" to the tunneling effect.

But the tunnelling effect *really does happen*. The entire semiconductor industry depends on it, through the p-n-p and n-p-n junctions in the legion of transistors embedded in computer chips. By varying the barrier height U_o, the electron tunnelling current coming through the junction can be precisely controlled. When this empirical observation is put into the language of complex de Broglie wave functions, the oscillatory quantum state $\psi \sim e^{ipx/\hbar}$ becomes, with imaginary momentum, $e^{-|p|x/\hbar}$. With the quantum mechanical probability density given by $\psi^*\psi$, the classically forbidden imaginary momentum offers a sensible interpretation as an exponentially damped probability. What was absolutely forbidden in Newtonian physics is merely exponentially damped in quantum physics.[14]

Complex numbers fall readily to hand in particle physics for another reason. A complex number $z = x + iy = re^{i\theta}$ locates a point in the (x, y) or (r, θ) plane. Its complex conjugate $z^* = x - iy = re^{-i\theta}$ locates that point's mirror image across the x-axis. If we think of θ as proportional to, say, an electric charge, then evidently the state of matter described by $\psi \sim e^{i\theta}$ and the state described by $\psi^* \sim e^{-i\theta}$ carry opposite charges. Antiparticles carry opposite charges from their corresponding particles. Therefore, complex numbers seem made to order for representing particles that carry a nonzero electric charge, or other discete quantum numbers that carry either sign.

Because in physics applications the Lagrangian density $\mathcal{L}$ is an energy density, it must be a real number. Therefore, whenever a complex quantum state ψ appears in $\mathcal{L}$, it must appear multiplicatively alongside its complex conjugate ψ^*. The two dependent variables $\psi^*(t, x, y, z)$ and $\psi(t, x, y, z)$ are to be considered distinct. There will be two Euler-Lagrange equations, one for ψ and another for ψ^*.

---

[14]What does it mean to say that the electrons in the beam "carry 1 eV of kinetic energy"? Thanks to quantum uncertainty, the energy cannot be known exactly. There is an intrinsic spread of energies about a "one 1 eV" beam of particles, and likwise an uncertainty in the barrier height.

Example (Quantum Wave Functions as Fields): When a body of mass m interacts with the real world through some potential energy function U, quantum theory requires us to mentally dance between the complementary concepts of particles and waves. The wave function $\psi(t, x)$ corresponds to a particle moving along the x-axis. In nonrelativistic quantum mechanics, the wave function evolves in time according to the Schrödinger equation,

$$-\frac{\hbar^2}{2m}\frac{\partial^2 \psi}{\partial x^2} + U\psi = -\frac{\hbar}{i}\frac{\partial \psi}{\partial t}. \tag{6.103}$$

For it to be an Euler-Lagrange equation derived from

$$\frac{\partial \mathcal{L}}{\partial \psi^*} = \partial_\mu \left(\frac{\partial \mathcal{L}}{\partial \psi_\mu^*} \right) \tag{6.104}$$

(where $\psi_\mu^* \equiv \partial \psi^*/\partial x^\mu$ and μ is summed over 0 and 1), the Lagrangian density must be

$$\mathcal{L}(t, x, \psi, \psi^*, \psi_0, \psi_1, \psi_0^*, \psi_1^*)$$
$$= -\frac{\hbar^2}{2m}\psi_1^*\psi_1 + \tfrac{1}{2}i\hbar(\psi^*\psi_0 - \psi_0^*\psi) - \psi^*U\psi. \tag{6.105}$$

The other Euler-Lagrange equation

$$\frac{\partial \mathcal{L}}{\partial \psi} = \partial_\mu \left(\frac{\partial \mathcal{L}}{\partial \psi_\mu} \right) \tag{6.106}$$

gives the complex conjugate of eq. (6.103).

With this Lagrangian density come four canonical momenta:

$$p^0 = \frac{\partial \mathcal{L}}{\partial \psi_0} = \frac{i\hbar}{2}\psi^*, \tag{6.107}$$

$$p^{*0} = \frac{\partial \mathcal{L}}{\partial \psi_0^*} = -\frac{i\hbar}{2}\psi, \tag{6.108}$$

$$p^1 = \frac{\partial \mathcal{L}}{\partial \psi_1} = -\frac{\hbar^2}{2m}\psi_1^*, \tag{6.109}$$

$$p^{*1} = \frac{\partial \mathcal{L}}{\partial \psi_1^*} = -\frac{\hbar^2}{2m}\psi_1. \tag{6.110}$$

From these we build the Hamiltonian tensor,

$$\mathcal{H}^\mu{}_\nu = p^\mu \psi_\nu + p^{*\mu}\psi_\nu^* - \delta^\mu{}_\nu \mathcal{L}. \tag{6.111}$$

Should the functional be invariant under an infinitesimal transformation and be extremal, then Noether's theorem gives the conserved current:

$$\partial_\nu j^\nu = 0, \tag{6.112}$$

where

$$j^\nu = p^\nu \zeta + p^{*\nu} \zeta^* - \mathcal{H}^\nu{}_\mu \tau^\mu. \tag{6.113}$$

Examples follow in section 6.8.

6.8   Global Gauge Transformations

An important class of transformations arises in the so-called *gauge transformation*, where the phase of a complex field is changed but the spacetime coordinates are left unchanged:

$$x'^\mu = x^\mu, \tag{6.114}$$

$$\psi' = e^{ig\varepsilon}\varphi, \tag{6.115}$$

$$\psi'^* = e^{-ig\varepsilon}\psi^*, \tag{6.116}$$

where g is a constant. With ε uniform throughout space, this is a global gauge transformation.[15] For infinitesimal ε the field transformations become

$$\psi' = \psi + ig\varepsilon\psi + \dots, \tag{6.117}$$

$$\psi'^* = \psi^* - ig\varepsilon\psi^* + \dots. \tag{6.118}$$

From the coefficients of ε we read off the generators:

$$\tau^\mu = 0, \tag{6.119}$$

$$\zeta = ig\psi, \tag{6.120}$$

$$\zeta^* = -ig\psi^*. \tag{6.121}$$

Because all the τ^μ vanish, the Jacobian equals unity, so the definition of invariance of the functional now reduces, by eq. (6.64), to $\mathcal{L}' - \mathcal{L} \sim \varepsilon^s$, where $s > 1$. Thus, as $\varepsilon \to 0$, the defintion of functional invariance further reduces, for a gauge transformation, to the invariance of the Lagrangian density,

$$\mathcal{L}' = \mathcal{L}. \tag{6.122}$$

Let us apply this gauge transformation to the complex scalar field of nonrelativistic quantum mechanics, using the generators just mentioned.

---

[15]Local gauge transformations, where $\varepsilon = \varepsilon(x^\mu)$, greatly occupy us later.

When the generators and potential energy are such that the functional is both invariant and extremal, Noether's conservation law gives the equation of continuity

$$\partial_0 j^0 + \partial_1 j^1 = 0, \tag{6.123}$$

with current density

$$j^\nu = ig(p^\nu \psi - p^{*\nu} \psi^*) \tag{6.124}$$

having time component

$$j^0 = -g\hbar \psi^* \psi \tag{6.125}$$

and space component

$$j^1 = \left(\frac{g\hbar^2}{2mi}\right)(\psi_1^* \psi - \psi_1 \psi^*). \tag{6.126}$$

A normalizable quantum wave function must vanish at spatial infinity, so integration of eq. (6.123) over all space gives

$$g \int_{-\infty}^{+\infty} \psi^* \psi \, dx = \text{const.} \tag{6.127}$$

Because $\psi^* \psi$ represents the probability density of locating the particle, if g denotes a quantum number such as the particle's charge, then $g\psi^*\psi$ represents the charge density, and its integral in eq. (6.127) states the conservation of charge. Thereby global gauge invariance leads, through Noether's theorem, to charge conservation.

Digression (Notes on Relativistic Scalar Fields): The Klein-Gordon equation (see appendix C),

$$\hbar^2(\nabla^2 - \partial_t^2)\varphi = m^2\varphi \tag{6.128}$$

describes the evolution of the quantum mechanical wave function φ for a relativistic free particle of mass m and integer spin, a "boson." In terms of the four-dimensional gradient in the spacetime of special relativity (see appendix B), the Klein-Gordon equation may be written

$$\hbar^2 \partial^\mu \partial_\mu \varphi = m^2 \varphi. \tag{6.129}$$

Notice how this becomes the homogeneous wave equation when $m = 0$.

Because $\partial^\mu \partial_\mu$ and m^2 are already Lorentz scalars, the transformation properties of the Klein-Gordon equation will be those of φ, such as a scalar field (spin 0), the component of a vector field (spin 1), or the component of a second-rank tensor field (spin 2). In this section let φ denote the wave function for a spin-0 boson, the quantum of a scalar field. As a scalar field, $\varphi(x^\mu)$ is unchanged when the coordinates are changed—this is what it means to be a scalar! Therefore, if you want to get a nonzero generator ζ, you have to put it in explicitly (see discussion question Q6.e. on "induced" field generators).

The object of interest here is the Klein-Gordon equation as an Euler-Lagrange equation. If the quanta are uncharged, then φ is a real field, and the Lagrangian density that produces the Klein-Gordon equation for a free particle may be written (with $\hbar = 1$ or, equivalently, the $\hbar$ absorbed into other factors)

$$\mathcal{L}(\varphi, \varphi_\mu) = \tfrac{1}{2}\varphi_\mu\varphi^\mu - \tfrac{1}{2}m^2\varphi^2. \tag{6.130}$$

where $\partial_\mu\varphi \equiv \varphi_\mu$. If the particle carries a specific nonzero value of a charge from a set of quantized possibilities (such as electric charge), then the Lagrangian will be that of a complex scalar field,

$$\mathcal{L}(\varphi, \varphi_\mu, \varphi^*, \varphi_\mu^*) = \varphi_\mu^*\varphi^\mu - m^2\varphi^*\varphi. \tag{6.131}$$

Notice the factor of $\tfrac{1}{2}$ in the Lagrangian density disappears with the transition from the real scalar field to the complex scalar field. This may seem puzzling, but there's a simple reason for it.

Consider a real scalar field α and a second real scalar field β, whose quanta have the same mass. The Lagrangian density of the combined α and β system is the sum of their separate Lagrangian densities,

$$\mathcal{L}^{(\alpha+\beta)} = \tfrac{1}{2}(\alpha_\mu\alpha^\mu + \beta_\mu\beta^\mu) - \tfrac{1}{2}m^2(\alpha^2 + \beta^2). \tag{6.132}$$

Suppose the wave functions of both fields have been normalized, $\int \alpha^2\, dV = 1$ and $\int |\beta|^2\, dV = 1$, as they would be if their respective quanta maintain their existence somewhere throughout space. If the quanta of the β field are the antiparticles of the quanta of the α field, as found for example in the positive and negative pions, instead of considering them as separate species we might think of the particle and its antiparticle as two states of one species. Their fields can be combined into one complex field $\varphi \sim \alpha + i\beta$. Since the Lagrangian density consists of energy density terms that must be real, then φ and its complex conjugate φ^* must always appear together multiplicatively, in the combination $\varphi^*\varphi \sim \alpha^2 + \beta^2$. This must normalize to 1 if the number if quanta is conserved. But if the positive and negative quanta were conserved when apart, then $\int(\alpha^2 + \beta^2)\, dV = 2$. Therefore, for φ to be properly normalized, when defined in terms of α and β a square root of 2 must be included:

$$\varphi \equiv \frac{1}{\sqrt{2}}(\alpha + i\beta), \quad \varphi^* \equiv \frac{1}{\sqrt{2}}(\alpha - i\beta). \tag{6.133}$$

Conversely,

$$\alpha = \sqrt{2}(\varphi^* + \varphi), \quad \beta = i\sqrt{2}(\varphi^* - \varphi). \tag{6.134}$$

When these expressions for α and β are substituted into the Lagrangian density of eq. (6.132), we find that the coefficients of $\frac{1}{2}$ go away, resulting in eq. (6.131).[16] **End of digression.**

To summarize: If φ is a real scalar field, its Lagrangian density is given by eq. (6.130). But if φ is a complex field made of two real fields whose quanta respect normalization separately, then eq. (6.131) holds. In the latter case the number of degrees of freedom is the same whether we use α and β or, instead, φ and φ^*. That is why φ and φ^* are considered distinct dependent variables in the Euler-Lagrange equation. To describe the evolution of the entire system requires two Euler-Lagrange equations, one with respect to φ, the other with respect to φ^*, which echoes the fact that had we stayed with α and β, two Euler-Lagrange equations would be required to describe them. The quanta of the field φ^* are the antiparticles of the quanta of φ.

Let us now apply Noether's theorem to the complex scalar field under a combined global spacetime and gauge transformation,

$$x'^{\mu} = x^{\mu} + \tau^{\mu}\varepsilon + \ldots, \tag{6.135}$$

$$\varphi' = e^{ig\varepsilon}\varphi = \varphi + ig\varphi\varepsilon + \ldots, \tag{6.136}$$

$$\varphi'^* = e^{-ig\varepsilon}\varphi^* = \varphi^* - ig\varphi^*\varepsilon + \ldots. \tag{6.137}$$

All terms in $\mathcal{L}$ proportional to $\varphi^*\varphi$ are clearly invariant. The same can be said for the kinetic energy term $\varphi^{*\mu}\varphi_{\mu}$ because $e^{\pm ig\varepsilon}$ does not depend on any of the spacetime coordinates, and therefore the derivative $\partial^{\mu}\varphi^*$ transforms the same as φ^* itself, and likewise for $\partial_{\mu}\varphi$ and φ. If the functional happens to be invariant, then is made an extremal, Noether's conservation law says

$$\partial_{\mu}j^{\mu} = 0, \tag{6.138}$$

where

$$j^{\mu} = p^{\mu}\zeta + p^{*\mu}\zeta^* - \mathcal{H}^{\mu}{}_{\nu}\tau^{\nu}. \tag{6.139}$$

For a pure global gauge transformation all the τ^{ν} vanish, so the surviving piece of the Noether conservation law may be written in terms of a current of the scalar field φ,

$$j^{\mu}_{(\varphi)} = ig(\varphi^{*\mu}\varphi - \varphi^*\varphi^{\mu})$$

$$\equiv ig(\varphi^* \overleftrightarrow{\partial}^{\mu}\varphi) \tag{6.140}$$

---

[16]The $\frac{1}{2}$ survives in eq. (6.101), even though χ is complex, because the y and z of eq. (6.95) are not normalized.

where we have introduced the notation

$$A \overset{\leftrightarrow}{\partial} B \equiv (\partial A)B - A(\partial B) \tag{6.141}$$

so characteristic of scalar field currents.

This $j^{\mu}_{(\varphi)}$ is the conserved current of a charged scalar field. Integrating over all space and using the divergence theorem, we are left with the conserved charge

$$g \int_{V} (\varphi^{*0}\varphi - \varphi^{*}\varphi^{0}) \, dV = \text{const.} \tag{6.142}$$

Here is an interesting aside: Unlike the corresponding case with the non-relativistic quantum wave function, the present integrand, which "ought" to be a probability density, and thus with the charge g "ought" to be the charge density, has a problem: The integrand is not positive-definite! Probability densities must be non-negative, because a probability, as a fraction of occurrences, exists on the closed interval [0,1]. This was an infamous problem with the Klein-Gordon equation, and Noether's theorem reveals it.[17] In the early days of relativistic quantum theory, the failure of the Klein-Gordon probability current to be positive-definite was a motivation for abandoning it and developing the Dirac equation as an alternative. But after the Dirac equation came along, it was realized that the Klein-Gordon equation could be rescued after all, with some reinterpretation—but that's another story [see, e.g., Bjorken & Drell (1964) 8].

Let us press on to *local* gauge invariance, using the scalar field for illustration, in chapters 7 and 8. There the distinction will be made between what Emmy Noether called her "first" and "second" theorems. Chapter 7 pushes local gauge invariance as far as possible with Noether's first theorem. In chapter 8 we bring out the big guns of Noether's second theorem.

Questions for Reflection and Discussion

Q6.a. How would the wave equation for the wave on the string be modified to include (a) the string's weight, (b) stretching of the string, or (c) the amplitude of oscillation not necessarily small? [See Holmes (1994).]

Q6.b. What modifications occur for the invariance identity to accommodate the complex scalar field?

Q6.c. Rotations, displacements in space and time, and velocity boosts, when done together, require more than one parameter ε in writing the transformation equations. Generalize the transformations, the definition of

---

[17]This is a problem for scalar field theory, not for Noether's theorem.

invariance, the invariance identity, and the Noether conservation law when there are a set of M transformation parameters $\{\varepsilon^k, k = 1, 2, \ldots, M\}$. In four-dimensional spacetime, what is the maximum number of transformation parameters necessary for velocity boost between inertial reference frames *and* rotations of axes *and* displacing the origins of both time and space coordinates? Find out whatever you can about the "Poincaré group."

Q6.d. If $\lambda = 0$ in the equation $\partial_\mu K^\mu = \lambda$, we have an equation of continuity. How do we interpret the equation if $\lambda \neq 0$? For a specific example, consider Poynting's theorem of electrodynamics,

$$\nabla \cdot \mathbf{S} + \frac{\partial \eta}{\partial t} = -\mathbf{j} \cdot \mathbf{E} \tag{6.143}$$

where $\mathbf{E}$ is the electric field, $\mathbf{B}$ the magnetic field, $\mathbf{S} = (\mathbf{E} \times \mathbf{B})/\mu_o$ denotes Poynting's vector, $\mathbf{j}$ the density of conduction current, and η the electromagnetic energy density,

$$\eta \equiv \tfrac{1}{2}\epsilon_o \mathbf{E}^2 + \frac{1}{2\mu_0}\mathbf{B}^2. \tag{6.144}$$

Can this apparent nonconservation law (nonconservation of *what*) be interpreted as a conservation law (conservation of *what*)?

Q6.e. Should we worry about "induced" transformations of the field $\varphi(x^\mu)$ due to transformations of the spacetime coordinates x^μ? We have been assuming that τ^μ and ζ are given separately. But these generators themselves depend on the coordinates. We might wonder: Can a spacetime generator τ induce a field transformation generator ζ, even when there is no explicit transformation of the fields? To see what I mean, consider the spacetime transformation

$$x'^\mu = x^\mu + \tau^\mu \varepsilon, \tag{6.145}$$

with no ζ given explicitly. In terms of the new spacetime coordinates, the field becomes

$$\varphi(x^\mu) \to \varphi(x'^\mu). \tag{6.146}$$

We might want to interpret this as a transformed field and write

$$\varphi' = \varphi(x'^\mu), \tag{6.147}$$

then expand it in a Taylor series about $\varepsilon = 0$, that is to say, about $x'^{\mu} = x^{\mu}$, to find some "induced" generator ζ:

$$\varphi(x'^{\mu}) = \varphi(x^{\mu}) + (x'^{\mu} - x^{\mu}) \left(\frac{\partial \varphi}{\partial x'^{\mu}} \right)_{x'^{\mu}=x^{\mu}} + \cdots$$

$$= \varphi^{\mu} + \varepsilon \tau^{\mu} \left(\frac{\partial \varphi}{\partial x'^{\mu}} \right)_{0} + \cdots$$

(6.148)

where it looks like a generator ζ proportional to τ^{μ} has been produced. But not so fast! The field φ has something to say about this. Suppose φ is a scalar field. What does it mean to say that φ is a scalar field? It means φ is invariant under a transformation of the spacetime coordinates. In other words, at any event in spacetime, the numerical value of φ at that event does not depend on the choice of coordinate system; under a change of coordinates, $\varphi' = \varphi$. In that case, $\zeta = 0$. If you want a nonzero ζ for a scalar field, you have to put it there, as a global gauge transformation (for example) does. The Lorentz transformation, by itself, generates no ζ for a *scalar* field.

However, if φ is not a scalar field, then τ^{μ} *may* induce field generators ζ^{ν}. We revisit this issue when we take up vector and higher-rank tensor fields in chapters 7 and 8. There we see that a $\tau^{\mu}(x^{\nu})$ may induce a nonzero ζ^{ν}. As a warm-up exercise, show that if $x'^{\mu} = x^{\mu} + \varepsilon \tau^{\mu}$, and if A'^{μ} means $A^{\mu}(x'^{\nu})$, then a Taylor series expansion of A'^{μ} about $x'^{\mu} = x^{\mu}$ gives, to first order in ε,

$$A'^{\mu} = A^{\mu} + \varepsilon \tau^{\nu} (\partial_{\nu} A^{\mu}).$$

(6.149)

Exercises

6.1. Consider the functional

$$\Gamma = \int_{\Re} \sqrt{1 + \varphi_x^2 + \varphi_y^2 + \varphi_z^2} \, dxdydz,$$

(6.150)

where the subscripts denote partial derivatives.
a. Find the canonical momenta and construct the Hamiltonian tensor components; and
b. write the Euler-Lagrange equations.

6.2. Consider the real scalar field $\varphi(t, x)$. The Lagrangian density, including a potential energy density u, is

$$\mathcal{L} = \frac{1}{2} \varphi_{\mu} \varphi^{\mu} - u(\varphi),$$

(6.151)

where the subscripts and superscripts denote partial derivatives.
a. Write the Euler-Lagrange equation of motion.

b. Construct the canonical momenta and the Hamiltonian tensor.

c. Consider a time translation $t \to t' = t + \varepsilon$. Show that the functional is invariant and find the conservation law.

d. Consider a space translation $x \to x' = x + \varepsilon$. Show that the functional is invariant and find the conservation law.

6.3. Consider the Lagrangian density of ex. 6.2, and the Lorentz boost described in appendix B, eqs. (B.20)–(B.22).

a. Show the functional to be invariant under the infinitesimal transformation.

b. Determine the conservation law.

6.4 Poisson's equation arises in static field theories, notably in Newtonian gravitation and electrostatics:

$$\frac{\partial^2 \Phi}{\partial x^2} + \frac{\partial^2 \Phi}{\partial y^2} + \frac{\partial^2 \Phi}{\partial z^2} = -\rho \qquad (6.152)$$

where Φ denotes the gravitational or electrostatic potential, and ρ denotes the density of mass or electric charge (including necessary constants). Notice that, in Euclidian space, superscripts versus subscripts on the coordinates and gradients makes no difference because the metric tensor is the Kronecker delta, $\delta_{\mu\nu} = 1$ if $\mu = \nu$, and $\delta_{\mu\nu} = 0$ if $\mu \neq \nu$.

a. Construct a Lagrangian density that gives Poisson's equation as the Euler-Lagrange equation.

b. Construct the canonical momenta.

c. Show that the Hamiltonian tensor components may be written

$$\mathcal{H}_{jk} = T_{jk} + \delta_{jk}\rho\Phi, \qquad (6.153)$$

for $j, k = 1$, 2, or 3, and where $T_{jk} \equiv E_j E_k - \delta_{jk}\eta$, with $E_j \equiv -\partial\Phi/\partial x_j$ and $\eta \equiv \frac{1}{2}(E_1^2 + E_2^2 + E_3^2)$.

d. Consider a rotation about the z-axis, with rotation matrix (for coordinates $(x^0, x^1, x^2, x^3) = (t, x, y, z)$)

$$\{\Lambda^\mu{}_\nu\} = \begin{bmatrix} 1 & 0 & 0 & 0 \\ 0 & \cos\varepsilon & \sin\varepsilon & 0 \\ 0 & -\sin\varepsilon & \cos\varepsilon & 0 \\ 0 & 0 & 0 & 1 \end{bmatrix}. \qquad (6.154)$$

Find the generators and show from the invariance identity, or by any other valid method, that the functional is invariant under the infinitesimal transformation.

e. Show that the Noether equation of continuity becomes

$$\mathbf{r} \times \mathbf{S} = \text{const.}, \qquad (6.155)$$

where **r** denotes the position vector. Identify the vector **S**. What would the same transformation yield for $\partial_\mu \partial^\mu \Phi = -\rho$, where $\mu = 0, 1, 2, 3$?

6.5. Consider the Lagrangian

$$L = \frac{1}{2}m\mathbf{v} \cdot \mathbf{v} + q\mathbf{A} \cdot \mathbf{v} - qV \tag{6.156}$$

for a nonrelativistic particle of mass m and electric charge q moving with velocity **v** in a reference frame that sees electromagnetic potentials **A** and V. The electric and magnetic fields **E** and **B** are given, respectively, by $\mathbf{E} = -\nabla V - \partial \mathbf{A}/\partial t$ and $\mathbf{B} = \nabla \times \mathbf{A}$.

a. Show that the Euler-Lagrange equations give Newton's second law with the Lorentz force.

b. Find the canonical momentum. Is it the same as the $m\mathbf{v}$ momentum?

c. Construct the Hamiltonian. Although the Hamiltonian is formally a function of the canonical momenta and not the velocities, write it in two ways: in terms of canonical momentum, and in terms of velocity. Is the Hamiltonian numerically equal to the particle's kinetic energy plus potential energy? *What* energy is recorded by this Hamiltonian?

d. Consider a Lorentz transformation (written here with c absorbed into t, measuring time in meters), where ε denotes the relative velocity between two reference frames:

$$\begin{aligned} t' &= \gamma(t - \varepsilon x), \\ x' &= \gamma(x - \varepsilon t), \\ y' &= y, \\ z' &= z, \end{aligned} \tag{6.157}$$

with $\gamma = (1 - \varepsilon^2)^{-1/2}$. Find the generators of the infinitesimal transformation and see under what circumstances the functional would be invariant.

e. If such invariances exist, find the Noether conservation laws that result when the functional is also made an extremal.

6.6. Generalize the invariance identity from the scalar field, eq. (6.65), to the complex scalar field, where

$$\mathcal{L} = \mathcal{L}(x^\mu, \varphi, \varphi^*, \varphi_\mu, \varphi_\mu^*). \tag{6.158}$$

6.7. a. Consider a nonrelativistic particle of mass m moving in three spatial dimensions, under the influence of the potential energy function U. Show that the Lagrangian density

$$\mathcal{L}(\psi, \psi^*, \psi_k, \psi_k^*) = -\frac{\hbar^2}{2m}\nabla\psi^* \cdot \nabla\psi + \frac{1}{2}i\hbar(\psi^*\psi_0 - \psi_0^*\psi) - \psi^*U\psi \tag{6.159}$$

gives the three-dimensional time-dependent Schrödinger equation,

$$-\frac{\hbar^2}{2m}\nabla^2\psi + U\psi = i\hbar\frac{\partial\psi}{\partial t} \tag{6.160}$$

by using the Euler-Lagrange equation for ψ^*. Here $\psi = \psi(\mathbf{r}, t)$, $\psi_0 \equiv \partial\psi/\partial t$, $\psi_k \equiv \partial\psi/\partial x^k$.

b. Show that the Euler-Lagrange equation for ψ gives the complex conjugate of the Schrödinger equation.

c. Write the canonical momenta and Hamiltonian densities in terms of ψ^* and ψ. How does the canonical momentum compare with the usual quantization rule, that momentum is $(\hbar/i)\nabla$?

d. Consider a global phase transformation $\psi' = e^{i\delta}\psi$, where δ is a real constant. Is the functional invariant? If so, what quantity is conserved?

e. Investigate whether this Lagrangian density is invariant under a time translation, a space translation, and a Lorentz boost (done separately). If it is invariant, derive the Noether conservation law. Check the Galilean relativity limit.

6.8. In the system of ex. 6.7, separate the real and imaginary parts of ψ by writing $\psi = X + iY$, where $X^* = X$ and $Y^* = Y$. In this exercise, let $\partial_\mu X$ be denoted X_μ, and similarly for Y_μ.

a. Write the Lagrangian density $\mathcal{L}(X, Y, X_\mu, Y_\mu)$ and write out the Euler-Lagrange equations for X and Y. Show the results to be the pair of coupled real equations

$$-\frac{\hbar^2}{2m}\nabla^2 X + UX = -\hbar\frac{\partial Y}{\partial t} \tag{6.161}$$

and

$$-\frac{\hbar^2}{2m}\nabla^2 Y + UY = \hbar\frac{\partial X}{\partial t}. \tag{6.162}$$

b. Assume X and Y are both expressible as superpositions of harmonic waves, each harmonic having wavenumber $\mathbf{k} = (k_x, k_y, k_z)$, so that

$$X(t, \mathbf{r}) = \sum_{\mathbf{k}} a_{\mathbf{k}} e^{i(\mathbf{k}\cdot\mathbf{r} - \omega t)} \tag{6.163}$$

and similarly for Y with coefficients $b_{\mathbf{k}}$. For each pair $a_{\mathbf{k}}$ and $b_{\mathbf{k}}$, show that $\mathbf{p}^2/2m + U = \pm i\hbar\omega$, where $\mathbf{p} = \hbar\mathbf{k}$. It appears we have a choice in doing quantum mechanics: We may have have complex wave functions and real energies, or real wave functions and complex energies. Why is the former preferred?

c. Consider the two-parameter transformation

$$x'^\mu = x^\mu,$$
$$X' = X + \varepsilon_1, \qquad (6.164)$$
$$Y' = Y + \varepsilon_2.$$

Determine whether $\mathcal{L}$ is divergence invariant under the given transformation. If so, what is the conserved quantity when the functional is also an extremal?

6.9. For the Lagrangian densities of the real and complex fields, whose quanta are electrically neutral and charged respectively,

$$\mathcal{L}^{(\text{neutral})} = \frac{1}{2}(\partial_\mu \varphi)(\partial^\mu \varphi) - \frac{1}{2}m^2\varphi^2, \qquad (6.165)$$

$$\mathcal{L}^{(\text{charged})} = (\partial_\mu \varphi^*)(\partial^\mu \varphi) - m^2\varphi^*\varphi, \qquad (6.166)$$

make a list of the transformations that lead to Noether conservation laws. Compare the conserved quantities on your two lists.

6.10. Show that the invariance identity for fields may also be written

$$-(\zeta - \varphi_\nu \tau^\nu)\left(\frac{\partial \mathcal{L}}{\partial \varphi} - \partial_\rho p^\rho\right) = \partial_\rho[p^\rho\zeta - \mathcal{H}_\mu{}^\rho \tau^\mu]. \qquad (6.167)$$

Hint: Consider the total derivatives

$$\frac{d\mathcal{L}}{dx^\mu}, \qquad \frac{d}{dx^\mu}\left(\zeta\frac{\partial \mathcal{L}}{\partial \varphi_\mu}\right), \qquad \frac{d}{dx^\mu}\left(\varphi_\rho \tau^\rho \frac{\partial \mathcal{L}}{\partial \varphi_\mu}\right) \qquad (6.168)$$

and make some substitutions in the invariance identity, eq. (6.65).

6.11. Prove the assertion that the charge and current densities $\{j^\nu\} = (\rho, \mathbf{j}) = (\rho, \rho\mathbf{v})$ ($\mathbf{v}$ is the charged particle's velocity) form the components of a four-vector, namely, that they transform the same way as the spacetime coordinate differentials dx^μ under a Lorentz transformation. Suggestion: Begin by going to the charge distribution's rest frame.

6.12. A wave travels down a string of linear density μ that has been tightened to tension T. Let the undisturbed string define the x axis, and consider a wave plane-polarized in the y-direction. Let a damping force per unit mass $-b\dot{y}$ act on the string, where $b = $ const. Including the string's weight, the y-component of Newton's second law gives, for a little string segment of mass dm,

$$-(dm)g + T(\sin\theta_2 - \sin\theta_1) - b(dm)\dot{y} = (dm)\ddot{y}, \qquad (6.169)$$

where θ is the angle of the string above the x-axis. Assuming small displacements and neglecting any stretching of the string (so that μ and T can be assumed constant), with some rearrangement the equation for the damped wave on the string becomes

$$\frac{\partial^2 y}{\partial x^2} - \frac{\partial^2 y}{\partial s^2} - \lambda \frac{\partial y}{\partial s} = \gamma, \tag{6.170}$$

where $s = vt$, $v^2 = T/\mu$, $\lambda = b/v$, and $\gamma = g/v^2$.

a. Show that the Lagrangian density, whose Euler-Lagrange equation gives the foregoing inhomogeneous wave equation, is

$$\mathcal{L}(s, y, y_0, y_1) = \frac{1}{2} \left(y_1^2 - y_0^2 + \gamma y \right) e^{\lambda s} \tag{6.171}$$

where $y_0 \equiv \partial y/\partial s$ and $y_1 \equiv \partial y/\partial x$.

b. Find the canonical momenta, and show the Hamiltonian tensor to be

$$\mathcal{H}^\mu{}_\nu = \begin{pmatrix} -(\xi + \gamma y) & -y_0 y_1 \\ y_0 y_1 & \xi - \gamma y \end{pmatrix} \tag{6.172}$$

where $\xi \equiv \frac{1}{2}(y_0^2 + y_1^2)$.

c. Show that

$$\xi + \gamma y = \frac{1}{T} \frac{dE}{dx} \tag{6.173}$$

where E is the sum of kinetic energy and the potential energies of the restoring force and gravity.

d. Consider the infinitesimal transformation

$$s' = s + \varepsilon \tau^0,$$

$$x' = x + \varepsilon \tau^1, \tag{6.174}$$

$$y' = y + \varepsilon \zeta.$$

Let the derivatives of the generators be denoted with subscripts, such as $\tau_0^1 \equiv \partial \tau^1/\partial s$ and so on. Impose the invariance identity and show the Killing equations to be

$$\zeta + y(\lambda \tau^0 + \tau_0^0 + \tau_1^1) = 0,$$

$$\zeta_0 = 0,$$

$$\zeta_1 = 0,$$

$$\lambda \tau^0 - \tau_1^1 + \tau_0^0 = 0, \tag{6.175}$$

$$-\lambda \tau^0 - \tau_1^1 + \tau_0^0 = 0,$$

$$\tau_1^0 - \tau_0^1 = 0.$$

e. Show that if $\tau^0 \neq 0$ then $\lambda = 0$, and the Killing equations become

$$\tau^0 = Cs + \kappa x + f(y),$$
$$\tau^1 = Cx + \kappa s + g(y), \tag{6.176}$$
$$\zeta = -2Cy$$

where C and κ are separation constants, and $f(y)$ and $g(y)$ are integration "constants" that arise when integrating partial derivatives over s or x.

e. Suppose we set $f(y) = f_o = \text{const.}$ and $g(y) = g_o = \text{const.}$ Show that the transformations become

$$s' = s(1 + \varepsilon C) + \varepsilon \kappa x + \varepsilon f_o,$$
$$x' = x(1 + \varepsilon C) + \varepsilon \kappa s + \varepsilon g_o, \tag{6.177}$$
$$y' = (1 - 2\varepsilon C)y.$$

Describe in words what kind of transformation results if, among the four arbitrary constants C, f_o, g_o, and κ, one at a time are taken to be nonzero. Can the Galilean transformation be seen as a special case of these transformations?

f. When C, κ, and g_o are all zero but $f_o \neq 0$, what is the Noether conservation law?

6.13 a. Show that the Lagrangian density

$$\mathcal{L} = \alpha \psi_1^* \psi_1 + \beta(\psi^* \psi_0 - \psi_0^* \psi) - \psi^* U \psi, \tag{6.178}$$

where $\alpha \equiv -\hbar^2/2m$ and $\beta \equiv i\hbar/2$, $\psi_0 \equiv \partial \psi / \partial t$ and $\psi_1 \equiv \partial \psi / \partial x$, gives the time-dependent Schrödinger equation in one spatial dimension,

$$-\frac{\hbar^2}{2m} \frac{\partial^2 \psi}{\partial x^2} + U\psi = i\hbar \frac{\partial \psi}{\partial t}. \tag{6.179}$$

b. Find the canonical momenta and show the Hamiltonian tensor to be

$$\mathcal{H}^\mu{}_\nu = \begin{pmatrix} -\alpha \psi_1^* \psi_1 + \psi^* U \psi & \beta(\psi^* \psi_1 - \psi_1^* \psi) \\ \alpha(\psi_1^* \psi_0 + \psi_0^* \psi_1) & \alpha \psi_1^* \psi_1 - \beta(\psi^* \psi_0 - \psi_0^* \psi) + \psi^* U \psi \end{pmatrix}. \tag{6.180}$$

c. Under the infinitesimal transformation

$$t' = t + \varepsilon \tau^0,$$
$$x' = x + \varepsilon \tau^1,$$
$$\psi' = \psi + \varepsilon \zeta, \tag{6.181}$$
$$\psi'^* = \psi^* + \varepsilon \zeta^*,$$

impose the invariance identity and derive the Killing equations:

$$-(\psi^*\zeta + cc)U + (\beta\psi^*\dot{\zeta} + cc) - \psi^*\psi\partial_\mu(U\tau^\mu) = 0,$$

$$\alpha\zeta' + \beta\dot{\tau}^1\psi = 0,$$

$$\zeta + \tau'^1\psi = 0, \tag{6.182}$$

$$\tau'^0 = 0,$$

$$\dot{\tau}^0 - \tau'^1 = 0$$

where cc means complex conjugate, $\dot{\zeta} \equiv d\zeta/dt$, $\zeta' = d\zeta/dx$, and likewise for derivatives of τ^μ.

d. Show that the Killing equations imply

$$\tau^0 = At + t_o$$
$$\tau^1 = Ax + g(t) \tag{6.183}$$

where A is a separation constant, t_o an integration constant, and $g(t)$ an integration "constant" coming from an integration over x.

e. Show that the result for τ^1 of part d, when inserted into part c, implies $\zeta = -A\psi$ and

$$\frac{\hbar}{m}\zeta' = i\dot{g}\psi, \tag{6.184}$$

and that consistency between these expressions for ζ requires

$$\psi \sim e^{-i\dot{g}mx/\hbar A} \equiv e^{ikx} \tag{6.185}$$

so that a wave number $k = -\dot{g}m/\hbar A$ may be identified. This suggests that ψ may be expanded in a superposition of harmonics (a Fourier transform). Show also from part c that τ^1 may be written

$$\tau^1 = Ax - A(\hbar k/m)t + x_o \tag{6.186}$$

where x_o is an integration constant.

f. If $A = 0$ and $x_0 = 0$ in the preceding results, so that

$$\tau^0 = t_o,$$
$$\tau^1 = 0, \tag{6.187}$$
$$\zeta = 0,$$

show that the functional is invariant and the Noether conservation law becomes

$$\int_{-\infty}^{+\infty} \psi^* \left(\frac{p^2}{2m} + U\right)\psi \, dx = \text{const.} \tag{6.188}$$

and interpret it in words.

Chapter 7

Local Gauge
Transformations of Fields

The idea of gauge transformations stems from the old observation that to every continuous symmetry of the Lagrangian there corresponds a conservation law.... Continuing along these lines it is easy to see that in a Lorentz invariant theory, the energy, momentum, and angular momentum can be defined and are conserved....

Here we will be interested in conservation laws that are not consequences of classical space-time symmetries. For every conserved quantum number one can construct *a transformation on the fields which leaves* L *invariant.*
—Ernest S. Abers and Benjamin W. Lee, "Gauge Theories," 1973

7.1   Local Gauge Invariance and Minimal Coupling

In chapter 6 we considered a global gauge transformation on real and complex scalar fields. A gauge transformation alters the phase of the complex field,

$$\varphi' = e^{ig\varepsilon}\varphi. \tag{7.1}$$

With all the $\tau^\mu = 0$, the definition of invariance for the functional requires only the invariance of the Lagrangian density,

$$\mathcal{L}' = \mathcal{L}. \tag{7.2}$$

When the functional is also an extremal, Noether's theorem gives the conservation law in the form of an equation of continuity,

$$\partial_\nu j^\nu = 0. \tag{7.3}$$

In the example of the noninteracting complex field, the current is

$$j^\nu_{(\varphi)} = ig(\varphi^{*\nu}\varphi - \varphi^*\varphi^\nu) \equiv ig(\varphi^* \overset{\leftrightarrow}{\partial}{}^\nu \varphi) \tag{7.4}$$

where $\varphi^\nu \equiv \partial\varphi/\partial x_\nu$. Integration over the spatial volume leads to the conserved charge[1]

$$g \int (\varphi^{*0}\varphi - \varphi^*\varphi^0)dV = \text{const.} \tag{7.5}$$

Two kinds of gauge transformations exist: global and local. In the 1918 publication of her theorem that forms the subject of this book, Emmy Noether distinguished two cases that she called the first and second theorems. The second theorem extends the first theorem, from global to local gauge invariance. In a global gauge transformation, ε has the same value throughout the entire space or spacetime. Imagine an array of analog gauges distributed throughout space. The needle on each gauge reads the value of the field's phase at that point in the space, or at each event in spacetime. Under a global gauge transformation, all the pointer needles rotate in concert by the same amount. We saw in chapter 6 that a global gauge transformation leads to conservation laws expressed with an equation of continuity, such as local charge conservation.

In this chapter, we study local gauge invariance, which in the analogy corresponds to rotating the pointer needles by different amounts at different locations in the space (or at different events in spacetime). It may be difficult to see how such a transformation can leave anything invariant. But under a local gauge transformation, when a gauge needle rotates, if it also emits some kind of signal to compensate for its rotation, the coupled needle-plus-signal system might exhibit some kind of invariance.

In this section we begin with the complex scalar field, whose quanta are spin-0 charged particles carrying mass m. The spacetime coordinates are labeled x^μ with $\mu = 0$ for time and $\mu = 1, 2, 3$ for the space dimensions. The Lagrangian density reads

$$\mathcal{L}^{(\varphi)} = \varphi^*_\mu\varphi^\mu - m^2\varphi^*\varphi. \tag{7.6}$$

Now suppose the field's phase is allowed to change by different amounts from one event in spacetime to another, so that in eq. (7.1) we allow $\varepsilon \to \varepsilon(x^\mu)$. The definition of invariance does not prohibit ε from being a function of the coordinates. Because we are considering purely gauge

---

[1] Again, this is not positive-definite, but that is a problem for scalar fields, not for Noether's theorem.

transformations, the spacetime coordinates are not transformed. But the transformation of the field $\varphi(t, x)$ and its conjugate says[2]

$$\varphi' = e^{ig\varepsilon(x^\mu)}\varphi,$$
$$\varphi'^* = e^{-ig\varepsilon(x^\mu)}\varphi^*. \tag{7.7}$$

In the Lagrangian density, the square of the field, $\varphi^*\varphi$, is nicely invariant, but the derivative term $\varphi_\mu^*\varphi^\mu$ is not, because

$$\partial_\mu\varphi' = e^{ig\varepsilon}[\partial_\mu\varphi + ig(\partial_\mu\varepsilon)\varphi]. \tag{7.8}$$

The first term displays the condition that, when put with its complex conjugate, would leave the kinetic energy term invariant. But the $\partial_\mu\varepsilon$ term spoils the invariance. Instead of abandoning invariance under a local change of phase, consider the possibility of enlarging the Lagrangian density.

Let's enlarge the definition of the derivative to compensate for the phase change when $\varepsilon = \varepsilon(x^\mu)$. In particular, add to the gradient ∂_μ another vector field with components A_μ. In this new kind of derivative, the vector field components A_μ hook onto φ with the coupling constant, or charge g:

$$\partial_\mu\varphi \to (\partial_\mu + igA_\mu)\varphi. \tag{7.9}$$

This algorithm is called "minimal coupling." The modified definition of the derivative is analogous to the covariant derivative of Riemannian geometries (see chapter 8 and appendix F). Because they arise from analogous motivations,[3] in this chapter we call the minimal coupling algorithm a covariant derivative, keeping in mind the distinction between it and the covariant derivative of Riemannian geometry.

The reason for introducing the vector field with components A_μ is the requirement that their transformation guarantees the invariance of the coupled $\varphi - A_\mu$ system under a local gauge transformation.

The A field components carry subscripts (or superscripts, $A^\mu = g^{\mu\nu}A_\nu$) even without being differentiated. Therefore our notation must distinguish between subscripts or superscripts that denote field components, and those that denote partial derivatives.[4] Several notations are used in the

---

[2]For simplicity, we maintain only one spatial dimension for now, $\{x^\mu\} = (t, x)$. This can be generalized to more dimensions by summing over more indices.

[3]The covariant derivative of chapter 8 and appendix F arises from taking derivatives of basis vectors; the minimal coupling algorithm of this chapter comes from heuristically enlarging the system to maintain local gauge invariance. Hence from different motivations come two contexts for covariant derivatives.

[4]Of course, one could do away with subscript notation for derivatives entirely, but then the Euler-Lagrange equation and invariance identity become cluttered with monstrosities such as

$$p_\mu = \frac{\partial\mathcal{L}}{\partial\left(\frac{\partial\varphi}{\partial x^\mu}\right)}. \tag{7.10}$$

literature to denote partial derivatives of A_μ and A^μ with respect to coordinates:

$$\frac{\partial A_\mu}{\partial x^\nu} \equiv \partial_\nu A_\mu \equiv A_{\mu,\nu}$$

$$\frac{\partial A^\mu}{\partial x^\nu} \equiv \partial_\nu A^\mu \equiv A^\mu{}_{,\nu}$$

$$\frac{\partial A_\mu}{\partial x_\nu} \equiv \partial^\nu A_\mu \equiv A_\mu{}^{,\nu}$$ (7.11)

$$\frac{\partial A^\mu}{\partial x_\nu} \equiv \partial^\nu A^\mu \equiv A^{\mu,\nu}$$

where the coordinate with respect to which the derivative is evaluated follows the comma. For the partial derivatives of the scalar field φ, to be consistent with the above usage on the A_μ we should write

$$\frac{\partial \varphi}{\partial x^\nu} \equiv \partial_\nu \varphi \equiv \varphi_{,\nu}$$

$$\frac{\partial \varphi}{\partial x_\nu} \equiv \partial^\nu \varphi \equiv \varphi^{,\nu}.$$ (7.12)

Because φ is a scalar field, it carries a subscript or superscript only when it has been differentiated, so for the usual partial derivative we may write $\partial_\nu \varphi$ either as $\varphi_{,\nu}$ or φ_ν. The φ_ν notation used in chapter 6 is continued here.

The covariant derivative may be dignified with the symbol D_μ:

$$D_\mu \varphi \equiv (\partial_\mu + ig A_\mu)\varphi.$$ (7.13)

Another notation in the literature replaces the comma for partial derivative with a semicolon for covariant derivative. For φ,

$$\varphi_{;\nu} \equiv D_\nu \varphi = \partial_\nu \varphi + ig A_\nu \varphi$$ (7.14)

and for φ^*,

$$\varphi^*_{;\nu} = (D_\nu \varphi)^* \equiv \partial_\nu \varphi^* - ig A_\nu \varphi^*.$$ (7.15)

Similar expressions hold for covariant derivatives with respect to upper indices.[5] Note that in the covariant derivative of φ^*, the entire covariant derivative is complex conjugated. The "gauge fields" A_μ are real, $A_\mu^* = A_\mu$.

---

[5]To keep the covariant derivative of this chapter distinct from the covariant derivative of Riemannian geometry, I use ∇_μ for the latter; see appendix F and chapter 8. The covariant derivative of Riemannian geometry is also denoted with the "comma to semicolon rule." To complicate the story further, the D notation is used in Riemannian geometry for the closely related "intrinsic derivative," which applies when a vector is differentiated along a trajectory with respect to a scalar parameter that labels events on the trajectory; see chapter 8.

Whatever the notation, the reason for introducing $D_\mu\varphi$ is to require the covariant derivative of φ to have the same transformation rule as φ itself under a local gauge transformation. Thus we impose

$$(D_\mu\varphi)' = e^{ig\varepsilon(x)}(D_\mu\varphi). \tag{7.16}$$

This is an essential part of the definition of the A^μ or A_μ, the reason they were introduced in the first place. This will make invariant the kinetic energy term in the Lagrangian density when ∂_μ is replaced with D_μ. Let us see what requirement this puts on the A_μ field components.

When we use eq. (7.13) to write out what eq. (7.16) says, we confront

$$(\partial_\mu + igA'_\mu)\varphi' = e^{ig\varepsilon}(\partial_\mu\varphi + igA_\mu\varphi). \tag{7.17}$$

Writing $\varphi' = e^{ig\varepsilon}\varphi$, several terms cancel, leaving the required transformation property of the vector field component A_μ:

$$A'_\mu = A_\mu - \partial_\mu\varepsilon. \tag{7.18}$$

Likewise,

$$A'^\mu = A^\mu - \partial^\mu\varepsilon. \tag{7.19}$$

The transformation of the gauge fields A_μ or A^μ do not have ε itself, but its derivative!

To modify the Lagrangian density so it is invariant under a local gauge transformation, replace the partial derivatives in the original $\mathcal{L}$ with covariant derivatives: $\partial_\mu\varphi \to D_\mu\varphi$, as illustrated with the complex scalar field, taking $\mathcal{L}^{(\varphi)}$ into $\mathcal{L}^{(\varphi-A)}$:

$$\mathcal{L}^{(\varphi)} = (\partial_\mu\varphi^*)(\partial^\mu\varphi) - m^2\varphi^*\varphi$$
$$\to \quad \mathcal{L}^{(\varphi-A)} = (D_\mu\varphi)^*(D^\mu\varphi) - m^2\varphi^*\varphi + \ldots \tag{7.20}$$

where $+\ldots$ denotes pure A-field terms that describe whatever contributions this new gauge field brings into the system, such as its kinetic energy.[6] After $(D_\mu\varphi)^*(D^\mu\varphi)$ is multiplied out explicitly, the Lagrangian density for the coupled $\varphi - A$ system becomes

$$\mathcal{L}^{(\varphi-A)} = (\partial_\mu\varphi^*)(\partial^\mu\varphi) - m^2\varphi^*\varphi + igA_\mu(\varphi^*{}^\mu\varphi - \varphi^*\varphi^\mu) + g^2 A_\mu A^\mu\varphi^*\varphi + \ldots$$
$$= \mathcal{L}^{(\varphi)} + j^\mu_{(\varphi)}A_\mu + g^2(A_\mu A^\mu)(\varphi^*\varphi) + \ldots \tag{7.21}$$

---

[6]Notice the distinction between $(D_\mu\varphi)^*$, which is the correct term in the gauged Lagrangian, versus $D_\mu\varphi^*$. In $\mathcal{L}^{(\varphi)}$ it made no difference, because $(\partial_\mu\varphi)^* = \partial_\mu\varphi^*$. But in the gauged Lagrangian the distinction matters. Why does it matter? One reason is clearly the i in the covariant derivative, but at a deeper level, an operator such as D_μ turns the state φ into a new state $\Phi_\mu \equiv D_\mu\varphi$. Since the Lagrangian must have $\Phi^{*\mu}$ and Φ_μ appearing together, the kinetic energy term becomes, after gauging, $\Phi^{*\mu}\Phi_\mu = (D_\mu\varphi)^*(D^\mu\varphi)$.

where $j^{\mu}_{(\varphi)}$ is the current of charged scalar particles we met in chapter 6:

$$j^{\mu}_{(\varphi)} \equiv ig(\varphi^* \overleftrightarrow{\partial}^{\mu} \varphi). \tag{7.22}$$

The terms in the Lagrangian density proportional to g and g^2, $j^{\mu}_{(\varphi)} A_{\mu}$ and $g^2(A_{\mu} A^{\mu})(\varphi^* \varphi)$ respectively, describe interactions between the φ and A_{μ} fields. If the coupling constant g is small, then as an approximate treatment the term proportional to g^2 is sometimes neglected [Aitchison & Hey (1982) 25]. We will retain it.

Using the Lagrangian density of the coupled fields, let us evaluate the Euler-Lagrange equation for φ^*,

$$\frac{\partial \mathcal{L}}{\partial \varphi^*} = \partial_{\rho} \left(\frac{\partial \mathcal{L}}{\partial \varphi^*_{\rho}} \right). \tag{7.23}$$

This gives the inhomogeneous Klein-Gordon equation, now featuring $\varphi - A$ interaction terms. After evaluating the derivatives in eq. (7.23) and doing some transposing, we come across the terms

$$\partial_{\mu} \partial^{\mu} \varphi - m^2 \varphi = -2ig A_{\mu} \varphi^{\mu} - ig(\partial^{\mu} A_{\mu})\varphi + g^2(A_{\mu} A^{\mu})\varphi. \tag{7.24}$$

Recognizing that the covariant derivative may operate on any quantity Λ, so that $D_{\mu}\Lambda = \partial_{\mu}\Lambda + ig A_{\mu}\Lambda$, the equation of motion for φ, when interacting with the gauge fields, may also be written

$$D_{\mu}\varphi^{\mu} - m^2 \varphi = -ig D_{\mu}(A^{\mu}\varphi), \tag{7.25}$$

and even more succinctly as

$$D_{\mu} D^{\mu} \varphi - m^2 \varphi = 0. \tag{7.26}$$

The free-particle Klein-Gordon equation, $\partial_{\mu}\partial^{\mu}\varphi - m^2\varphi = 0$, has been neatly made to describe interactions with the gauge fields by replacing $\partial_{\mu}\partial^{\mu}$ with the covariant derivatives $D_{\mu}D^{\mu}$.

The presentation of the A_{μ} dynamics through its Euler-Lagrange equation will have to await the missing "$+\ldots$" terms noted above. We infer them in section 7.2 by taking as our model the dynamics of a vector field we know, the electromagnetic field.

The trail from the field of the charged spin-0 boson to the requirements of local gauge invariance has brought us into a direct encounter with A_{μ}. Now that trail temporarily stops, because we need to know more about this vector field. In section 7.2 we start from a different trail—the electromagnetic field—and look at gauge invariance from its perspective, including the field's interactions with charged matter such as the φ quanta. Then we will be able to connect the two trails together.

7.2 Electrodynamics as a Gauge Theory, Part 1: Field Tensors

Maxwell's equations in differential form are four first-order partial differential equations that relate the electric field $\mathbf{E}$ and magnetic field $\mathbf{B}$ to each other and to their sources. Those sources are the electric charge density ρ and current density $\mathbf{j}$. If the source charges are the only matter present in what is otherwise a vacuum, the Maxwell equations are:

Gauss' law for $\mathbf{E}$:

$$\nabla \cdot \mathbf{E} = \frac{\rho}{\epsilon_o} \tag{7.27}$$

Gauss's law for $\mathbf{B}$:

$$\nabla \cdot \mathbf{B} = 0 \tag{7.28}$$

the Faraday-Lenz law,

$$\nabla \times \mathbf{E} = -\frac{\partial \mathbf{B}}{\partial t} \tag{7.29}$$

and the Amperé-Maxwell law,

$$\nabla \times \mathbf{B} = \mu_o \left(\mathbf{j} + \epsilon_o \frac{\partial \mathbf{E}}{\partial t} \right) \tag{7.30}$$

where ϵ_o and μ_o are, respectively, the electric permittivity and magnetic permeability of empty space, and

$$\mu_o \epsilon_o = \frac{1}{c^2}, \tag{7.31}$$

where c denotes the speed of light in vacuum. The four Maxwell equations can be combined into two second-order partial differential equations with the $\mathbf{E}$ and $\mathbf{B}$ fields themselves,[7] or, as we do here, in terms of their potentials V and $\mathbf{A}$. In the latter strategy local gauge invariance plays a pivotal role, so we pursue it here.

Because the divergence of a curl vanishes identically, from Gauss's law for $\mathbf{B}$, eq. (7.28), we may introduce a vector potential $\mathbf{A}$, where

$$\mathbf{B} = \nabla \times \mathbf{A}. \tag{7.32}$$

Insert this into Faraday's law, eq. (7.29), and recall that the curl of a gradient vanishes identically, so that we may introduce the scalar[8] potential V, to obtain

$$\mathbf{E} = -\nabla V - \frac{\partial \mathbf{A}}{\partial t}. \tag{7.33}$$

---

[7]Take the curl of each of the curl equations, then use vector identity $\nabla \times (\nabla \times \mathbf{V}) = \nabla(\nabla \cdot \mathbf{V}) - \nabla^2 \mathbf{V}$ and the other Maxwell equations.

[8]V is a scalar and $\mathbf{A}$ a vector under rotations and translations in Euclidean space. In special relativity, V and $\mathbf{A}$ are the time and space components respectively of a four-vector, a feature we will soon exploit.

Upon substituting eqs. (7.32) and (7.33) into the Maxwell equations with the source terms, eqs. (7.27) and (7.30), with the assistance of some vector identities we obtain a pair of second-order partial differential equations,

$$\nabla^2 \mathbf{A} - \frac{1}{c^2} \frac{\partial^2 \mathbf{A}}{\partial t^2} = -\mu_o \mathbf{j} + \nabla \left(\nabla \cdot \mathbf{A} + \frac{1}{c^2} \frac{\partial V}{\partial t} \right) \qquad (7.34)$$

and

$$\nabla^2 V + \frac{\partial (\nabla \cdot \mathbf{A})}{\partial t} = -\frac{\rho}{\epsilon_o}. \qquad (7.35)$$

However, these equations are coupled: to solve for V we need $\mathbf{A}$, and to get $\mathbf{A}$ we need V! From elementary calculus, we know that a function $f(x)$ and $f(x) + $ const. have the same derivative. Similarly, because V and $\mathbf{A}$ are introduced through derivatives, $\mathbf{E}$ and $\mathbf{B}$ are unchanged under a local gauge transformation of the form

$$\mathbf{A} \to \mathbf{A}' = \mathbf{A} + \nabla \varepsilon, \qquad (7.36)$$

and

$$V \to V' = V - \frac{\partial \varepsilon}{\partial t} \qquad (7.37)$$

for a function ε whose derivatives exist but is otherwise arbitrary. Under these transformations of the potentials, $\mathbf{E}' = \mathbf{E}$ and $\mathbf{B}' = \mathbf{B}$ (check it out). Note the resemblence between the transformations of eqs. (7.36) and (7.37) and those of eqs. (7.18) or (7.19).

A theorem in vector calculus, called the Helmholtz theorem [Panofsky & Phillips (1962) 2–6] proves that if the curl and the divergence of a vector field are known, then so is the vector field itself. The curl of $\mathbf{A}$ is $\mathbf{B}$, but the divergence of $\mathbf{A}$ has not been specified. Because of gauge invariance, we can choose $\nabla \cdot \mathbf{A}$ to be whatever we like. It works like this: suppose I give you some specific expression for $\mathbf{A}$ and you evaluate its divergence and come up with some number or function α. But you say, "I don't want the divergence of $\mathbf{A}$ to be α, I want it to be β." No problem! Merely invoke gauge invariance and require the divergence of $\mathbf{A}'$ to equal β. The gauge-shifting function ε that makes this happen will no longer be arbitrary, of course, and one speaks of "fixing the gauge." To find out what ε has to be to fix the gauge, merely take the divergence of $\mathbf{A}' = \mathbf{A} + \nabla \varepsilon$, which gives you a Poisson-like equation to solve for the necessary ε:

$$\nabla^2 \varepsilon = \beta - \alpha. \qquad (7.38)$$

With this ε in hand, you now have the new $\mathbf{A}'$ and the new V' which gives you the divergence of $\mathbf{A}'$ that you want. Notice that, even if you are happy with the divergence of $\mathbf{A}$ being α, you can still do a gauge transformation within the α-gauge; in other words $\boldsymbol{\nabla} \cdot \mathbf{A} = \alpha$ forms a family of gauges, because you can shift the potentials in a gauge transformation for which $\nabla^2 \varepsilon = 0$ and maintain $\boldsymbol{\nabla} \cdot \mathbf{A}' = \alpha$.

The coupled differential equations for $\mathbf{A}$ and V, eqs. (7.34) and (7.35), neatly uncouple if we choose the Lorenz gauge,[9]

$$\boldsymbol{\nabla} \cdot \mathbf{A} = -\frac{1}{c^2} \frac{\partial V}{\partial t}. \tag{7.39}$$

This Lorenz condition turns eqs. (7.34) and (7.35) into uncoupled inhomogeneous wave equations for V and $\mathbf{A}$:

$$\nabla^2 \mathbf{A} - \frac{1}{c^2} \frac{\partial^2 \mathbf{A}}{\partial t^2} = -\mu_o \mathbf{j} \tag{7.40}$$

and

$$\nabla^2 V - \frac{1}{c^2} \frac{\partial^2 V}{\partial t^2} = -\frac{\rho}{\epsilon_o}. \tag{7.41}$$

Next let us write these equations covariantly, consistent with special relativity, treating space and time in a unified way. The potentials V and $\mathbf{A}$ may be gathered into a spacetime four-vector,[10]

$$(A^0, A^1, A^2, A^3) \equiv (V, \mathbf{A}), \tag{7.42}$$

along with the four-vector of the charge density and the current density,

$$(j^0, j^1, j^2, j^3) \equiv (\rho, \mathbf{j}). \tag{7.43}$$

The metric tensor of Minkowskian spacetime may be employed (see appendix A), whose nonzero components are

$$\eta_{00} = 1 = -\eta_{11} = -\eta_{22} = -\eta_{33} \tag{7.44}$$

and its multiplicative inverse, with nonzero components

$$\eta^{00} = 1 = -\eta^{11} = -\eta^{22} = -\eta^{33}. \tag{7.45}$$

---

[9]James Maxwell derived the wave equation for electromagnetic fields using the Coulomb gauge, $\boldsymbol{\nabla} \cdot \mathbf{A} = 0$. The Lorenz gauge was introduced by the Danish physicist and mathematician Ludvig Lorenz (1829–1891) to incorporate into his solution of the wave equations the effects of the so-called retarded time, where the signal received by the observer at time t was emitted from the source the time R/c earlier, where R is the distance between source and observer. Because of the similar names, the Lorenz gauge is frequently called the "Lorentz gauge," for Dutch physicist Hendrick A. Lorentz (1853–1928), after whom the Lorentz transformation, Lorentz invariance, and the Lorentz force are rightly named.

[10]The c has been absorbed into the t-coordinate to turn a time increment into a distance. Thus $1s = 3 \times 10^8$m, or equivalently $c = 1$, which implies that $1/\epsilon_o = \mu_o$. In SI units the four-vector potential reads $(V/c, \mathbf{A})$, the velocity-four vector $\gamma(c, \mathbf{v})$, the current density four-vector $(c\rho, \mathbf{j})$, and the four-gradient $(\partial_{ct}, \boldsymbol{\nabla})$.

Spacetime four-vectors with upper or lower indices are related to each other through the metric tensor by $A^\mu = \eta^{\mu\nu} A_\nu$ and $A_\mu = \eta_{\mu\nu} A^\nu$, where repeated indices are summed. Explicitly,

$$(A_0, A_1, A_2, A_3) = (V, -\mathbf{A}). \tag{7.46}$$

Partial derivatives with respect to spacetime coordinates are prototypes of lower-index four-vectors,

$$\{\partial_\mu\} = \left(\frac{\partial}{\partial t}, \mathbf{\nabla}\right), \tag{7.47}$$

which can also be altered into upper-index versions with the aid of the metric tensor, to give

$$\{\partial^\mu\} = \left(\frac{\partial}{\partial t}, -\mathbf{\nabla}\right). \tag{7.48}$$

Returning to electrodynamics, in terms of the A^μ a local gauge transformation may be written covariantly as

$$A'^\mu = A^\mu - \partial^\mu \varepsilon \tag{7.49}$$

(compare to eq. [7.19]), and the Lorenz gauge condition, eq. (7.39), becomes compactly expressed as

$$\partial_\mu A^\mu = 0. \tag{7.50}$$

In the Lorenz gauge the wave equations, eqs. (7.40) and (7.41), for $\mathbf{A}$ and V can be combined and written covariantly as a single expression:

$$\partial_\mu \partial^\mu A^\nu = \mu_o j^\nu. \tag{7.51}$$

If $\nu = 0$, then we get eq. (7.41), and if $\nu = 1, 2$, or 3, we get a component of eq. (7.40).

Notice that upon taking the divergence of eq. (7.51), consistency with the conservation law $\partial_\nu j^\nu = 0$ follows only in the Lorenz gauge, because on the left-hand side the divergence becomes $\partial_\nu (\partial_\mu \partial^\mu A^\nu) = \partial_\mu \partial^\mu (\partial_\nu A^\nu)$. Since the theory is supposed to be gauge-invariant, how can a point of physics as fundamental as local electric charge conservation depend on the choice of gauge? The answer is, of course, that it doesn't. To see this, let us write a gauge-independent version of eq. (7.51) and reevaluate the situation. Return to the equations for the potentials as they appeared before a gauge was selected. The equation for V, eq. (7.35), may be written in terms of the A^μ according to

$$\partial^k \partial_k A^0 - \partial^0 (\partial_k A^k) = \mu_o j^0, \tag{7.52}$$

where k gets summed over 1, 2, 3. This can be regrouped into

$$\partial_k (\partial^k A^0 - \partial^0 A^k) = \mu_o j^0. \tag{7.53}$$

The equation for the A^n ($n = 1$, 2, or 3), eq. (7.34), may be transcribed as

$$\partial_k \partial^k A^n + \partial_0 \partial^0 A^n = \mu_o j^n + \partial^n (\partial_\nu A^\nu) \tag{7.54}$$

and rearranged into

$$\partial_\nu (\partial^\nu A^n - \partial^n A^\nu) = \mu_o j^n. \tag{7.55}$$

In view of eqs. (7.53) and (7.55), we seem beholden to introduce the Faraday tensor with components $F^{\mu\nu}$ defined by

$$F^{\mu\nu} \equiv \partial^\mu A^\nu - \partial^\nu A^\mu, \tag{7.56}$$

or in terms of the comma notation for the partial derivatives,

$$F^{\mu\nu} \equiv A^{\nu,\mu} - A^{\mu,\nu}. \tag{7.57}$$

Notice that the Faraday tensor is antisymmetric:

$$F^{\mu\nu} = -F^{\nu\mu}. \tag{7.58}$$

By eqs. (7.32) and (7.33), in rectangular spatial coordinates the tensor's elements can be expressed as the 4×4 matrix

$$\{F^{\mu\nu}\} = \begin{bmatrix} 0 & -E_1 & -E_2 & -E_3 \\ E_1 & 0 & -B_3 & B_2 \\ E_2 & B_3 & 0 & -B_1 \\ E_3 & -B_2 & B_1 & 0 \end{bmatrix}. \tag{7.59}$$

$F^{\mu\nu}$ is manifestly gauge invariant, as we can quickly see: Under a gauge transformation $A'^\mu = A^\mu - \partial^\mu \varepsilon$, we find at once that $F'^{\mu\nu} = F^{\mu\nu}$.

Now equations (7.53) and (7.55) can be written collectively as

$$\partial_\mu F^{\mu\nu} = \mu_o j^\nu. \tag{7.60}$$

No gauge has been specified. Now when we take the divergence of eq. (7.60), the left-hand-side becomes zero, because $(\partial_\nu \partial_\mu)(F^{\mu\nu}) = (+\partial_\mu \partial_\nu)(-F^{\nu\mu})$. The divergence of $\partial_\mu F^{\mu\nu}$ equals its negative, consistent with $\partial_\nu j^\nu = 0$, and this is true regardless of the gauge.

In the Lagrangian density for the electromagnetic field to follow, we need Faraday tensor components with lower indices. They are produced by lowering the upper indices with the metric tensor, in this case the Minkowskian metric tensor,

$$F_{\mu\nu} = \eta_{\mu\rho} \eta_{\nu\sigma} F^{\rho\sigma}, \tag{7.61}$$

which changes the signs of the components of **E** but leaves **B** alone:

$$\{F_{\mu\nu}\} = \begin{bmatrix} 0 & E_1 & E_2 & E_3 \\ -E_1 & 0 & -B_3 & B_2 \\ -E_2 & B_3 & 0 & -B_1 \\ -E_3 & -B_2 & B_1 & 0 \end{bmatrix}. \tag{7.62}$$

Although we won't need them in what follows, for completeness I should mention that the source-free Maxwell equations,

$$\nabla \cdot \mathbf{B} = 0, \tag{7.63}$$

$$(\nabla \times \mathbf{E}) + \frac{\partial \mathbf{B}}{\partial t} = 0, \tag{7.64}$$

can be conveniently written in terms of the $F_{\mu\nu}$ through a cyclic permutation, as you can readily verify:

$$\partial_\rho F_{\mu\nu} + \partial_\mu F_{\nu\rho} + \partial_\nu F_{\rho\mu} = 0. \tag{7.65}$$

Back to sources: the current j^ν that produces the electromagnetic field must be given before eq. (7.60) can be solved for the $F^{\mu\nu}$. For instance, spin-0 particles that carry electric charge e as sources of the electromagnetic field have for their current

$$j^\mu_{(\varphi)} = ie(\varphi^* \overleftrightarrow{\partial^\mu} \varphi). \tag{7.66}$$

Having emphasized the importance of sources, let's set them aside for the moment and consider the electromagnetic field all by itself, in spacetime away from all charged matter, where $j^\nu = 0$.[11] Let's call this "pure electrodynamics," just light, radio waves, and so on, in a vacuum. We will restore charged matter back into the problem later, after we get a feel for the mathematics of the electromagnetic field all by itelf, in the language of a Lagrangian density. These fields, according to eq. (7.60), are controlled locally by the homogeneous equation

$$\partial_\mu F^{\mu\nu} = 0. \tag{7.67}$$

We require eq. (7.67) to be the Euler-Lagrange equation of some Lagrangian $\mathcal{L}^{(A)}$, a function of the A^μ, or rather their derivatives with respect to spacetime coordinates. The Euler-Lagrange equation for the vector field A^μ looks like this:

$$\frac{\partial \mathcal{L}}{\partial A^\mu} = \partial^\nu \left(\frac{\partial \mathcal{L}}{\partial A^{\mu,\nu}} \right). \tag{7.68}$$

---

[11]The sources have to exist somewhere, but differential equations describe things locally. The electromagnetic field *here*, where no sources reside, is due to charges and their currents *over there* where $j^\nu \neq 0$.

You can verify by direct substitution that the Lagrangian density which turns eq. (7.68) into eq. (7.67) for pure electrodynamics, where $j^\nu = 0$, is[12]

$$\mathcal{L}^{(A)} = -\frac{1}{4\mu_o} F_{\mu\nu} F^{\mu\nu}. \tag{7.69}$$

For the purpose of evaluating $\partial \mathcal{L}^{(A)}/\partial A^{\mu,\nu}$ in eq. (7.68), it's a good idea to express $\mathcal{L}^{(A)}$ entirely in terms of the $A^{\mu,\nu}$ by using the metric tensor:[13]

$$\begin{aligned}
\mathcal{L}^{(A)} &= -\frac{1}{4\mu_o} \eta_{\mu\rho} \eta_{\nu\sigma} F^{\rho\sigma} F^{\mu\nu} \\
&= -\frac{1}{4\mu_o} \eta_{\mu\rho} \eta_{\nu\sigma} (A^{\sigma,\rho} - A^{\rho,\sigma})(A^{\nu,\mu} - A^{\mu,\nu}).
\end{aligned} \tag{7.70}$$

In terms of $\mathbf{E}$ and $\mathbf{B}$, this Lagrangian density is numerically equal to[14]

$$\mathcal{L}^{(A)} = \frac{1}{2\mu_o}(\mathbf{E}^2 - \mathbf{B}^2). \tag{7.71}$$

Throughout section 7.3 we consider pure electrodynamics with no sources. With the powerful tool of Noether's theorem, we investigate the conservation laws that hold for pure electromagnetism. In section 7.4 we restore the charged particles of matter that serve as sources of the electromagnetic field, coupling the particles and fields together, through the minimal coupling algroithm of local gauge invariance.

7.3 Pure Electrodynamics, Spacetime Invariances, and Conservation Laws

Now that we have the Lagrangian density for the electromagnetic field, we can define its canonical momenta and Hamiltonian. According to the usual prescription, the canonical momentum component $p_{\rho\sigma}$ is

$$p_{\rho\sigma} \equiv \frac{\partial \mathcal{L}^{(A)}}{\partial A^{\rho,\sigma}}. \tag{7.72}$$

Making use of

$$\frac{\partial A^{\sigma,\rho}}{\partial A^{\nu,\mu}} = \delta^\sigma{}_\nu \delta^\rho{}_\mu, \tag{7.73}$$

---

[12]The $1/\mu_o$ is included because it will be needed when local sources are restored, according to eq. (7.60).

[13]I say this because the Euler-Lagrange equation, eq. (7.68), was written in terms of $A^{\mu,\nu}$. One could as well write the Euler-Lagrange equation exclusively in terms of $A_{\mu,\nu}$.

[14]In conventional units, $\mathcal{L}^{(A)} = \frac{1}{2}\epsilon_o \mathbf{E}^2 - \frac{1}{2\mu_o}\mathbf{B}^2$.

with careful attention to superscript and subscript manipulations, you can show that

$$p_{\rho\sigma} = \frac{1}{\mu_o} F_{\rho\sigma}. \tag{7.74}$$

Now the Hamiltonian tensor can be worked out for pure electromagnetism:

$$\mathcal{H}^{(A)\mu\nu} = A^{\rho,\mu} p_\rho{}^\nu - \delta^{\mu\nu} \mathcal{L}^{(A)}$$

$$= \frac{1}{\mu_o} (A^{\rho,\mu}) F_\rho{}^\nu - \delta^{\mu\nu} \mathcal{L}^{(A)}.$$

The definition of invariance requires

$$\mathcal{L}' J - \mathcal{L} \sim \varepsilon^s, \tag{7.75}$$

where $s > 1$ and J denotes the Jacobian of the spacetime transformation. Recall that, when this definition is differentiated with respect to ε, after which ε is set equal to zero, we derive the vector field version of the invariance identity:

$$-(\zeta^\mu - A^\mu{}_{,\sigma}\tau^\sigma)\left(\frac{\partial\mathcal{L}^{(A)}}{\partial A^\mu} - \partial_\nu p_\mu{}^\nu\right) = \partial_\nu[p_\rho{}^\nu\zeta^\rho - \mathcal{H}^{(A)}{}_\rho{}^\nu\tau^\rho]. \tag{7.76}$$

The left-hand side becomes zero when the Euler-Lagrange equation also holds, that is, when the functional is an extremal. Then Noether's theorem gives a conservation law in the form of an equation of continuity:

$$\partial_\nu S^\nu = 0, \tag{7.77}$$

where

$$S^\nu \equiv p_\rho{}^\nu\zeta^\rho - \mathcal{H}^{(A)}{}_\rho{}^\nu\tau^\rho. \tag{7.78}$$

The conserved quantities S^ν form a current of electromagnetic field energy and momentum, which we now investigate.

Our attention has been turning to gauge invariance, but let's check the familiar conservation laws of energy and momentum in pure electrodynamics, using Noether's theorem with only the spacetime coordinate transformations,

$$x'^\mu = x^\mu + \varepsilon\tau^\mu. \tag{7.79}$$

Suppose the spacetime transformation is the translation of one coordinate, $\tau^\rho = 1$ with all other $\tau^\mu = 0$. This presents four possible conservation laws, one each for $\rho = 0, 1, 2, 3$:

$$\partial_\nu\mathcal{H}^{(A)}{}_\rho{}^\nu = 0. \tag{7.80}$$

Now recall the pure electromagnetic Hamiltonian, which with the help of $A^{\mu,\rho} = A^{\rho,\mu} + F^{\mu\rho}$ may be written in a way that allows a straightforward interpretation:

$$
\begin{aligned}
\mathcal{H}^{(A)\mu\nu} &= \frac{1}{\mu_o}(A^{\mu,\rho} + F^{\mu\rho})F_\rho{}^\nu - \delta^{\mu\nu}\mathcal{L}^{(A)} \\
&= \frac{1}{\mu_o}F^{\mu\rho}F_\rho{}^\nu - \delta^{\mu\nu}\mathcal{L}^{(A)} + \frac{1}{\mu_o}A^{\mu,\rho}F_\rho{}^\nu.
\end{aligned}
\tag{7.81}
$$

Define the energy tensor for the electromagnetic field, according to

$$
U^{\mu\nu} \equiv \frac{1}{\mu_o}F^{\mu\rho}F_\rho{}^\nu - \delta^{\mu\nu}\mathcal{L}^{(A)}.
\tag{7.82}
$$

In terms of it the Hamiltonian tensor becomes

$$
\mathcal{H}^{(A)\mu\nu} = U^{\mu\nu} + \frac{1}{\mu_o}A^{\mu,\rho}F_\rho{}^\nu.
\tag{7.83}
$$

Now eq. (7.77) becomes

$$
\begin{aligned}
0 &= \partial_\nu[U^{\mu\nu} + \frac{1}{\mu_o}(\partial^\rho A^\mu)F_\rho{}^\nu] \\
&= \partial_\nu U^{\mu\nu} + \frac{1}{\mu_o}(\partial_\nu\partial^\rho A^\mu)F_\rho{}^\nu + \frac{1}{\mu_o}(\partial^\rho A^\mu)\partial_\nu F_\rho{}^\nu.
\end{aligned}
\tag{7.84}
$$

These steps are useful because, by virtue of the Euler-Lagrange equation (7.67), the last term vanishes in source-free regions.[15] Furthermore, the second term vanishes identically because it equals its negative under the exchange of ρ and ν; the Faraday tensor is antisymmetric and the second derivatives are symmetric. Therefore, Noether's conservation law for space-time transformation invariance in source-free electrodynamics reduces to

$$
\partial_\nu U^{\mu\nu} = 0,
\tag{7.85}
$$

where $\mu = 0, 1, 2, 3$. If $\mu = 0$ for a time translation, this conservation law becomes the matter-free version of Poynting's theorem, an equation of continuity for field energy in source-free regions:

$$
\nabla \cdot \mathbf{S} + \frac{\partial\eta}{\partial t} = 0,
\tag{7.86}
$$

where $\eta \equiv U^{00} = \frac{1}{2\mu_o}(\mathbf{E}^2 + \mathbf{B}^2)$ denotes the energy density of the electromagnetic field, and $U^{0k} = S_k$, the kth component of Poynting's vector,[16]

---

[15]Instead of vanishing, the last term becomes $(\partial^\rho A^\mu)j_\rho$ when $j_\rho \neq 0$.

[16]John H. Poynting (1852–1914) was an English physicist who in 1884 introduced the vector that bears his name. For another way to derive Poynting's theorem, $\nabla \cdot \mathbf{S} + \partial\eta/\partial t = -\mathbf{j} \cdot \mathbf{E}$, evaluate $\nabla\cdot(\mathbf{E} \times \mathbf{B})$, then use some vector identities and Maxwell's equations.

$\mathbf{S} = (\mathbf{E} \times \mathbf{B})/\mu_o$, where $k = 1, 2$ or 3 denotes x, y and z coordinates. Integrate the source-free Poynting's theorem over a volume $\mathcal{V}$ bounded by the closed surface $\mathcal{S}$, and use Gauss's divergence theorem, to obtain

$$\oint_{\mathcal{S}} \mathbf{S} \cdot \hat{\mathbf{n}} \, da = -\frac{d}{dt} \int_{\mathcal{V}} \eta \, d\mathcal{V} \qquad (7.87)$$

where $\hat{\mathbf{n}}$ denotes an outwardly pointing unit vector normal to the patch of surface area da that resides on the surface $\mathcal{S}$ that encloses volume $\mathcal{V}$. This integral form of Poynting's theorem says that the rate at which the electromagnetic energy decreases within $\mathcal{V}$ equals the flux of $\mathbf{S}$ through the surface $\mathcal{S}$. In field theory, as in particle mechanics, invariance of the system under a time translation implies energy conservation.

If $\mu = 1$, which describes a spatial translation parallel to the x-axis, the equation of continuity for the Hamiltonian density gives the electromagnetic analog of Newton's second law. For then eq. (7.85) becomes

$$-\partial_0 S_1 + \partial_1 T_{11} + \partial_2 T_{21} + \partial_3 T_{31} = 0, \qquad (7.88)$$

where the electromagnetic energy-momentum tensor components[17] are

$$T_{kn} = -\frac{1}{\mu_o}\left(F_{k\rho}F^\rho{}_n - \frac{1}{4}\delta_{kn}F_{\rho\sigma}F^{\rho\sigma}\right)$$
$$= \frac{1}{\mu_o}(E_k E_n + B_k B_n) - \delta_{kn}\eta. \qquad (7.89)$$

Therefore, the equation of continuity for the Hamiltonian density with $\mu = k$ becomes

$$\frac{\partial T_{kn}}{\partial x^k} = \frac{\partial S_n}{\partial t}. \qquad (7.90)$$

After integration over a volume bounded by a closed surface and the use of the divergence theorem, this says the decrease of the electromagnetic field's momentum density (the role of $\mathbf{S}$ in this context) within the volume equals the flux of the Maxwell stress tensor through the surface that bounds the volume:

$$\oint_{\mathcal{S}} T_{kj}\hat{n}^j \, da = -\frac{d}{dt} \int_{\mathcal{V}} S_k \, d\mathcal{V}, \qquad (7.91)$$

where $\hat{n}^j$ denotes the jth component of the unit normal vector $\hat{\mathbf{n}}$.

As an aside, if the electromagnetic field quantum—the photon—were to somehow acquire a mass M, the Lorenz-gauge version of the equation of motion, eq. (7.51), would (without the source term) have to fit the free-particle Klein-Gordon equation,

$$\partial_\mu \partial^\mu A^\nu - M^2 A^\nu = 0. \qquad (7.92)$$

---

[17]Also called the "Maxwell stress tensor."

For this to be an Euler-Lagrange equation, the required Lagrangian density would be

$$\mathcal{L}^{(A,M)} = -\frac{1}{4\mu_o} F_{\mu\nu} F^{\mu\nu} + \frac{1}{2} M^2 A_\nu A^\nu. \tag{7.93}$$

Under a local gauge transformation $A'^\nu = A^\nu - \partial^\nu \varepsilon$, the $\partial^\nu \varepsilon$ term cancels from the $F'_{\mu\nu} F'^{\mu\nu}$ term, but not from the $M^2 A'_\nu A'^\nu$ term. If we want local gauge invariance in electrodynamics, a freely propagating photon must carry zero mass.

What is the measured value of the photon mass? Measurements always include uncertainties, and the measured photon mass is less than 4×10^{-48} g, about 21 orders of magnitude smaller than the electron's mass [Jackson (1975) 6] Such measurements support the concept of the photon mass being exactly zero. The requirement of local gauge invariance gives a profound theoretical reason for thinking so.

7.4 Electrodynamics as a Gauge Theory, Part 2: Matter-Field Interactions

We now have everything we need to couple charged matter to radiation. The main features of minimal coupling, local gauge invariance, and conservation laws are illustrated with the electromagnetic interactions of charged spin-0 particles, such as the charged pion.[18]

Because we are considering local gauge invariance, where $\varepsilon = \varepsilon(x^\mu)$, when examining this system through Noether's theorem we must turn to Emmy Noether's second theorem. Her first theorem is ideally equipped to deal with spacetime coordinate transformations and global gauge transformations like the ones we considered earlier, transformations of the type[19]

$$x'^\mu = x^\mu + \varepsilon^k \tau_k^\mu \tag{7.94}$$

$$\varphi' = \varphi + \varepsilon^k \zeta_k^\mu \tag{7.95}$$

where each ε^k is uniform across the space, and k is summed over the various ε^k parameters, for instance 10 of them in the case of the Poincaré group.[20]

---

[18]Electrons are more ubiquitous charged particles, but to discuss them properly would require digressions into the Dirac equation and the transformation properties of Dirac spinors [Bjorken & Drell (1964)]. While that would be fun, it would not deepen our understanding of Noether's theorem, other than to provide another important illustration of its use: same principles, different Lagrangian density. Rudimentary introductions to the Dirac Lagrangian density appear in the exercises and in appendix C.

[19]Since the Greek and Latin indices on τ_k^μ and ζ_k^μ are not in the same space, it's okay for the upper and lower indices to stand in the same column.

[20]The 10-parameter Poincaré group of special relativity springs to mind, allowing for a time translation; a spatial transformation each for x, y, and z; rotations about

Such a countable set of discrete ε^k was the topic of Emmy Noether's first theorem, which yields the familiar conservation laws of mechanics and electrodynamics: energy, linear and angular momentum, and electric charge. She denoted such a set of transformations as $\underline{G}_\rho$ where ρ (our k) was a countable index. She wrote,

> The group [of transformations] will be called a finite continuous group $\underline{G}_\rho$ if its transformations are contained in a most general (transformation) depending analytically on ρ essential parameters ε.

In this passage, the discrete ε^k do not depend on the spacetime coordinates. But when the ε^k are functions of the spacetime coordinates, so that $\varepsilon^k = \varepsilon^k(x^\mu)$, then for each k the range of possible transformations varies continuously from one event to another, an uncountable set of parameters. Noether denoted this group of transformations $\underline{G}_{\infty\rho}$, and continued

> Correspondingly, an infinite continuous group $\underline{G}_{\infty\rho}$ is understood to be a group whose most general transformations depend on ρ essentially arbitrary functions $p(x)$ and their derivatives analytically, or at least in a continuous and finite-fold continuously differentiable manner. [Noether (1918); Noether & Tavel (1971)]

The group $\underline{G}_{\infty\rho}$ carried out by our $\varepsilon^k(x^\mu)$, Noether's $p(x)$, allows for local gauge invariance.

Before turning to Noether's second theorem in chapter 8, let's see how far we can push her first theorem in describing the coupled matter-field system that exhibits local gauge invariance. Through Noether's first theorem a vanishing divergence, like an equation of continuity, does indeed emerge. But it includes covariant derivatives, which raises a problem of interpretation: What, precisely, is conserved? Noether's second theorem shows the question to be a genuine one, affirming that the problem of interpretation is intrinsic to local gauge invariance itself.

In the remainder of this chapter, let's forge ahead with Noether's first theorem as far as possible, applying it to systems featuring local gauge invariance. In the mathematical formalism, the interactions are turned on by invoking the covariant derivative of minimal coupling. We have an algorithm:

1. Write the Lagrangian density $\mathcal{L}^{(\varphi)}$ for the φ field;
2. Replace its partial derivatives $\partial_\mu \varphi$ with covariant derivatives $D_\mu \varphi$;

---

the three spatial axes; and a three-component relative velocity vector that describes a Lorentz boost from a rest frame to a moving frame, 10 parameters in all, $\varepsilon^1, \varepsilon^2, \ldots, \varepsilon^{10}$, in the Poincaré group.

3. Add the Lagrangian density $-\frac{1}{4\mu_o}F_{\mu\nu}F^{\mu\nu} \equiv \mathcal{L}^{(A)}$ for the pure electromagnetic gauge field A_μ whose quanta have zero mass.[21]

Let's try it![22]

The Lagrangian density for the noninteracting complex (charged) spin-0 field whose quanta carry mass m reads

$$\mathcal{L}^{(\varphi)} = (\partial_\mu\varphi)^*(\partial^\mu\varphi) - m^2\varphi^*\varphi. \tag{7.96}$$

To introduce the electromagnetic interaction to the scalar field, replace ∂_μ with D_μ and add $\mathcal{L}^{(A)}$, whence the gauged Lagrangian density becomes

$$\mathcal{L}^{(\varphi-A)} = (D_\mu\varphi)^*(D^\mu\varphi) - m^2\varphi^*\varphi - \frac{1}{4\mu_o}F_{\mu\nu}F^{\mu\nu}, \tag{7.97}$$

where

$$D_\mu\varphi = (\partial_\mu + ieA_\mu)\varphi. \tag{7.98}$$

When everything gets multiplied out, an avalanche of terms results (recall $\partial_\mu\varphi \equiv \varphi_\mu$):

$$\mathcal{L}^{(\varphi-A)} = \varphi_\mu^*\varphi^\mu - m^2\varphi^*\varphi + ie[\varphi^\mu\varphi - \varphi^*\varphi^\mu]A_\mu$$
$$+ e^2(A_\mu A^\mu)(\varphi^*\varphi) - \frac{1}{4\mu_o}F_{\mu\nu}F^{\mu\nu} \tag{7.99}$$

$$= \varphi_\mu^*\varphi^\mu - m^2\varphi^*\varphi + j^\mu_{(\varphi)}A_\mu + e^2(A_\mu A^\mu)(\varphi^*\varphi) - \frac{1}{4\mu_o}F_{\mu\nu}F^{\mu\nu}$$

where we recognize the charged particle current density,

$$j^\mu_{(\varphi)} = ie(\varphi^{*\mu}\varphi - \varphi^*\varphi^\mu) = ie(\varphi^* \overset{\leftrightarrow}{\partial}{}^\mu \varphi). \tag{7.100}$$

The $\varphi_\mu^*\varphi^\mu - m^2\varphi^*\varphi$ terms are the kinetic and mass energy densities of the scalar particles, and $-\frac{1}{4\mu_o}F_{\mu\nu}F^{\mu\nu}$ is the kinetic energy density of the electromagnetic field (photons). The $j^\mu_{(\varphi)}A_\mu$ and $g^2(A_\mu A^\mu)(\varphi^*\varphi)$ terms describe electromagnetic interactions between the scalar charged particles and the electromagnetic field.

To simplify the expressions, in the remainder of this chapter let's neglect the mass m of the quanta of the scalar field. While mass is an important

---

[21]Although there is no *harm* in doing so, if we replace the partial derivatives in $\mathcal{L}^{(A)}$ with covariant derivatives by rewriting $F^{\mu\nu}$ as $D^\mu A^\nu - D^\nu A^\mu$, the newly introduced terms merely cancel out, leaving $\partial^\mu A^\nu - \partial^\nu A^\mu$, as you can readily verify. However, there is no *reason* to do this, because the electromagnetic field does not "minimally couple" to itself. The minimal coupling is of the charged matter to the electromagnetic field.

[22]This algorithm is an *ansatz* that usually works, not a fundamental principle. The minimal coupling procedure for electrodynamics is violated in some instances, as shown for example by the anamolous magnetic moments of neutrons and protons; see Huang (1982) 245–247.

part of dynamics, the mass terms in our case study here merely get carried along without being a central point in our study of Noether's theorem.

There are three Euler-Lagrange equations that describe the dynamics: one for φ, one for φ^*, and one for the electromagnetic fields A^μ. Let's examine each of them in turn. Dropping the $(\varphi - A)$ superscript on $\mathcal{L}^{(\varphi-A)}$, let an unadorned $\mathcal{L}$ denote the Lagrangian density for the coupled $\varphi - A$ system. For φ^*,

$$\frac{\partial \mathcal{L}}{\partial \varphi^*} = \partial^\beta \left(\frac{\partial \mathcal{L}}{\partial \varphi^{*\beta}} \right). \tag{7.101}$$

which produces

$$\partial_\mu \partial^\mu \varphi = -ieA_\mu \varphi^\mu - ie[\partial^\mu (A_\mu \varphi) + ieA^\mu (A_\mu \varphi)]. \tag{7.102}$$

This we have seen before (compare eq. 7.24). This may be written

$$D_\mu D^\mu \varphi = 0. \tag{7.103}$$

The interaction terms of φ with the field are buried in the covariant derivatives. The Euler-Lagrange equation for φ,

$$\frac{\partial \mathcal{L}}{\partial \varphi} = \partial^\beta \left(\frac{\partial \mathcal{L}}{\partial \varphi^\beta} \right). \tag{7.104}$$

gives the complex conjugate of the Euler-Lagrange equation for φ, $(D_\mu D^\mu \varphi)^* = 0$.

The Euler-Lagrange equation for A^μ reads

$$\frac{\partial \mathcal{L}}{\partial A^\mu} = \partial^\nu \left(\frac{\partial \mathcal{L}}{\partial A^{\mu,\nu}} \right). \tag{7.105}$$

Recalling that $p_{\mu\nu}^{(A)} \equiv \partial \mathcal{L} / \partial A^{\mu,\nu} = F_{\mu\nu}/\mu_o$, writing $F_{\mu\nu} = g_{\mu\rho} g_{\nu\sigma} F^{\rho\sigma}$ to facilitate the bookkeeping, and noting that $\partial A^{\mu,\nu}/\partial A^{\rho,\sigma} = \delta^\mu{}_\rho \delta^\nu{}_\sigma$, the equation of motion for the electromagnetic field becomes

$$\partial_\mu F^{\mu\nu} = \mu_o[ie(\varphi^* \overleftrightarrow{\partial^\mu} \varphi) + 2e^2 A^\mu \varphi^* \varphi]. \tag{7.106}$$

The "current" on the right-hand side—call it $J^\nu_{(\varphi)}$—picks up a covariant derivative:

$$J^\mu_{(\varphi)} = ie(\varphi^* \overleftrightarrow{\partial^\mu} \varphi) + 2(ie)(-ie)A^\mu \varphi^* \varphi \tag{7.107}$$

$$= ie[\varphi^{*\mu}\varphi - \varphi^* \varphi^\mu - ieA^\mu \varphi^* \varphi - ieA^\mu \varphi^* \varphi]$$

$$= ie[(\varphi^{*\mu} - ieA^\mu \varphi^*)\varphi - \varphi^*(\varphi^\mu + ieA^\mu \varphi)]$$

$$= ie[(D^\mu \varphi)^* \varphi - \varphi^*(D^\mu \varphi)]$$

$$= ig(\varphi^* \overleftrightarrow{D^\mu} \varphi)$$

where $(A\overleftrightarrow{D}B) \equiv (DA)B - A(DB)$. Now the field equations for electromagnetism coupled to the charged scalar particles, eq. (7.106), may be elegantly written as

$$\partial_\mu F^{\mu\nu} = \mu_o J^\nu_{(\varphi)}. \tag{7.108}$$

Looking over the set of Euler-Lagrange equations for the scalar particle and the electromagnetic field, we see they are coupled by virtue of sharing the covariant derivative. These results are the primary elements in describing the interaction between two charged particles. The interaction proceeds through the exchange of photons. One charged particle emits a photon as its wave function's phase changes locally. That photon is absorbed by the second charged particle, which also gets the phase of its wave function changed. Meanwhile, the whole show is locally gauge invariant. In this context photons are called "gauge bosons" because they enforce local gauge invariance.

For the sake of completeness, let's collect the canonical momenta and the Hamiltonian tensor for the coupled matter-radiation system. The momentum conjugate to φ is

$$p^{(\varphi)}_\mu = \frac{\partial \mathcal{L}}{\partial \varphi^\mu} = \varphi^*_\mu - ieA_\mu\varphi^* = (D_\mu\varphi)^*. \tag{7.109}$$

Similarly, we find for the momentum conjugate to φ^*,

$$p^{(\varphi^*)}_\mu = \frac{\partial \mathcal{L}}{\partial \varphi^{*\mu}} = \varphi_\mu + ieA_\mu\varphi = D_\mu\varphi. \tag{7.110}$$

For the electromagnetic field, we have already seen that

$$p^{(A)}_{\mu\nu} = \frac{1}{\mu_o}F_{\mu\nu}. \tag{7.111}$$

Now the Hamiltonian tensor can be constructed:

$$\mathcal{H}_{\mu\nu} = p^{(\varphi)}_\mu\varphi_\nu + p^{(\varphi^*)}_\mu\varphi^*_\nu + p^{(A)\rho}_\mu A_{\nu,\rho} - \delta_{\mu\nu}\mathcal{L}. \tag{7.112}$$

After some work this becomes

$$\mathcal{H}_{\mu\nu} = (\varphi^*_\mu\varphi_\nu + \varphi^*_\nu\varphi_\mu - \delta_{\mu\nu}\varphi^*_\rho\varphi^\rho) \tag{7.113}$$
$$+ A_\mu j_{\nu(\varphi)} - \delta_{\mu\nu}A_\rho j^\rho_{(\varphi)}$$
$$+ \frac{1}{\mu_o}F_\mu{}^\rho A_{\nu,\rho} - \delta_{\mu\nu}\left(e^2 A_\rho A^\rho\varphi^*\varphi - \frac{1}{4\mu_o}F_{\rho\sigma}F^{\rho\sigma}\right).$$

We turn next to the Noether conservation law for the coupled $\varphi - A$ fields and see how far we can push Noether's first theorem.

7.5   Local Gauge Invariance and Noether Currents

With the canonical momenta and Hamiltonian tensor in hand for the charged spin-0 particle coupled to the electromagnetic field, let's see how far we can go with the conservation law of Noether's first theorem under a local gauge transformation. The Lagrangian density for a massless scalar particle, coupled to the electromagnetic field, is

$$\mathcal{L}(\varphi, \varphi^*, \varphi_\mu, \varphi_\mu^*, A_\mu, A^{\mu,\nu}) = (D_\mu\varphi)^*(D^\mu\varphi) - \frac{1}{4\mu_o}F_{\mu\nu}F^{\mu\nu} \qquad (7.114)$$

where

$$D_\mu = \partial_\mu + ieA_\mu \qquad (7.115)$$

and

$$F_{\mu\nu} = \partial_\mu A_\nu - \partial_\nu A_\mu. \qquad (7.116)$$

Here we are not transforming the spacetime coordinates, $x'^\mu = x^\mu$, so all the spacetime generators vanish, $\tau^\mu = 0$. For a local gauge transformation of the fields, where $\varepsilon = \varepsilon(x^\mu)$, the field transforms according to

$$\varphi' = e^{ie\varepsilon}\varphi = \varphi + ie\varepsilon\varphi + \ldots, \qquad (7.117)$$

$$\varphi'^* = e^{-ie\varepsilon}\varphi^* = \varphi^* - ie\varepsilon\varphi^* + \ldots, \qquad (7.118)$$

$$A'^\mu = A^\mu - \partial^\mu\varepsilon \qquad (7.119)$$

so that $\zeta = ie\varphi$ and $\zeta^* = -ie\varphi^*$ are the generators of the scalar fields.

The definition of invariance requires

$$\mathcal{L}'J - \mathcal{L} \sim \varepsilon^s \qquad (7.120)$$

where $s > 1$ and J is the Jacobian of the spacetime transformation, for which $J = 1 + \varepsilon\partial_\mu\tau^\mu$. In a pure gauge transformation $J = 1$. To derive Noether's theorem as done before we proceed through the invariance identity, which follows by differentiating the definition of invariance with respect to ε, and afterward setting $\varepsilon = 0$. Since $J = 1$, we need to evaluate only

$$\left[\frac{d\mathcal{L}'}{d\varepsilon}\right]_0 = 0 \qquad (7.121)$$

where $\mathcal{L}'$ means

$$\mathcal{L}' = (\varphi_\mu'^* - ieA_\mu'\varphi'^*)(\varphi'^\mu + ieA'^\mu\varphi') - \frac{1}{4\mu_o}(A_{\mu,\nu}' - A_{\nu,\mu}')(A'^{\mu,\nu} - A'^{\nu,\mu}). \qquad (7.122)$$

Differentiating the invariance definition with respect to ε (after putting all the indices either upstairs or downstairs using the metric tensor $\eta_{\mu\nu}$ or $\eta^{\mu\nu}$) and working through the derivatives of the Lagrangian density with the chain rule, we have to decide what to do with the transfomation of the gauge field, which is not proportional to ε but to its derivative. In general, ε and $\partial^\mu \varepsilon$ are different functions. Let us proceed in that manner (see ex. 7.9 for another option) by setting $\partial A'^\mu / \partial \varepsilon = 0$.[23] We find

$$\frac{\partial \mathcal{L}}{\partial \varphi}\left[\frac{d\varphi'}{d\varepsilon}\right]_0 + \frac{\partial \mathcal{L}}{\partial \varphi^*}\left[\frac{d\varphi'^*}{d\varepsilon}\right]_0 + \frac{\partial \mathcal{L}}{\partial \varphi^\mu}\left[\frac{d\varphi'^\mu}{d\varepsilon}\right]_0 + \frac{\partial \mathcal{L}}{\partial \varphi^{*\mu}}\left[\frac{d\varphi'^{*\mu}}{d\varepsilon}\right]_0 = 0, \quad (7.123)$$

or in terms of the generators,

$$\frac{\partial \mathcal{L}}{\partial \varphi}\zeta + \frac{\partial \mathcal{L}}{\partial \varphi^*}\zeta^* + \frac{\partial \mathcal{L}}{\partial \varphi^\mu}\zeta^\mu + \frac{\partial \mathcal{L}}{\partial \varphi^{*\mu}}\zeta^{*\mu} = 0. \quad (7.124)$$

If the functional is both invariant and extremal, then we may invoke the Euler-Lagrange equations for φ and φ^*. With $p_\mu = \partial \mathcal{L}/\partial \varphi^\mu$ and $p_\mu^* = \partial \mathcal{L}/\partial \varphi^{*\mu}$, we obtain

$$(\partial^\mu p_\mu)\zeta + (\partial^\mu p_\mu^*)\zeta^* + p_\mu \zeta^\mu + p_\mu^* \zeta^{*\mu} = 0. \quad (7.125)$$

Recalling that $p_\mu = (D_\mu \varphi)^*$ and $p_\mu^* = D_\mu \varphi$, the invariance identity becomes

$$0 = [\partial^\mu(\varphi_\mu^* - ieA_\mu \varphi^*)]\zeta + [\partial^\mu(\varphi_\mu + ieA_\mu \varphi)]\zeta^*$$
$$+ (\varphi_\mu^* - ieA_\mu \varphi^*)\zeta^\mu + (\varphi_\mu + ieA_\mu \varphi)\zeta^{*\mu} \quad (7.126)$$
$$= \partial^\mu[(\varphi_\mu^* \zeta + \varphi_\mu \zeta^*) + ieA_\mu(\varphi \zeta^* - \varphi^* \zeta)].$$

For the scalar field, the generators are $\zeta = ie\varphi$ and $\zeta^* = -ie\varphi^*$, which turns this into the Noether conservation law,

$$\partial_\mu J^\mu_{(\varphi)} = 0 \quad (7.127)$$

where

$$J^\mu_{(\varphi)} \equiv ie(\varphi^{*\mu}\varphi - \varphi^* \varphi^\mu) + 2e^2 \varphi^* \varphi A_\mu = ie\varphi^* \overleftrightarrow{D}^\mu \varphi, \quad (7.128)$$

which we met previously,[24] and thus the conservation law is an equation of continuity, but with a current that includes a covariant derivative,

$$ie\partial_\mu \left(\varphi^* \overleftrightarrow{D}^\mu \varphi\right) = 0. \quad (7.129)$$

---

[23]Recall that the gauge boson sector of the Lagrangian density is designed to be gauge invariant.

[24]The conservation law $\partial_\nu J^\nu_{(\varphi)} = 0$ is consistent with the inhomogeneous Maxwell equation $\partial_\mu F^{\mu\nu} = \mu_o J^\nu$ because $F^{\mu\nu} = -F^{\nu\mu}$ but $\partial_\nu \partial_\mu = \partial_\mu \partial_\nu$.

Integrating over all space, and assuming the fields vanish at spatial infinity, gives the conserved quantity,

$$ie \int_{\mathcal{R}} \left(\varphi^* \overset{\leftrightarrow}{D}{}^0 \varphi \right) \, d\mathcal{V} = const. \qquad (7.130)$$

But what is this the conservation of? Writing out the covariant derivative, this volume integral has two pieces:

$$\int_{\mathcal{R}} j^0_{(\varphi)} \, d\mathcal{V} + 2e \int_{\mathcal{R}} \varphi^* (eA^0) \varphi \, d\mathcal{V} = const. \qquad (7.131)$$

The first term represents the electric charge carried by the matter, the quanta of the scalar field. The second term is the expectation value of an interaction of the electromagnetic potential with the matter fields. Thus this conservation law is not simply a statement about a charge distribution that can be cleanly isolated within a region $\mathcal{R}$; it also includes an *interaction* within $\mathcal{R}$ between an electromagnetic potential energy eA^0 and a charge density $2e\varphi^*\varphi$.

A connection can be made to an analogous situation in electrodynamics where the language of covariant derivatives is not routinely used. Recall Poynting's theorem with the possibility of charged matter present,

$$\boldsymbol{\nabla} \cdot \mathbf{S} + \frac{\partial \eta}{\partial t} = -\mathbf{j} \cdot \mathbf{E}, \qquad (7.132)$$

where $\mathbf{S} \equiv \mathbf{E} \times \mathbf{B} / \mu_o$ denotes Poynting's vector, η the energy density of the electric and magnetic fields, and $\mathbf{j}$ the conduction current density. If $\mathbf{j} = \mathbf{0}$, then the electromagnetic field energy is locally conserved (see section 7.3), and we write

$$\boldsymbol{\nabla} \cdot \mathbf{S} + \frac{\partial \eta}{\partial t} = 0 \qquad (7.133)$$

as

$$\partial_\mu U^\mu = 0. \qquad (7.134)$$

where $\{U^\mu\} = (\eta, \mathbf{S}).$[25] But when $\mathbf{j} \cdot \mathbf{E} \neq 0$, the electromagnetic field energy is no longer locally conserved, because the field exchanges energy with charged matter. However, the energy of the total matter+fields system *is* conserved. If we were clever we might transpose the $\mathbf{j} \cdot \mathbf{E}$ term to the other side of Poynting's theorem,

$$\boldsymbol{\nabla} \cdot \mathbf{S} + \frac{\partial \eta}{\partial t} + \mathbf{j} \cdot \mathbf{E} = 0, \qquad (7.135)$$

---

[25]Modulo factors of c.

then invent a new "covariant derivative"[26]

$$\mathcal{D}_\mu U^\mu \equiv \partial_\mu U^\mu + \mathbf{j} \cdot \mathbf{E}, \tag{7.136}$$

so that eq. (7.135) looks like an equation of continuity,

$$\mathcal{D}_\mu U^\mu = 0. \tag{7.137}$$

In chapter 8 we revisit conservation laws under local gauge invariance, examined through the exquisite microscope of Noether's second theorem.

7.6 Internal Degrees of Freedom

The problems in variation here concerned are such as to admit a continuous group (in Lie's sense).... What is to follow, therefore, represents a combination of the methods of the formal calculus of variations and those of Lie's group theory. —Emmy Noether, "Invariante Variationsprobleme," 1918

This section lies out of the main line of development for Noether's theorem itself. It is included as an introduction to the jargon of a field where Noether's theorem is routinely applied. The relation between symmetries and conservation laws, which Emmy Noether pioneered, has become an essential part of theories of fundamental forces, and local gauge invariance plays a central role. I do what I can here to reduce barriers between the student initiate and the main ideas behind gauge invariance as practiced in elementary particle physics.

Theorists go into this business relying on conservation laws from the outset. They start from what is known or hypothesized to be conserved, then build their theories around those conservation laws. In other words, they start with the conservation laws then reverse engineer to the Lagrangian density. Noether's theorem provides the consistency check between the assumed conservation laws and the demonstration that, under a set of well-defined transformations, the proposed Lagrangian density does, indeed, produce them. Gauge invariance in particle physics is applied this way to so-called internal degrees of freedom. Internal to what?

The quantum state of a particle typically factors into an "external" piece that depends on the particle's spacetime coordinates, and an "internal" piece that describes a set of options the particle carries around with it. To

---

[26]I am not proposing this as a serious candidate covariant derivative; I am only making a point about a system consisting of two interacting parts, in this case matter coupled to radiation. Neither one has an equation of continuity all to itself; the conservation laws must extend to the coupled system. The point is, a vanishing divergence with a covariant derivative looks like an equation of continuity, but it has to be interpreted in terms of the coupled system.

illustrate the distinction between external and internal degrees of freedom, let's try an analogy. Imagine cruising through the mountains in a convertible sports car. To map the car's location at a given time requires a coordinate system external to the car itself. That information is carried by some function $\Psi(t, \mathbf{r})$. But the roadster also carries its own internal variables around with it. For instance, think of the convertible top as a two-state system. When you drive the car the top will be in one of two states: up or down. Which one is realized may depend on the car's interactions with the external environment (raining $\rightarrow$ top up, sunny $\rightarrow$ top down). An arbitrary state of the top can be represented as a two-component column matrix χ,

$$\chi = \begin{pmatrix} a_{up} \\ a_{down} \end{pmatrix} \tag{7.138}$$

where $|a_{up}|^2$ is the probability that the top will be found up, and $|a_{down}|^2$ the probability of it being found in the down position. You don't drive the car with the top in between up or down; when you drive it, the top is either up or it is down. On a cloudy day, let us say two-thirds of the drivers of such roadsters put their top down and the other one-third leave it up. The "state of a roadster top" has $a_{down} = \sqrt{2/3}$ and $a_{up} = \sqrt{1/3}$. Such a variable forms an "internal space" because it is a set of options carried with each car, no matter where it goes. Taking into account both the geographical and the top up/down information, the "roadster state" is $\Psi(t, \mathbf{r})\chi$.

Similarly, as an elementary particle moves through spacetime, its quantum state may carry along with it some quantized "internal coordinate" options, such as one of a set of possible discrete values for the z-component of spin.

The numerical values of a quantized internal degree of freedom can be expressed as the product of a fundamental physical constant rescaled by a dimensionless rational number. For instance, the electric charge Q carried by any particle can be written $Q = qe$, where $e = 1.6 \times 10^{-19}$ C denotes the elementary unit of charge, and $q = -1$ for electrons, $+1$ for positrons, and $= \pm\frac{1}{3}$ or $\pm\frac{2}{3}$ for quarks.

Another example is found in a particle's intrinsic angular momentum, or spin, described by a vector $\mathbf{S}$. In classical mechanics, $|\mathbf{S}|$ and all three of its components can, in principle, be precisely known simultaneously. But in quantum mechanics, to have a spin vector $\mathbf{S}$ means the particle always carries around with it a quantum number s that can have only one of the values $0, \frac{1}{2}, 1, \frac{3}{2}, 2, \frac{5}{2}, \ldots$, where $|\mathbf{S}|^2 = \hbar^2 s(s+1)$. Furthermore, in quantum theory one and only one component, say the z-component, can be known. Its quantized values may be one of $\hbar m_s$ where $m_s = s, s-1, \ldots, -s$.[27]

---

[27]These rules about quantized spin follow from the definition of angular momentum, $\mathbf{L} = \mathbf{r} \times \mathbf{p}$, and the quantum rule that momentum is the operator $(\hbar/i)\boldsymbol{\nabla}$. Consequently,

Particles with integer values of s are bosons. The name comes from Bose-Einstein statistics, where no limit exists on how many identical bosons may be in the same quantum state. Their relativistic quantum equation of motion is the Klein-Gordon equation (see appendix C). Particles with half odd integer values of s are called fermions, from Fermi-Dirac statistics, where no two identical fermions can be in the same quantum state. The relativistic quantum equation of motion for fermions is the Dirac equation (appendix C). For example, the electron, having spin $s = \frac{1}{2}$, has two options for its z-component of spin, $m_s = +\frac{1}{2}$ ("spin up") and $m_s = -\frac{1}{2}$ ("spin down").

Instead of treating spin-up and spin-down electrons as two distinct particles, we see them as different states of one particle. Similarly, despite their differences, for some purposes the proton and neutron can be conceptualized as two basis states of the "nucleon" that carries, by analogy to spin-$\frac{1}{2}$, a quantum number $I = \frac{1}{2}$ called "isotopic spin," or "isospin." The "isospin up" state with $m_I = +\frac{1}{2}$ is identified with the neutron, leaving $m_I = -\frac{1}{2}$ assigned to the proton. The proton gets selected as ground state of the nucleon, because the neutron, being slightly more massive of the two, can spontaneously beta decay into a proton. When the weak nuclear force changes protons and neutrons into one another, the nucleon flips its isospin.

If this talk of internal spaces seems contrived, let us remember what we are doing here. A bewildering array of structures and interactions exist in nature, and to make sense of them we have to create concepts in terms of which their doings can be organized into patterns. Concepts such as electric charge, spin, and isospin help us classify particle species into families whose members can turn into one another in various interactions.

Why should we, as Noether theorem appreciators, care about internal degrees of freedom? Because in the transformations between them, invariances and conservation laws must be considered. Furthermore, internal transformations such as isospin flip can be gauged! The bosons introduced to enforce local gauge invariance among these internal states have been

---

r and p do not commute, which leads to the result that each component of angular momentum commutes with L^2 but none of the components commute with each other. Therefore, the most we can know simultaneously is L^2 and one component, typically taken to be L_z. The z-axis is defined by some other vector that breaks the symmetry of space, such as a magnetic field or the direction of the particle's linear momentum. Since a component of a vector cannot be larger than the entire vector, this leads to the conditions that L^2 can have the values $\hbar^2 \ell(\ell+1)$, where $\ell = 0, \frac{1}{2}, 1, \frac{3}{2}, 2, \frac{5}{2}, 3, \ldots$, and L_z may have the values $\hbar m_\ell$ where $m_\ell = \ell, \ell-1, \ldots, -\ell$. (An important exception exists in massless spin-1 and spin-2 particles, such as the photon and graviton respectively. They lack the $m_s = 0$ state, and have Lorentz transformation properties slightly different from those of massive spin-1 particles that can realize all their spin options.) As an angular momentum, spin respects the algebra just described. To distinguish spin from orbital angular momentum, for spin the quantum number ℓ is denoted s.

observed in the laboratory. Let us examine the mathematical infrastructure used to describe such internal spaces.

An illustrative example of scalar fields grouped into families, or "multiplets" through internal quantum numbers, presents itself with the two states of the kaon, K^+ and the K^0. The superscripts denote their electric charges. Rather than thinking of the K^+ and K^0 as independent creatures, we may think of them as basis states of a generic kaon state, represented as a vector in abstract two-dimensional kaon space (Figure 7.1). To carry out this concept, write the generic kaon state as

$$|K\rangle = \begin{pmatrix} a_+ \\ a_0 \end{pmatrix} \tag{7.139}$$
$$= a_+|K^+\rangle + a_0|K^0\rangle,$$

where

$$|K^+\rangle \equiv \begin{pmatrix} 1 \\ 0 \end{pmatrix} \tag{7.140}$$

denotes the pure K^+ state, and

$$|K^0\rangle \equiv \begin{pmatrix} 0 \\ 1 \end{pmatrix} \tag{7.141}$$

denotes the pure K^0 state. When "asked" in a measurement what kind of kaon it is, the particle has the probability $|a_+|^2$ of answering "I'm a positive kaon," and the probability $|a_0|^2$ of responding "I'm a neutral kaon." Meanwhile, when not asked, the kaon moves merrily along in some unknown mixture of these two eigenstates. So long as the kaon exists, the probabilities for both alternatives must sum to unity:

$$|a_+|^2 + |a_0|^2 = 1. \tag{7.142}$$

The objectives of quantum mechanics include predicting probabilities for a system to be in various states and estimating probabilities for making transitions between those states. Probabilities are real numbers on the closed interval $[0,1]$, computed in the quantum context as the absolute square of the quantum state, or in the case of transitions, absolute square of an interaction matrix element connecting two quantum states. Because we represent the kaon state as an abstract column vector $|K\rangle$, we also need its "adjoint" row vector $\langle K|$:

$$\langle K| \equiv \begin{pmatrix} a_+^* & a_0^* \end{pmatrix} \tag{7.143}$$

where * denotes complex conjugate. Now eq. (7.142) can be expressed compactly as

$$\langle K|K\rangle = 1. \tag{7.144}$$

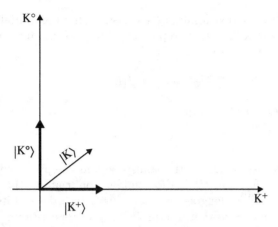

Figure 7.1: *Kaon isospin state space, where the positive and neutral kaons are basis eigenstates for the generic kaon state. An arbitrary kaon state is a superposition of these two eigenstates. In the laboratory, when an experiment detects a K^+ kaon turning into a K^0 kaon, that process gets represented in the isospin space by the state vector rotating from $|K^+\rangle$ to $|K^0\rangle$.*

As an isospin doublet, quantum numbers may be assigned to the K^+ and K^0 eigenstates as follows:

$$
\begin{array}{cccc}
& I & I_z & q \\
K^+ : & \frac{1}{2} & +\frac{1}{2} & +1 \\
K^0 : & \frac{1}{2} & -\frac{1}{2} & 0.
\end{array}
\tag{7.145}
$$

Transitions between the K^+ and K^0 states can be represented with isospin flips. Given the conservation of electric charge, such a notion suggests the reactions

$$K^+ \to K^0 + W^+, \tag{7.146}$$

$$K^0 \to K^+ + W^-. \tag{7.147}$$

The proposed $W^\pm$ particles carry ± 1 units of electric charge because electric charge is conserved. They also carry ± 1 unit of I_z, respectively, if isospin is also conserved. Isospin may or may not be conserved in all interactions, but little would be gained by assigning quantum numbers that are never conserved.[28] Noether's theorem will have something to say about this!

---

[28]Some quantum numbers, such as parity, are not conserved in the weak interaction, but are conserved in the other fundamental forces.

When an interaction results in a transition from an initial state represented by $|K\rangle$ to a final state represented by $|K'\rangle$, the interaction can be represented by a transition matrix U:

$$|K'\rangle = U|K\rangle. \tag{7.148}$$

By the rules of matrix transposition,

$$\langle K'| = \langle K|U^\dagger \tag{7.149}$$

where the row vector $\langle K|$ is the transpose and complex conjutate of the column vector $|K\rangle$, and $U^\dagger$ is the complex conjugate of the transpose of U, its "adjoint."[29] So long as the kaon exists, regardless of its mixture as a K^+ and K^0 superposition, the probability of its existence must remain unity, after the transition as well as before. This means that

$$\langle K'|K'\rangle = \langle K|K\rangle = 1, \tag{7.150}$$

so that

$$\langle K|U^\dagger U|K\rangle = \langle K|K\rangle, \tag{7.151}$$

which requires

$$U^\dagger U = 1, \tag{7.152}$$

where 1 denotes the unit matrix. Therefore, the adjoint of U equals its multiplicative inverse,

$$U^\dagger = U^{-1}. \tag{7.153}$$

Such an operator is said to be "unitary." Any unitary operator can always be written

$$U = e^{i\Lambda}, \tag{7.154}$$

where $i^2 = -1$ and the operator Λ is self-adjoint,

$$\Lambda^\dagger = \Lambda \tag{7.155}$$

and said to be "Hermitian."[30]

If $|K\rangle$ is a solitary complex number (1×1 matrix), then U merely produces a phase shift, $U = e^{i\theta}$, with θ a real number. Such transformations taken as a set form the "unitary group" in one dimension, U(1). We already met U(1); it appeared as the transformation $\varphi' = e^{ig\varepsilon}\varphi$. But when $|K\rangle$

---

[29]The precise meaning of the term "adjoint" depends on the context. In the context of matrices with complex entires, "adjoint" means transpose and complex conjugate. The Hermitian adjoint appears in quantum mechanics. In chapter 8 another adjoint appears in the context of Noether's second theorem.

[30]The bizarre notion of making an operator the argument of a transcendental function makes sense when defined in terms of a Taylor series, so that $e^{i\Lambda} = 1 + i\Lambda + \frac{1}{2!}(i\Lambda)^2 + \dots$.

denotes an abstract vector in an internal space of N dimensions, U and Λ will be represented as $N \times N$ matrices. If the determinant of U is set equal to 1, the trace of Λ vanishes, leaving $N^2 - 1$ independent components. Such transformation matrices are elements of the "special unitary group" in N dimensions, SU(N). This condition on the determinant of U is not as arbitrary as it sounds. Evaluate the determinant of eq. (7.152). Since $|U^\dagger| = |U|$, it follows that $|U|^2 = 1$, and therefore $|U| = \pm 1$. Whether the determinant is $+1$ or -1 is our choice to make.

We can spread the matrix Λ out into a sum of $N^2 - 1$ independent basis matrices denoted $\frac{1}{2}\lambda^k$ where $k = 1, 2, \ldots, N^2 - 1$.[31] Each λ^k matrix has numerical constants for all its matrix elements. A superposition of the $\frac{1}{2}\lambda^k$ matrices forms any matrix Λ needed for SU(N). Thus we can introduce, in the abstract SU(N) space, a vector of transformation coefficients $\varepsilon \equiv (\varepsilon_1, \varepsilon_2, \ldots, \varepsilon_{N^2-1})$, and write

$$\Lambda = \sum_{k=1}^{N^2-1} \varepsilon_k \frac{\lambda^k}{2}, \tag{7.156}$$

denoted more compactly as a scalar product of vectors in the internal space:

$$\Lambda = \frac{1}{2}(\boldsymbol{\varepsilon} \cdot \boldsymbol{\lambda}). \tag{7.157}$$

Let's apply these ideas to a scalar field φ whose quanta, the K^+ and K^0 particles, carry spin-0 and internal degrees of freedom such as isospin. Thus the field φ becomes, in isospin space, a two-component column vector denoted φ or $|K\rangle$,

$$|K\rangle \equiv \begin{pmatrix} \varphi^+ \\ \varphi^0 \end{pmatrix} \equiv \varphi. \tag{7.158}$$

Its corresponding adjoint is

$$\langle K| = \begin{pmatrix} \varphi^{*+} & \varphi^{*0} \end{pmatrix} \equiv \varphi^\dagger. \tag{7.159}$$

Under a gauge transformation, these two-component fields transform under the group SU(2). The field transformation $\varphi' = e^{i\Lambda}\varphi$ becomes, to first order in the ε_k,

$$\varphi' = \varphi + \frac{i}{2}\varepsilon_k \lambda^k \varphi + \ldots. \tag{7.160}$$

For the nth component of the new isospin state, in terms of individual matrix elements this would be written

$$\varphi'_n = \varphi_n + \frac{i}{2}\varepsilon_k \lambda^k_{nm} \varphi_m \tag{7.161}$$

---

[31] The $\frac{1}{2}$ is traditional, coming I assume from the legacy of SU(2) in describing spin-$\frac{1}{2}$ states.

(sum over m from 1 to N; sum over k from 1 to $N^2 - 1$). We are looking at a set of $N^2 - 1$ generators of a gauge transformation, matrices operating on the internal degrees of freedom

$$\zeta^k \equiv \frac{i}{2}\lambda^k \varphi. \tag{7.162}$$

Matrix representations of the λ^k matrices can be constructed explicitly from the requirements that they be Hermitian and traceless. The procedure is easily illustrated with SU(2). Start with a generic 2×2 matrix Λ,

$$\Lambda = \begin{pmatrix} A & B \\ C & D \end{pmatrix}. \tag{7.163}$$

Now construct its adjoint:

$$\Lambda^\dagger = \begin{pmatrix} A^* & C^* \\ B^* & D^* \end{pmatrix}. \tag{7.164}$$

Because Λ is Hermitian, $A = A^*, D = D^*$, and $C^* = B$. Requiring Λ to be traceless means $D = -A$. Therefore we can write

$$\Lambda = \begin{pmatrix} 0 & B \\ B^* & 0 \end{pmatrix} + A\begin{pmatrix} 1 & 0 \\ 0 & -1 \end{pmatrix}. \tag{7.165}$$

Separate the real from the imaginary components of B by setting $B = \alpha - i\beta$, where α and β are real, so that

$$\Lambda = \alpha\begin{pmatrix} 0 & 1 \\ 1 & 0 \end{pmatrix} + \beta\begin{pmatrix} 0 & -i \\ i & 0 \end{pmatrix} + A\begin{pmatrix} 1 & 0 \\ 0 & -1 \end{pmatrix}. \tag{7.166}$$

Denote $\alpha = \varepsilon_1, \beta = \varepsilon_2$, and $A = \varepsilon_3$, and define

$$\sigma^1 \equiv \begin{pmatrix} 0 & 1 \\ 1 & 0 \end{pmatrix}, \quad \sigma^2 \equiv \begin{pmatrix} 0 & -i \\ i & 0 \end{pmatrix}, \quad \sigma^3 \equiv \begin{pmatrix} 1 & 0 \\ 0 & -1 \end{pmatrix}. \tag{7.167}$$

These λ^k matrices for SU(2) are known as the Pauli matrices $\sigma^k (k = 1, 2, 3)$. Traditionally we use the Pauli matrices σ^k for SU(2) and λ^k for all other SU(N) for which $N \neq 2$. Recalling the customary $\frac{1}{2}$, we have shown that, for any Λ in SU(2),

$$\Lambda = \Lambda(\varepsilon) = \varepsilon_1 \frac{\sigma^1}{2} + \varepsilon_2 \frac{\sigma^2}{2} + \varepsilon_3 \frac{\sigma^3}{2} \equiv \frac{1}{2}(\varepsilon \cdot \boldsymbol{\sigma}). \tag{7.168}$$

The Pauli matrix σ^3 has $+1$ and -1 along its diagonal, and zeroes for all its off-diagonal entries. Therefore, $\frac{1}{2}\sigma^3$ has for its eigenstates the basis states $|K^+\rangle$ and $|K^0\rangle$ (eqs. [7.140] and [7.141]), with their m_I values as the eigenvalues:

$$\tfrac{1}{2}\sigma^3|K^+\rangle = +\tfrac{1}{2}|K^+\rangle \tag{7.169}$$

$$\tfrac{1}{2}\sigma^3|K^0\rangle = -\tfrac{1}{2}|K^0\rangle. \tag{7.170}$$

Introducing an isospin coupling constant g, a kaon in the K^+ state is assigned the isospin charge $+\frac{g}{2}$, and a kaon in the K^0 state carries isospin charge $-\frac{g}{2}$.[32] Thus an arbitrary kaon state

$$|K\rangle = \begin{pmatrix} a \\ b \end{pmatrix} = a|K^+\rangle + b|K^0\rangle \tag{7.171}$$

has the probability $|a|^2$ of having isospin charge $+\frac{g}{2}$, and the probability $|b|^2$ of having isospin charge $-\frac{g}{2}$, as shown by evaluating $\langle K|\frac{g}{2}\sigma^3|K\rangle$, the operator $\frac{g}{2}\sigma^3$ that picks out isospin charge. The other operators, $\frac{g}{2}\sigma^1$ and $\frac{g}{2}\sigma^2$, produce transitions between the eigenstates. For example, the operator $\frac{g}{2}\sigma^1$ changes the isospin state according to

$$\frac{g}{2}\sigma^1|K\rangle = \frac{g}{2}\begin{pmatrix} 0 & 1 \\ 1 & 0 \end{pmatrix}\begin{pmatrix} a \\ b \end{pmatrix}$$
$$= \frac{g}{2}\begin{pmatrix} b \\ a \end{pmatrix}. \tag{7.172}$$

This operator changes the probability of the kaon being in the K^+ state from $|a|^2$ to $|b|^2$, and changes its probability of being in the K^0 state from $|b|^2$ to $|a|^2$.[33]

This example with SU(2) illustrates a general feature of SU(N): of the N^2-1 matrices λ^k, $N-1$ of them are diagonal. For example, SU(3) has two diagonal matrix generators, so in a family of particles that transform under SU(3), every particle requires two quantum numbers to specify its "charge." An application arises in the "color charge" of quantum chromodynamics.

These transformations form a group (which can be represented by matrix multiplication): an identity transformation exists, each transformation has an inverse, three transformations are associative, and the group elements respect closure—if e^A and e^B are group elements, then so is $e^A e^B$. In SU(N),

$$e^A e^B = \exp\left(A + B + \frac{1}{2}(A, B) + \dots\right), \tag{7.173}$$

---

[32]Although this language is not commonly used, one could also say that spin "couples to $\hbar$." To participate in any interaction that involves angular momentum a particle must carry some $\hbar$, the fundamental unit of angular momentum.

[33]Recall that the superscripts on K^+ and K^0 describe their electric charges, which interact under the group $U(1)$. So far as isospin is concerned, electric charge just goes along for the ride. In a transition between K^0 and K^+, the isospin changes by ± 1.

where $(A, B) \equiv AB - BA$ denotes the commutator of A and B. The terms farther out in the argument of the exponential are commutators of commutators. To preserve closure in $SU(N)$, the commutator algebra is determined by "structure constants," f^{abc}:

$$\left(\frac{\lambda^a}{2}, \frac{\lambda^b}{2}\right) = i f^{abc} \frac{\lambda^c}{2} \tag{7.174}$$

(sum over c).[34] The structure constants are all zero in $U(1)$. For $SU(2)$ they are the Levi-Civita symbols: $\epsilon^{ijk} = +1(-1)$ when ijk is an even (odd) permutation of 1-2-3, and $\epsilon^{ijk} = 0$ when any two indices are equal. If the structure constants are all zero, so that all elements commute, the group is called "Abelian," after Norwegian mathematician Niels Henrik Abel (1802–1829). When an Abelian group is gauged, the result is isomorphic to electrodynamics. When a non-Abelian group is gauged, the spin-1 gauge bosons are "Yang-Mills fields." Such is the case with our $W^\pm$ gauge bosons, which come from gauging isospin under $SU(2)$ and to which we turn next.

For Emmy Noether, her theorem about invariance was an exercise in Lie algebra.

7.7   Noether's Theorem and Gauged Internal Symmetries

To illustrate Noether's theorem applied to gauged transformations within an internal space, we return to the kaon system, with its $|K^+\rangle$ and $|K^0\rangle$ eigenstates of the isospin charge operator $\frac{g}{2}\sigma^3$. As a two-state system, in analogy to quantized angular momentum, we postulated that the generic kaon isospin state respects the same algebra as the angular momentum of spin-$\frac{1}{2}$ particles. Continuous transformations of the fields within this isospin space are described by $SU(2)$ acting on the kaon state vector $|K\rangle$ in isospin space, or, in alternative notation, φ,

$$|K\rangle = \left(\begin{array}{c} \varphi^+ \\ \varphi^0 \end{array} \right) \equiv \varphi \tag{7.175}$$

and its adjoint

$$\langle K| = (\begin{array}{cc} \varphi^{*+} & \varphi^{*0} \end{array}) \equiv \varphi^\dagger. \tag{7.176}$$

Without interactions, the kaon Lagrangian density is[35]

$$\mathcal{L}^{(\varphi)} = (\partial_\mu \varphi)^\dagger (\partial^\mu \varphi) - m^2 \varphi^\dagger \varphi. \tag{7.177}$$

---

[34]The up or down placement of repeated indices for internal degrees of freedom is not crucial.

[35]Notice this is matrix multiplication, with $\varphi^\dagger$ a row vector and φ a column vector.

The choice of SU(2) transformation will preserve unitarity, as implemented by

$$\varphi' = \exp\left(\frac{ig}{2}\varepsilon_{\mathbf{k}}\sigma^{\mathbf{k}}\right)\varphi, \qquad (7.178)$$

where $k = 1, 2, 3$, the matrices $\sigma^{\mathbf{k}}$ are the Pauli matrices, and g denotes the coupling constant (a "charge" of some kind) necessary for a particle to "hook onto" rotations in isospin space.

To first order in the ε_k, the transformation of the fields becomes the matrix equation

$$\varphi' = \varphi + \frac{ig}{2}\varepsilon_k\sigma^k\varphi, \qquad (7.179)$$

where $k = 1, 2, 3$ and repeated indices are summed. For the component φ_a (where a denotes $+$ or 0) of the φ vector in isospin space, eq. (7.179) says[36]

$$\varphi'_a = \varphi_a + \frac{ig}{2}\varepsilon_k(\sigma^k)_{ba}\varphi_b, \qquad (7.180)$$

where in the last term b is summed over the $+$ and 0 components. The adjoint transforms as

$$\varphi'^*_a = \varphi^*_a - \frac{ig}{2}\varepsilon_k(\sigma^k)^*_{ba}\varphi^*_b. \qquad (7.181)$$

Therefore the generators that concern us in the context of Noether's theorem are

$$\zeta^k{}_a = \frac{ig}{2}(\sigma^k)_{ba}\varphi_b \qquad (7.182)$$

and its adjoint.[37]

If all the parameters ε_k are independent of spacetime coordinates, then a global gauge transformation results, and the covariant derivative is unnecessary. Noether's theorem gives the equation of continuity

$$\partial_\mu j^\mu = 0, \qquad (7.183)$$

with

$$j^\mu = ig[(\partial^\mu\varphi)^\dagger\varphi - \varphi^\dagger(\partial_\mu\varphi)] \equiv ig(\varphi^\dagger \overleftrightarrow{\partial^\mu} \varphi). \qquad (7.184)$$

Integrating over all space, if the fields vanish at infinity then the conservation law results,

$$g\int j^0 \, dV = \text{const.} \qquad (7.185)$$

---

[36]φ is a scalar in spacetime but a vector in isospin space. It is unchanged under a spacetime transformation, but the isospin components get rearranged under an internal SU(2) transformation.

[37]Recall that in the gauge transformations considered here, the spacetime coordinates are left alone, $x'^\mu = x^\mu$, so that $\tau^\mu = 0$.

Therefore the isospin charge is conserved, consistent with the original assumption to that effect. As Howard Georgi said, "Symmetry is a tool that should be used to determine underlying dynamics, which must in turn explain the success (or failure) of the symmetry arguments" [Georgi (1982) xix].

The situation grows more interesting when local gauge invariance is imposed, so that $\varepsilon_k = \varepsilon_k(x^\mu)$. The algorithm for locally gauging the system has us replace ∂_μ with the covariant derivative D_μ and add the Lagrangian density for the gauge fields. Let W_μ^a denote the isospin gauge fields coupled to the kaon doublet with coupling constant g. The Latin indices denote a component of the isospin state vector, and the Greek indices denote space-time components of vectors and tensors. The covariant derivative becomes

$$D_\mu \varphi \equiv \left[\partial_\mu + \frac{ig}{2} \left(\boldsymbol{\sigma} \cdot \mathbf{W}_\mu \right) \right] \varphi \qquad (7.186)$$

where the scalar product is in isospin space with φ a two-component isospin vector, although it is a scalar in spacetime. As before, we require the covariant derivative to have the same transformation rule as the field φ itself, so that

$$(D_\mu \varphi)' = \exp \left(\frac{ig}{2} \boldsymbol{\varepsilon} \cdot \boldsymbol{\sigma} \right) (D_\mu \varphi), \qquad (7.187)$$

which requires

$$W_\mu'^a = W_\mu^a - \partial_\mu \varepsilon^a + f^{abc} \varepsilon^b W_\mu^c, \qquad (7.188)$$

where in SU(2) the structure constants f^{abc} are the Levi-Civita coefficients. Notice the transformation of the non-Abelian gauge field has the $\partial_\mu \varepsilon^a$ similar to the Abelian case, but also picks up a generator $\zeta_\mu^{ab} = f^{abc} W_\mu^c$.

We saw that the Abelian gauge field's Lagrangian density was $-\frac{1}{4\mu_o} F_{\mu\nu} F^{\mu\nu}$. Let its analog for the gauge bosons of isospin be denoted $-\frac{1}{4} G_{\mu\nu}^a G^{a\mu\nu}$.[38] The $G_{\mu\nu}$ pick up a term quadratic in the W fields:

$$G^a{}_{\mu\nu} \equiv \partial_\mu W_\nu^a - \partial_\nu W_\mu^a + g f^{abc} W_\mu^b W_\nu^c. \qquad (7.189)$$

The term $g f^{abc} W^a{}_\mu W^b{}_\nu$ expresses self-interactions among the gauge fields.[39]

Assuming a common mass m for all kaon states and using the notation $\partial_\mu \varphi \equiv \varphi_\mu$, the gauged Lagrangian density becomes

---

[38]Sum over repeated Latin indices that denote internal space coordinates, for which the up or down position makes no difference.

[39]In contrast, the local gauge transformations of the electromagnetic interaction are described by the Abelian field U(1), for which all $f^{abc} = 0$. The photon carries no electric charge itself; photons do not interact *directly* with one another. See Q.7.h.

$$\mathcal{L} = (D_\mu \varphi)^\dagger (D^\mu \varphi) - m^2 \varphi^\dagger \varphi - \frac{1}{4} G^a_{\mu\nu} G^{a\mu\nu}$$

$$= \varphi^\dagger_\mu \varphi^\mu - m^2 \varphi^\dagger \varphi + \frac{ig}{2} \left[\varphi^{\dagger\mu} (\boldsymbol{\sigma} \cdot \mathbf{W}_\mu) \varphi - \varphi^\dagger (\boldsymbol{\sigma} \cdot \mathbf{W}_\mu)^\dagger \varphi^\mu \right] \quad (7.190)$$

$$+ \frac{g^2}{4} (\boldsymbol{\sigma} \cdot \mathbf{W}_\mu)(\boldsymbol{\sigma} \cdot \mathbf{W}^\mu)^\dagger (\varphi^\dagger \varphi) - \frac{1}{4} G^a_{\mu\nu} G^{a\mu\nu}.$$

Notice that the Pauli matrices are Hermitian and the gauge fields real, so that $(\boldsymbol{\sigma} \cdot \mathbf{W}_\mu)^\dagger = \boldsymbol{\sigma} \cdot \mathbf{W}_\mu$. Therefore the Lagrangian density may be written

$$\mathcal{L} = \varphi^\dagger_\mu \varphi^\mu - m^2 \varphi^\dagger \varphi + \frac{ig}{2} (\boldsymbol{\sigma} \cdot \mathbf{W}_\mu) \left(\varphi^{\dagger\mu} \varphi - \varphi^\dagger \varphi^\mu \right)$$

$$+ \frac{g^2}{4} (\boldsymbol{\sigma} \cdot \mathbf{W}^\mu)(\boldsymbol{\sigma} \cdot \mathbf{W}_\mu)(\varphi^\dagger \varphi) - \frac{1}{4} G^a_{\mu\nu} G^{a\mu\nu}. \quad (7.191)$$

Let's write the equations of motion for the matter and gauge fields. For the scalar particles, we need the Euler-Lagrange equation

$$\frac{\partial \mathcal{L}}{\partial \varphi^\dagger} = \partial_\mu p^{(\varphi^\dagger)}_\mu \quad (7.192)$$

where

$$p^{(\varphi^\dagger)}_\mu \equiv \frac{\partial \mathcal{L}}{\partial \varphi^{\mu\dagger}} = D_\mu \varphi. \quad (7.193)$$

After regrouping some terms, eq. (7.191) becomes

$$D_\mu D^\mu \varphi + m^2 \varphi = 0, \quad (7.194)$$

which resembles the Klein-Gordon equation, with the crucial difference that the derivatives are covariant derivatives, which contain interaction terms within them.

The equation of motion for the gauge boson field (i.e., the analog of Maxwell's equations) is the Euler-Lagrange equation

$$\frac{\partial \mathcal{L}}{\partial W^{\mu a}} = \partial^\nu p^{(W^a)}_{\mu\nu} \quad (7.195)$$

where

$$p^{(W^a)}_{\mu,\nu} \equiv \frac{\partial \mathcal{L}}{\partial W^{a\mu,\nu}}. \quad (7.196)$$

By writing $G^a_{\mu\nu} = g_{\mu\rho} g_{\nu\sigma} G^{a\rho\sigma}$, noting that

$$\frac{\partial W^{c\alpha,\beta}}{\partial W^{qx,\xi}} = \delta^c{}_q \, \delta^\alpha{}_x \, \delta^\beta{}_\xi \quad (7.197)$$

and $f^{abc} = -f^{acb}$, all needed to show that $G^{a\mu\nu} = -G^{a\nu\mu}$, with some patience one finds at the end of the day

$$p_{\mu\nu}^{(W^a)} = G_{\mu\nu}^a. \tag{7.198}$$

After some work the Euler-Lagrange equation for the gauge fields becomes

$$\partial_\nu G^{a\mu\nu} = ig\left(\varphi^\dagger \overleftrightarrow{D}^\mu \frac{\sigma^a}{2}\varphi\right) + gf^{aqs}\varphi^\dagger(G^{q\mu\nu}W_\nu^s)\varphi \equiv J_{(W)}^{a\mu}. \tag{7.199}$$

The kaon-gauge boson interaction "currents" $J_{(W)}^{a\mu}$ are conserved in the sense of Noether's first theorem,

$$\partial_\mu J_{(W)}^{a\mu} = 0. \tag{7.200}$$

As we saw in the interactions of charged scalar particles with the electromagnetic field under local gauge invariance, this is not simply a statement of a locally conserved isospin charge. It also includes an interaction between the isospin charge and the gauge fields, so other than enclosing the entire system and saying that everything inside it is conserved, the question, "This equation describes the conservation of *what?*" has no precise answer. Eq. (7.200) is not wrong, but ambiguous. Perhaps that ambiguity is necessary; perhaps not; we shall see. More is said about local gauge invariance and conservation laws in chapter 8.

Questions for Reflection and Discussion

Q7.a. Lie groups such as SO(N), the special orthogonal group in N dimensions, have their uses in several areas of physics. Rotation of rectangular axes from xy to $x'y'$ offers an example of SO(2). Write the generators of SO(2) for a rotation of the xy axes about the z-axis. Under what circumstances should one use SO(2) instead of SU(2)?

Q7.b. Recall the roller coaster problems of introductory physics, whose solutions depend on the conservation of mechanical energy. The zero of potential energy could be shifted by a constant without changing the dynamics. Why does this make no difference in the application of the conservation law $K + U = $ const.? Why does this make no difference in $\mathbf{F} = m\mathbf{a}$? Can we call this shift of the potential energy a gauge transformation?

Q7.c. Find out whatever you can about the Lagrangian density in the SU(3) model of quantum chromodynamics, or QCD [e.g., see Marciano & Pagels (1978)]. What plays the role of charge? What are the gauge bosons called? Why is the theory called "chromo"dynamics? What are some consequences of the non-Abelian nature of the theory? QCD has been enormously successful, accounting for many features of the strong nuclear force between quarks. What are some of its successes? What are some of its challenges?

Q7.d. Find a paper [e.g., Weinberg (1967)] or a textbook [e.g., Aitchison & Hey (1982)] that displays the Lagrangian density of the $SU(2) \times U(1)$ Weinberg-Salam model of the electroweak interaction. Identify the role of each term in the Lagrangian density. What are the groups of gauge transformations under which the system is invariant? Why does this Lagrangian density use scalar fields coupled to the fermion fields (see Q7.g.)? What are some consequences of the non-Abelian $SU(2)$ sector of the theory? What are the theory's gauge bosons? The Weinberg-Salam model successfully unifies electrodynamics with the weak interaction, and endows some of the gauge bosons with mass without spoiling gauge invariance—now there's a trick (consult the loophole called the "Higgs mechanism")!

Q7.e. Find out whatever you can about the Lagrangian density of the $SU(5)$ model of Howard Georgi and Sheldon Glashow [Georgi & Glashow (1974)], the simplest model interaction proposed so far that unifies the electroweak and chromodynamic interactions. This hypothetical interaction transforms quarks into leptons and vice versa. What new gauge bosons does the model predict that did not already exist in the $SU(3) \times SU(2) \times U(1)$ "standard model" of quarks and leptons (the union of the Weinberg-Salam and QCD models)? What plays the role of the unified charge? What does the $SU(5)$ model predict about proton decay? Proton decay as predicted by $SU(5)$ has been experimentally ruled out, but the model was important for being suggestive, inspiring more complicated gauge groups (e.g., superstrings) for unifying nature's fundamental forces.

Q7.f. Consider some musings on induced field transformation generators. Could it be possible that a transformation of the coordinates, without a transformation postulated explicitly for the fields, nevertheless produces a transformation of the fields? In other words, if we postulate a nonzero τ^ν, would a ζ^μ be produced that depends on τ^ν? To investigate this possibility, consider the component A^μ of a vector field in spacetime. As a vector component, it transforms the same as the coordinate displacements:

$$A'^\mu = \frac{\partial x'^\mu}{\partial x^\nu} A^\nu. \tag{7.201}$$

For the reverse transformation, interchanging the old and new coordinates, we would write

$$A^\mu = \frac{\partial x^\mu}{\partial x'^\nu} A'^\nu. \tag{7.202}$$

Differentiation of eq. (7.202) with respect to ε yields

$$0 = \frac{\partial}{\partial \varepsilon}\left(\frac{\partial x^{\mu}}{\partial x'^{\nu}}\right) A'^{\nu} + \frac{\partial x^{\mu}}{\partial x'^{\nu}} \frac{\partial A'^{\nu}}{\partial \varepsilon}. \tag{7.203}$$

The transformations of interest considered so far take the form, to first order in ε,

$$x'^{\mu} = x^{\mu} + \varepsilon \tau^{\mu}. \tag{7.204}$$

Noting that $\partial x'^{\mu}/\partial x'^{\nu} = \delta^{\mu}{}_{\nu}$,

$$\begin{aligned}
\frac{\partial x^{\mu}}{\partial x'^{\nu}} &= \partial_{\nu'}(x'^{\mu} - \varepsilon \tau^{\mu}) \\
&= \delta^{\mu}{}_{\nu} - \varepsilon(\partial_{\nu'}\tau^{\mu}) - (\partial_{\nu'}\varepsilon)\tau^{\mu}.
\end{aligned} \tag{7.205}$$

For the moment let us forget about $A'^{\mu} = A^{\mu} - \partial^{\mu}\varepsilon$ and write

$$A'^{\mu} = A^{\mu} + \varepsilon \zeta^{\mu} \tag{7.206}$$

so that $\partial A'^{\nu}/\partial \varepsilon = \zeta^{\mu}$. After making these substitutions and then setting $\varepsilon = 0$, eq. (7.203) becomes

$$\zeta^{\mu} = (\partial_{\nu}\tau^{\mu})A^{\nu}. \tag{7.207}$$

While this statement follows logically from the *ansatz* $A'^{\mu} = A^{\mu} + \varepsilon \zeta^{\mu}$, we do not make use of it here because the fields A^{μ} were invented to make the model gauge-invariant, which required $A'^{\mu} - A^{\mu}$ to equal $-\partial^{\mu}\varepsilon$, not something proportional to an undifferentiated ε.

However, we could merge eq. (7.206) with a local gauge transformation by postulating

$$A'^{\mu} = A^{\mu} + \varepsilon \zeta_0^{\mu} + (\partial^{\mu}\varepsilon)\zeta_1 \tag{7.208}$$

for some functions ζ_0^{μ} and ζ_1, and see what happens. This is the stuff of Noether's second theorem. Furthermore, in general relativity, when equations are transformed from one system of coordinates to another under transformations more general than the Lorentz transformation, the metric tensor components—the gravitational fields—pick up an induced transformation that is equivalent to a local gauge transformation. Stay tuned (see chapter 8).

Q7.g. The gauge boson of the weak interaction (the W particle) has a mass of about 80 proton masses—far from being massless!—so there is more to the story of local gauge invariance and mass than we have presented here. Find out whatever you can about spontaneous symmetry breaking and the Higgs mechanism, as used in theories with local gauge invariance.

Essentially, one introduces spin-0 complex fields, the Higgs fields, that couple to familiar particles, all taken to be initially massless. The Lagrangian density includes interaction terms like $g\varphi\psi^\dagger\psi$, where ψ is a field whose quanta are familiar particles, φ denotes the Higgs scalar field, and g is a coupling constant. We can postulate that the scalar field's vacuum expectation value assumes a nonzero value,[40] $\langle\varphi\rangle = const. \neq 0$. In that case, the terms describing interactions between Higgs bosons and familiar particles become mass terms for the latter, including gauge bosons. Overall gauge invariance is maintained because of the extra degrees of freedom offered by the Higgs fields. If you are especially ambitious, research the difference between spontaneous symmetry breaking and dynamical symmetry breaking [see Nambu & Jona-Lasinio (1961a); Nambu & Jona-Lasinio (1961b)].

Q7.h. A photon *couples* to electric charge, even though the photon, itself, does not *carry* electric charge. Therefore a photon only scatters directly off of electrically charged matter; a photon cannot scatter directly off of another photon. Nevertheless, photon-photon scattering can happen indirectly:

$$\gamma + \gamma \to X + Y \to \gamma + \gamma \tag{7.209}$$

where γ denotes a photon. What possibilities exist for the intermediate state $X+Y$ that provides an indirect channel for photon-photon scattering?

Q7.i In quantum mechanics, the wave function ψ is always undetermined up to an overall phase, because $\psi' = e^{i\delta}\psi$ gives the same probability density as ψ. It is also true in classical mechanics that the zero of potential energy can be adjusted to suit one's convenience—the dynamics are unchanged if the potential energy U is shifted everywhere by a constant, $U(\mathbf{r}) \to U(\mathbf{r}) + U_o$ where $U_o = $ const. Show how this freedom to choose the zero of potential energy in mechanics is reconciled with the Schrödinger equation through a global gauge transformation.

7.8.j Consider a refrence frame that rotates with angular velocity $\boldsymbol{\omega}$ relative to an inertial frame (recall ex. 5.4). Discuss whether a "rotational covariant derivative" can be defined that takes one from the inertial frame to the rotating frame by replacing $d\mathbf{r}/dt$ in the original Lagrangian, $L = \frac{1}{2}m(d\mathbf{r}/dt)^2 - U$, with [Dallen & Neuenschwander (2011)]

$$\frac{D\mathbf{r}}{Dt} \equiv \frac{d\mathbf{r}}{dt} + (\boldsymbol{\omega} \times \mathbf{r}). \tag{7.210}$$

Does the meaning of $\mathbf{r}$ change?

---

[40]Since $\langle\varphi\rangle \neq 0$ by assumption, this mechanism is called "spontaneous" symmetry breaking.

Exercises

7.1. Suppose we liberalize the definition of invariance in field theory, and allow an inhomogeneity in the form of a divergence proportional to ε:

$$\mathcal{L}'J - \mathcal{L} = \varepsilon(\partial_\mu V^\mu) + O(\varepsilon^s) \tag{7.211}$$

where $s > 1$ and the V^μ are given functions.

a. Show that, when the invariance identity is derived, the inhomogeneity survives,

$$-(\zeta^\mu - A^\mu{}_{,\nu}\tau^\nu)\left[\frac{\partial \mathcal{L}}{\partial A^\mu} - \partial_\rho\left(\frac{\partial \mathcal{L}}{\partial A^\mu{}_{,\rho}}\right)\right] \tag{7.212}$$
$$= \partial_\nu[p_\rho{}^\nu \zeta^\rho - \mathcal{H}_\rho{}^\nu \tau^\rho - V^\nu],$$

and invoking the Euler-Lagrange equation produces a liberalized expression for the conserved Noether current:

$$\partial_\nu(j^\nu - V^\nu) = 0. \tag{7.213}$$

b. Generalize these results for an M-parameter family of transformations ε^k and $V^{k\mu}$ where $k = 1, 2, \ldots, M$.

7.2. Recall Poynting's theorem in electrodynamics,

$$\nabla \cdot \mathbf{S} + \frac{\partial \eta}{\partial t} = -\mathbf{j} \cdot \mathbf{E}, \tag{7.214}$$

where $\mathbf{S}$ denotes Poynting's vector, η the energy density of electric and magnetic fields, and $\mathbf{j}$ the three-vector current density of charged particles. Show how this can be interpreted as an instance of the work-energy theorem.

7.3. Derive the λ matrices for SU(3). This group has $3^2 - 1 = 8$ independent λ's. In going through the same kind of analysis that led to the Pauli matrices of SU(2), you should find two of the SU(3) matrices to be diagonal. Proceeding as before, show that the eight λ-matrices of SU(3) may be written

$$\lambda^1 = \begin{pmatrix} 0 & 1 & 0 \\ 1 & 0 & 0 \\ 0 & 0 & 0 \end{pmatrix}, \quad \lambda^2 = \begin{pmatrix} 0 & -i & 0 \\ i & 0 & 0 \\ 0 & 0 & 0 \end{pmatrix}, \quad \lambda^3 = \begin{pmatrix} 1 & 0 & 0 \\ 0 & -1 & 0 \\ 0 & 0 & 0 \end{pmatrix},$$

$$\lambda^4 = \begin{pmatrix} 0 & 0 & 1 \\ 0 & 0 & 0 \\ 1 & 0 & 0 \end{pmatrix}, \quad \lambda^5 = \begin{pmatrix} 0 & 0 & -i \\ 0 & 0 & 0 \\ i & 0 & 0 \end{pmatrix}, \quad \lambda^6 = \begin{pmatrix} 0 & 0 & 0 \\ 0 & 0 & 1 \\ 0 & 1 & 0 \end{pmatrix},$$

$$\lambda^7 = \begin{pmatrix} 0 & 0 & 0 \\ 0 & 0 & -i \\ 0 & i & 0 \end{pmatrix}, \quad \lambda^8 = \frac{1}{\sqrt{3}} \begin{pmatrix} 1 & 0 & 0 \\ 0 & 1 & 0 \\ 0 & 0 & -2 \end{pmatrix}.$$

Notice that λ^3 and λ^8 are diagonal and yield distinct eigenvalues. How many independent charges can be accommodated by SU(3)? This group is used in quantum chromodynamics, where each quark carries "color charge." Quarks never appear in isolation, but within bound states of baryons (sets of three quarks) and mesons (quark-antiquark pairs) that have zero net color charge.

7.4. The proton and neutron are spin-$\frac{1}{2}$ fermions. Their Euler-Lagrange equation is the Dirac equation. The Lagrangian density for a noninteracting Dirac particle of mass m is

$$\mathcal{L} = \bar{\psi}(\gamma^\mu i \partial_\mu - m)\psi \tag{7.215}$$

where the γ^μ are the Dirac matrices and $\bar{\psi}$ is a kind of adjoint of ψ (see appendix D).

a. Show that the Dirac Lagrangian density, when inserted into the Euler-Lagrange equation for $\bar{\psi}$,

$$\frac{\partial \mathcal{L}}{\partial \bar{\psi}} = \partial_\mu \left(\frac{\partial \mathcal{L}}{\partial(\partial_\mu \bar{\psi})} \right), \tag{7.216}$$

gives the Dirac equation [Bjorken & Drell (1964)].

b. Through the weak interaction, neutrons and protons can transform into one another. For instance, an isolated neutron decays with a half-life of about 15 minutes into a proton plus an electron and an electron-type antineutrino: $n \rightarrow p + e^- + \bar{\nu}_e$. This reaction is spontaneous because the neutron's mass exceeds the proton's by a few parts out of a thousand. The reverse reaction can be induced by a collision: $e^- + p \rightarrow n + \nu_e$. Such reactions, and the near-equality of the proton and neutron mass, suggest that, for some processes, the proton and neutron may be thought of as two different states of the nucleon. Because the neutron has the larger mass, let it be the excited state of the nucleon, and let the proton be the ground state. An arbitrary nucleon state, or state vector, consists of a superposition of two basis states,

$$|\psi\rangle = \begin{pmatrix} \psi_n \\ \psi_p \end{pmatrix}$$

$$= \psi_n |n\rangle + \psi_p |p\rangle,$$

where

$$|n\rangle \equiv \begin{pmatrix} 1 \\ 0 \end{pmatrix} \tag{7.217}$$

denotes the pure neutron state, and

$$|p\rangle \equiv \begin{pmatrix} 0 \\ 1 \end{pmatrix} \tag{7.218}$$

the pure proton state. The set of all nucleon states forms an abstract space spanned by the two basis vectors $|n\rangle$ and $|p\rangle$, with their doings described by a 2×2 matrix M:

$$M = \begin{pmatrix} M_{nn} & M_{np} \\ M_{pn} & M_{pp} \end{pmatrix} \tag{7.219}$$

where $M_{nn} = m_n c^2$ for the neutron, $M_{pp} = m_p c^2 = m_n c^2(1 - \delta)$, where $\delta \ll 1$ comes from electromagnetic interactions. The off-diagonal terms in the matrix M describe weak interactions that convert protons into neutrons and vice versa.

Show that the Dirac equation for this system,

$$i\gamma^\mu \partial_\mu |\psi\rangle = M|\psi\rangle, \tag{7.220}$$

when written out by components, becomes the coupled system of equations for the neutron and proton;

$$i\gamma^\mu \partial_\mu \psi_n = M_{nn}\psi_n + M_{np}\psi_p, \tag{7.221}$$

$$i\gamma^\mu \partial_\mu \psi_p = M_{pn}\psi_n + M_{pp}\psi_p. \tag{7.222}$$

b. show that if the matrix M is diagonal and we set $\delta = 0$ (i.e., the electromagnetic interaction is neglected), then the equations uncouple and the proton and neutron do not change into one another.

c. If the off-diagonal elements of M are not zero, and if $\delta = 0$, what kinds of neutron-proton transformations would occur? Why isn't this a good model for the weak interaction?

d. Suppose $\delta \neq 0$ but the off-diagonal elements of M are equal, $M_{np} = M_{pn} \equiv V$. What are the eigenstates of the weak interaction in this model?

7.5. Put three quarks with different color charges, a positron, and the electron-type neutrino into a five-member multiplet, represented by a column matrix with five entries. Next, propose qualitatively a 5×5 transition matrix whose entries are gauge bosons. In other words, in the reaction $a \to b + G$ where a and b can be either quarks or leptons, and G a known or hypothetical gauge boson, fill in the matrix elements—the Gs—that occur at the intersection of the column labeled a and the row labeled b. What familiar gauge bosons appear and what new gauge bosons (perhaps with fractional electric charge) are predicted, that transform quarks into leptons and vice versa? Such a model was worked out in 1974 by Howard

Georgi and Sheldon Glashow, using the 5×5 matrices of SU(5) [Georgi & Glashow (1974)]. How many gauge bosons and how many diagonal generator matrices exist for SU(5)? The Georgi-Glashow model predicts proton decay with a half-life of $\sim 10^{32}$ years, or one decay annually in a sample of matter holding 10^{32} protons. How large a tank of water would be necessary to contain this many protons? Unfortunately for the SU(5) model, despite valiant efforts, proton decay has not been observed.

7.6. The Lagrangian density for a nonrelativistic quantum particle of mass m moving in three spatial dimensions, in the presence of a potential energy function U, is

$$\mathcal{L}(\psi, \psi^*, \psi_k, \psi_k^*) = -\frac{\hbar^2}{2m} \boldsymbol{\nabla}\psi^* \cdot \boldsymbol{\nabla}\psi +$$
$$\frac{1}{2}i\hbar \left(\psi^* \frac{\partial \psi}{\partial t} - \frac{\partial \psi^*}{\partial t} \psi \right) - \psi^* U \psi. \tag{7.223}$$

Its Euler-Lagrange equation for ψ^* becomes the three-dimensional time-dependent Schrödinger equation,

$$-\frac{\hbar^2}{2m} \nabla^2 \psi + U\psi = i\hbar \frac{\partial \psi}{\partial t}. \tag{7.224}$$

a. Derive the conserved current for the time-dependent Schrödinger equation

$$\frac{\partial \rho}{\partial t} + \boldsymbol{\nabla} \cdot \mathbf{j} = 0 \tag{7.225}$$

where $\rho = \psi^* \psi$ and

$$\mathbf{j} = \frac{\hbar}{2mi} [\psi^*(\boldsymbol{\nabla}\psi) - (\boldsymbol{\nabla}\psi)^* \psi] \tag{7.226}$$

using only the Schrödinger equation itself, without going through Noether's theorem.
b. Under what transformation does the result derived this way agree with the Noether conserved current?

7.7. Recall the Lagrangian density for the charged scalar particle coupled to the electromagnetic field,

$$\mathcal{L} = (D_\mu\varphi)^*(D^\mu\varphi) - \frac{1}{4\mu_o} F_{\mu\nu}F^{\mu\nu}. \tag{7.227}$$

Consider the infinitesimal transformation

$$x'^\mu = x^\mu,$$
$$\varphi' = \varphi + (ig\varphi)\varepsilon,$$
$$\varphi'^* = \varphi^* - (ig\varphi^*)\varepsilon, \tag{7.228}$$
$$A'^\mu = A^\mu + \eta^\mu \varepsilon.$$

In other words, instead of the local transformation of the gauge fields written as $A'^\mu = A^\mu - \partial^\mu \varepsilon$, treat the gauge field as just another field with generator η^μ. Allowing $\varepsilon = \varepsilon(x^\nu)$, show that

$$\mathcal{L}' = \mathcal{L} + J_{\mu(\varphi)}(\eta^\mu \varepsilon + \partial^\mu \varepsilon) - \frac{1}{4\mu_o} F_{\mu\nu} N^{\mu\nu} \varepsilon + \frac{1}{2\mu_o} F_{\mu\nu} E^{\mu\nu} \qquad (7.229)$$

where

$$J_{\mu(\varphi)} = ig(\varphi^* \overset{\leftrightarrow}{D}_\mu \varphi) \qquad (7.230)$$

is the current density we have met before,

$$N_{\mu\nu} \equiv \partial^\nu \eta^\nu - \partial^\nu \eta^\mu, \qquad (7.231)$$

and

$$E^{\mu\nu} \equiv \partial^\mu \partial^\nu \varepsilon - \partial^\nu \partial^\mu \varepsilon. \qquad (7.232)$$

Note that $E^{\mu\nu} = 0$ if the partial derivatives meet typical continuity conditions so that

$$\frac{\partial^2 \varepsilon}{\partial x^\mu \partial x^\nu} = \frac{\partial^2 \varepsilon}{\partial x^\nu \partial x^\mu}. \qquad (7.233)$$

Show that in order for $A'^\mu = A^\mu + \eta^\mu \varepsilon$ agree with $A'^\mu = A^\mu - \partial^\mu \varepsilon$, we must set $\eta^\mu \varepsilon = -\partial^\mu \varepsilon$—in other words $\eta^\mu = -\partial^\mu$ operating on ε. When this is done, show that $\mathcal{L}' = \mathcal{L}$.

7.8. Return to the derivation of the conserved current $J_{(\varphi)}^\mu$, and treat the covariant derivative as a single entity in the Lagrangian density, coupling matter to the electromagnetic field according to

$$\mathcal{L}(\varphi_{;\mu}, \varphi_{;\mu}^*, \partial_\mu A_\nu) = \varphi_{;\mu}^* \varphi^{;\mu} - \frac{1}{4\mu_o} F_{\mu\nu} F^{\mu\nu} \qquad (7.234)$$

where $\varphi_{;\mu} \equiv D_\mu \varphi \equiv \varphi_\mu + igA_\mu\varphi$. Under the U(1) local gauge transformation, show that, when partial derivatives are replaced by covariant derivatives, the Euler-Lagrange equation, derived for a variation of the charged scalar fields,

$$\left[\frac{d\Gamma(\varepsilon)}{d\varepsilon}\right]_0 = \int_a^b \left(\frac{\partial \mathcal{L}}{\partial \varphi_{;\mu}}\left[\frac{\partial \varphi'_{;\mu}}{d\varepsilon}\right]_0 + \frac{\partial \mathcal{L}}{\partial \varphi_{;\mu}^*}\left[\frac{\partial \varphi'^*_{;\mu}}{d\varepsilon}\right]_0\right) dt = 0, \qquad (7.235)$$

gives the same Noether conserved current as did our treatment of $\mathcal{L}(\varphi, \varphi^*, \varphi_\mu, \varphi_\mu^*, A^\mu, \partial_\mu A_\nu)$ that used only the usual derivatives, $\varphi_\mu \equiv \partial_\mu \varphi \equiv \varphi_{,\mu}$.

7.9. Consider a Lagrangian $L = L(t, x, \dot{x})$ where all indices have been stripped away, $x = x(t)$ and $\dot{x} \equiv dx/dt$. More independent and dependent variables can be included by sticking on indices and summing, and renaming

L a Lagrangian density. Consider a transformation that is more generic than the ones considered so far:

$$t' = t + T(\varepsilon)$$
$$x' = x + Z(\varepsilon)$$

$$(7.236)$$

where T and Z are linear in ε and its first k derivatives:

$$T(\varepsilon) \equiv \Sigma_{n=0}^{k} \tau_n \, \varepsilon^{(n)} \tag{7.237}$$

$$Z(\varepsilon) \equiv \Sigma_{n=0}^{k} \zeta_n \, \varepsilon^{(n)} \tag{7.238}$$

or we can use a summation convention and write $T = \tau_n \varepsilon^{(n)}$ and $Z = \zeta_n \varepsilon^{(n)}$ for short. In these definitions of T and Z, the coefficients τ_n and ζ_n may be functions of t, as is ε and its derivatives. The notation $\varepsilon^{(n)}$ means $d^n \varepsilon / dt^n$, where $\varepsilon^{(0)} \equiv \varepsilon$. Notice that τ_0 and ζ_0 are the familiar τ and ζ we have been working with all along. Now comes the statement of the problem. In going from the definition of invariance to the invariance identity, we differentiated

$$L\left(t', x'(t'), \frac{dx'(t')}{dt'}\right)\frac{dt'}{dt} - L\left(t, x(t), \frac{dx(t)}{dt}\right) \sim \varepsilon^s, \quad s > 1 \tag{7.239}$$

with respect to ε, then set $\varepsilon = 0$. Let's try something else this time. Using the transformations above with T and Z, suppose we return repeatedly to the invariance definition, eq. (7.239), and differentiate it with respect to $\varepsilon^{(n)}$, once for each n from 0 to k, thereby producing $k + 1$ identities.
a. Defining $\tau_{-1} = 0$ and $\zeta_{-1} = 0$, show that, for each n, the differentiated invariance identity becomes

$$\frac{\partial L}{\partial x}\zeta_n + \frac{\partial L}{\partial \dot{x}}(\dot{\zeta}_n + \zeta_{n-1}) + \frac{\partial L}{\partial t}\tau_n - H(\dot{\tau}_n + \tau_{n-1}) = 0 \tag{7.240}$$

where H is the Hamiltonian.
b. Show that if the Euler-Lagrange equations hold, each separate invariance identity that comes from differentiation with respect to $\varepsilon^{(n)}$ takes the form

$$\frac{d}{dt}(p\zeta_n - H\tau_n) + (p\zeta_{n-1} - H\tau_{n-1}) = 0. \tag{7.241}$$

c. Interpret this for $n = 0$ and then for $n = 1$. For successive n, does a chain of identities result?[41]

---

[41]See the discussion on conformal invariance by Logan (1974).

Chapter 8

Emmy Noether's Elegant (Second) Theorem

Passing over from these identities to the corresponding variation problem, i.e., putting $\psi = 0$,[1] Theorem I in the one-dimensional case... asserts the existence of ρ first integrals... often referred to of late as "laws of conservation."...

The simplest example of Theorem II... is afforded by the Weierstrass parametric representation.... Another example is presented by the "general theory of relativity" of the physicists. —Emmy Noether, "Invariante Variationsprobleme," 1918

8.1 Two Noether Theorems

In April 1915, Emmy Noether came to Göttingen University at the invitation of David Hilbert, to join him and Felix Klein in their ongoing studies of invariants, a field in which Noether was by then an expert. A few weeks later, in late June and early July, Albert Einstein visited Göttingen to deliver six lectures on his progress in developing the general theory of relativity. He later wrote, "To my great joy, I completely succeeded in convincing Hilbert and Klein" [Pais (1982) 524]. Hilbert was electrified by general relativity, and tore into it with enthusiasm. Although Einstein finished the theory first and presented its final form to the Prussian Academy of Sciences on November 25, 1915, Hilbert's great contribution to the theory

---

[1]By "ψ" Noether means $\frac{\partial L}{\partial q} - \frac{d}{dt}\left(\frac{\partial L}{\partial \dot{q}}\right)$.

Figure 8.1: *David Hilbert (1862–1943).* (American Institute of Physics)

was to construct the final version of the Lagrangian density that yielded Einstein's field equations (see section 8.6).

With the gravitational Lagrangian density in hand, Hilbert and his colleagues examined the conservation laws. They had trouble with interpretation—as did Einstein when he faced the same problem—in reconciling the general relativity field equations with the conservation of energy. Part of solving any problem is stating it usefully, and Hilbert made a distinction between what he called "proper" and "improper" conservation laws. A conservation law is "proper" if the conserved quantity can be tracked cleanly with an equation of continuity $\partial_\nu j^\nu = 0$, and done independent of gauge.

Proper conservation laws for field theories emerge when Noether's theorem is applied to global spacetime transformations and global gauge transformations. "Global" means the infinitesimal parameter ε has the same value throughout the functional's domain. Global spacetime transformations gave conservation of energy and momenta; global gauge invariance gave conservation of charge, such as $\partial_\nu j^\nu_{(\varphi)} = 0$ for noninteracting charged scalar particles.

But in local gauge invariance, the parameter ε varies across the space or spacetime: $\varepsilon = \varepsilon(x^\mu)$. When we tried local gauge invariance with charged spin-0 particles coupled to the electromagnetic field, we *did* obtain a vanishing divergence,

$$\partial_\nu J^\nu_{(\varphi)} = 0 \tag{8.1}$$

where

$$J_{(\varphi)}^\nu \equiv ie(\varphi^* \overset{\leftrightarrow}{D}{}^\nu \varphi)$$

$$= ie(\varphi^* \overset{\leftrightarrow}{\partial}{}^\nu \varphi) + 2e^2 \varphi^* A^\nu \varphi \tag{8.2}$$

$$\equiv j_{(\varphi)}^\nu + 2e^2 \varphi^* A^\nu \varphi.$$

But this is not a proper conservation law, because it contains a covariant derivative, and a covariant derivative means *coupled* systems. $J_{(\varphi)}^\nu$ includes interactions between the charged particles and the gauge fields, and thus $\partial_\nu J_{(\varphi)}^\nu = 0$ means $\partial_\nu j_{(\varphi)}^\nu = -2e^2 \partial_\nu(\varphi^* A^\nu \varphi) \neq 0$ in an arbitrary gauge. The term $2e^2 \varphi^* A^\nu \varphi$ is not a current of just electric charge or just field energy. It is an *interaction* between charges and fields. That's why $J_{(\varphi)}^\nu$ cannot be cleanly interpreted as a current—it's the current of *what?* In Hilbert's jargon, eq. (8.1) is an improper conservation law.

In the infinitesimal spacetime and global gauge transformations that led to proper conservation laws, the coordinates transformed according to $x'^\mu = x^\mu + \epsilon \tau^\mu$ and the fields as $\varphi' = \varphi + ig\varphi\varepsilon$. We read off the generators τ^μ and $\zeta = ig\varphi$. When we differentiate the invariance identity with respect to ε, then set $\varepsilon = 0$, the ε makes no appearance in the invariance identity's final form, which can be expressed in terms of canonical momenta and the Hamiltonian,

$$-(\zeta - \varphi_\mu \tau^\mu)\psi = \partial^\mu(p_\mu \zeta - H_{\mu\nu}\tau^\nu) \tag{8.3}$$

where ψ is the Euler-Lagrange operator,

$$\psi \equiv \frac{\partial \mathcal{L}}{\partial \varphi} - \partial_\mu p^\mu. \tag{8.4}$$

When the functional is made an extremal, so that $\psi = 0$, an equation of continuity follows. Everything is neat and clean, a respectable proper conservation law. The quantity $(p_\mu \zeta - H_{\mu\nu}\tau^\nu)$ is locally conserved.

However, in a local gauge transformation where $\varepsilon = \varepsilon(x^\nu)$, the transformation of the gauge fields have derivatives of the parameter, as we saw in electrodynamics, where

$$A'^\mu = A^\mu - \partial^\mu \varepsilon. \tag{8.5}$$

In deriving the invariance identity by differentiating the invariance definition with respect to ε, was it correct to ignore $\partial^\mu \varepsilon$, as we did in chapter 7? After all, ε and $\partial^\mu \varepsilon$ are different functions. Or should we derive a set of coupled invariance identities by also taking derivatives of the definition with respect to $\partial^\mu \varepsilon$, as considered in ex. 7.9? To address such questions unambiguously, through judicious use of integration by parts, the definition of invariance can be reworked by swapping the $\partial^\mu \varepsilon$ terms for terms proportional to ε, and still have this reworked definition of invariance reduce to the familiar one when $\partial^\mu \varepsilon = 0$. In so doing we re-create Noether's second theorem. In particular, we will consider transformations of the form $t' = t + T(\varepsilon)$, where $T(\varepsilon)$ is a linear differential operator (and similarly for the fields). Defining invariance as $\Gamma' - \Gamma = 0$ under this kind of transformation, and assuming the variation vanishes on the boundary of the functional's integration region, gives expressions of the form $\psi T(\varepsilon)$ appearing within the functional's integrand. Corresponding to T we construct its *adjoint* $\tilde{T}$ where, by definition,

$$\psi T(\varepsilon) = \varepsilon \tilde{T}(\psi) + divergence\ terms. \qquad (8.6)$$

Because this step occurs under the functional's integral, the divergence terms can be disposed of at the boundary.

In this way, through the adjoint operator, Noether's second theorem moves the derivatives of ε from "inside" the T operator to "outside" the $\tilde{T}$ operator. Then $\Gamma' - \Gamma = 0$ becomes

$$\int_a^b \left(L' \frac{dt'}{dt} - L \right) dt = \int_a^b \varepsilon \tilde{T}(\psi)\, dt = 0. \qquad (8.7)$$

Because the functional's boundaries and ε are all arbitrary, by the fundamental lemma of the calculus of variations, we set

$$\tilde{T}(\psi) = 0. \qquad (8.8)$$

This is all done without requiring the functional to be an extremal—in other words, without setting $\psi = 0$. The equation $\tilde{T}(\psi) = 0$ with $\psi \neq 0$ gives a set of "dependencies," as Noether put it, which in applications to field theories include relationships between field components not determined by the equations of motion themselves.[2] Once these dependencies are in place, conservation laws, whether proper or improper, follow at once. The improper conservation laws will be seen to be characteristic of local gauge invariance. Noether's second thorem drives home the point that a functional being an extremal and being invariant are separate considerations.

Let us return to Hilbert, Klein, and Einstein trying to make sense out of energy conservation in general relativity, back in 1915.

The equations of general relativity lead to a vanishing divergence. But the divergence has a covariant derivative, $\partial_\mu + other\ terms$. This means the

---

[2]Such "dependencies" include symmetry constraints such as Bianchi identities [e.g., Laugwitz (1965) 103–104] for the Riemann curvature tensor, and the Ward-Takahashi identitites [e.g., Roman (1968) 540] of quantum field theory. We will see such a "dependency" emerge for the Faraday tensor $F_{\mu\nu}$ of electrodynamics.

energy densities of matter and electromagnetic radiation are not conserved, because they are not the whole system; the "other terms" describe the *exchange* of energy between them and gravitational fields.

Hilbert and Klein asked Dr. Noether to use her expertise to help them interpret energy conservation in general relativity. She warmed up with transformations featuring discrete sets of global parameters, whereby she re-derived, in a unified formalism, the "proper" conservation laws of mechanics and electrodynamics (the "first" theorem). After that she allowed the parameters to become functions of the spacetime coordinates, $\varepsilon^\mu = \varepsilon^\mu(x^\nu)$. For Noether, the issue was one of group theory. The spacetime transformations, Lorentz boosts, and global gauge transformations are a finite group; in other words, the distinct generators are finite in number.[3] But they are a subgroup of an infinite group that includes local gauge transformations, where each point in spacetime has its own transformation generator. Since spacetime is presumably a continuum, local gauge transformations have an uncountably infinite number of generators.

Furthermore, these locally variable coordinate transformations induce transformations of the components of the metric tensor $g_{\mu\nu}$ that depend on derivatives of the parameters,

$$g_{\mu\nu}(x'^\rho) - g_{\mu\nu}(x^\rho) \sim \partial_\mu \varepsilon_\nu. \tag{8.9}$$

Changing the coordinates locally induces a local gauge transformation on the metric tensor, and thus on the gravitational field, since in general relativity the metric tensor components play the role of gravitational potentials.

In section 8.2 we develop the machinery of Noether's second theorem, along the lines outlined above, then illustrate its use. Section 8.3 looks at another application of the second theorem in the context of "parametric invariance." In the remainder of the chapter, we apply Noether's second theorem to general relativity, and, with Emmy Noether, return to Hilbert's assertion.

All physics majors take a course in electrodynamics and become famiilar with Maxwell's equations. Not all physics majors have a course in general relativity with tensors,[4] so in sections 8.4 and 8.5 we review its two main differential equations, and turn to their corresponding functionals in section 8.6. Section 8.7 describes local gauge invariance in general relativity. Once we have all the tools needed, we can see (in section 8.8) how Emmy Noether brought clarity to David Hilbert, Felix Klein, Albert Einstein—and the rest of us—on the important point of energy conservation in general relativity.

---

[3] A countable infinite set of generators would also work, where each generator could be put into a one-to-one correspondence with the integers.

[4] In recent years, more undergraduate physics departments have started offering courses in general relativity, but as of this writing it is not universally offered like electrodynamics.

8.2 Noether's Second Theorem

To approach Noether's "second theorem," consider the familiar functional

$$\Gamma = \int_a^b L(t, x^\mu, \dot{x}^\mu)\, dt, \tag{8.10}$$

where $\dot{x}^\mu \equiv dx^\mu/dt$, but generalize the transformation from $t' = t + \varepsilon\tau$ and $x'^\mu = x^\mu + \varepsilon\zeta^\mu$ that we considered before, to[5]

$$\begin{aligned} t' &= t + T(\varepsilon) \\ x'^\mu &= x^\mu + Z^\mu(\varepsilon) \end{aligned} \tag{8.11}$$

where T and Z^μ are linear functions of $\varepsilon = \varepsilon(t)$ and its derivatives. For T,

$$\begin{aligned} T(\varepsilon) &\equiv \tau_0\varepsilon + \tau_1\frac{d\varepsilon}{dt} \cdots + \tau_n\frac{d^k\varepsilon}{dt^k} \\ &\equiv \tau_n\varepsilon^{(n)} \end{aligned} \tag{8.12}$$

(note the summation convention over $n = 0$ to k) where $\varepsilon^{(n)} \equiv d^n\varepsilon/dt^n$ and $\varepsilon^{(0)} \equiv \varepsilon$. For Z^μ,

$$Z^\mu(\varepsilon) \equiv \zeta_n^\mu\varepsilon^{(n)}. \tag{8.13}$$

The leading coefficients τ_0 and ζ_0 are the familiar τ and ζ of Noether's first theorem, contained here as special cases. In these expressions for T and Z^μ, all the τ_n and ζ_n^μ may be functions of t.[6] Consider now the transformed Lagrangian

$$L' \equiv L\left(t', x'^\mu(t'), \frac{dx'^\mu(t')}{dt'}\right). \tag{8.14}$$

Along the lines of our previous experience with functional invariance, with Emmy Noether we make the following definition of invariance under $T(\varepsilon)$ and $Z^\mu(\varepsilon)$.

Definition: The functional

$$\Gamma = \int_a^b L(t, x^\mu, \dot{x}^\mu)\, dt \tag{8.15}$$

---

[5]To compare our notation with Noether's 1918 paper [Noether (1918), Noether & Tavel (1971)]: our Γ is her I, our L is her f, our t her x, our t' her y, our x^μ her u, our x'^μ her v, and our $\varepsilon(x^\mu)$ is her $p(x)$.

[6]The functional and transformations can be generalized to multiple integral problems by including more indices with summation conventions.

is said to be invariant under the transformation $t' = t + T(\varepsilon), x'^{\mu} = x^{\mu} + Z^{\mu}(\varepsilon)$ if and only if for every continuous $x(t)$ on $[a, b]$, $\Gamma' - \Gamma = 0$. Written out in detail this says

$$\int_a^b \left[L\left(t, x'(t'), \frac{dx'(t')}{dt'}\right) \frac{dt'}{dt} - L\left(t, x^{\mu}(t), \frac{dx^{\mu}(t)}{dt}\right) \right] dt = 0. \quad (8.16)$$

The parameter ε and its derivatives are considered infinitesimal.[7]

Perform a Taylor series expansion of the new Lagrangian about (t, x^{μ}), that is, about $\varepsilon = 0$:

$$L' = L + (t' - t)\left(\frac{\partial L'}{\partial t'}\right)_0 + (x'^{\mu} - x^{\mu})\left(\frac{\partial L'}{\partial x'^{\mu}}\right)_0$$

$$+ \left(\frac{dx'^{\mu}}{dt'} - \frac{dx^{\mu}}{dt}\right)\left(\frac{\partial L'}{\partial \dot{x}'^{\mu}}\right)_0 + \cdots \quad (8.17)$$

$$= L + T\frac{\partial L}{\partial t} + Z^{\mu}\frac{\partial L}{\partial x^{\mu}} + \left(\frac{dx'^{\mu}}{dt'} - \frac{dx^{\mu}}{dt}\right)p_{\mu} + \cdots$$

where $p_{\mu} \equiv \partial L/\partial \dot{x}^{\mu}$ is the canonical momentum. The dx'^{μ}/dt' term takes some additional work:

$$\frac{dx'^{\mu}}{dt'} = \frac{dx^{\mu} + dZ^{\mu}}{dt + dT}$$

$$= \frac{\dot{x}^{\mu} + \dot{Z}^{\mu}}{1 + \dot{T}} \quad (8.18)$$

$$= (\dot{x}^{\mu} + \dot{Z}^{\mu})(1 - \dot{T} + \cdots)$$

$$\approx \dot{x}^{\mu} + \dot{Z}^{\mu} - \dot{x}^{\mu}\dot{T}$$

to first order in ε and its derivatives. Returning to the Taylor series, we have

$$L' = L + T\frac{\partial L}{\partial t} + Z^{\mu}\frac{\partial L}{\partial x^{\mu}} + (\dot{Z}^{\mu} - \dot{x}^{\mu}\dot{T})p_{\mu} + \cdots. \quad (8.19)$$

The definition requires $L'(dt'/dt)$, which to first order in ε is

$$L'\frac{dt'}{dt} \approx L'(1 + \dot{T})$$

$$= L + L\dot{T} + T\frac{\partial L}{\partial t} + Z^{\mu}\frac{\partial L}{\partial x^{\mu}} + (\dot{Z}^{\mu} - \dot{x}^{\mu}\dot{T})p_{\mu}. \quad (8.20)$$

Integrating over t, we write out the integral that appears in the invariance definition, to first order in ε:

---

[7]We do not define invariance here as $\Gamma' - \Gamma \sim \varepsilon^s, s > 1$ because we will not differentiate this definition with respect to ε.

$$\int_a^b \left(L'\frac{dt'}{dt} - L\right) dt = \int_a^b \left(L\dot{T} + T\frac{\partial L}{\partial t} + Z^\mu \frac{\partial L}{\partial x^\mu}\right.$$
$$\left. + (\dot{Z}^\mu - \dot{x}^\mu \dot{T})p_\mu\right) dt. \tag{8.21}$$

Perform an integration by parts on $Z^\mu p_\mu$ and identify the Euler-Lagrange operator

$$\psi_\mu \equiv \frac{\partial L}{\partial x^\mu} - \dot{p}_\mu \tag{8.22}$$

which turns the integral into

$$\int_a^b \left(L'\frac{dt'}{dt} - L\right) dt = \int_a^b \left(L\dot{T} + T\frac{\partial L}{\partial t} + Z^\mu \psi_\mu - \dot{x}^\mu \dot{T} p_\mu\right) dt$$
$$+ \int_a^b \frac{d}{dt}\left(Z^\mu p_\mu\right) dt. \tag{8.23}$$

The partial derivative of the Lagrangian with respect to time can be written in terms of other variables by expanding the total derivative of the Lagrangian:

$$\frac{dL}{dt} = \frac{\partial L}{\partial t} + \frac{\partial L}{\partial x^\mu}\dot{x}^\mu + \frac{\partial L}{\partial \dot{x}^\mu}\ddot{x}^\mu$$
$$= \frac{\partial L}{\partial t} + \dot{x}^\mu \psi_\mu + \frac{d}{dt}(\dot{x}^\mu p_\mu), \tag{8.24}$$

and thus

$$\frac{\partial L}{\partial t} = -\dot{H} - \dot{x}^\mu \psi_\mu \tag{8.25}$$

with H the Hamiltonian,

$$H \equiv \dot{x}^\mu p_\mu - L. \tag{8.26}$$

Therefore,

$$\int_a^b \left(L'\frac{dt'}{dt} - L\right) dt = \int_a^b \left(L\dot{T} - (\dot{H} + \dot{x}^\mu \psi_\mu)T + \psi_\mu Z^\mu - \dot{x}^\mu \dot{T} p_\mu\right) dt$$
$$+ \int_a^b \frac{d}{dt}(Z^\mu p_\mu)dt. \tag{8.27}$$

In the $L\dot{T}$ term, write $L = p_\mu \dot{x}^\mu - H$. The $\dot{x}^\mu \dot{T} p_\mu$ neatly cancels out, and we obtain

$$\int_a^b \left(L'\frac{dt'}{dt} - L\right) dt = \int_a^b \psi_\mu \left[Z^\mu(\varepsilon) - \dot{x}^\mu T(\varepsilon)\right] dt$$
$$+ \int_a^b \frac{d}{dt}[Z^\mu(\varepsilon)p_\mu - T(\varepsilon)H] dt \tag{8.28}$$

where the product rule for derivatives has been used on TH. I have restored the ε argument in T and Z^μ to emphasize that they take derivatives of $\varepsilon(t)$. The last integral may be written as a boundary term, so that

$$\int_a^b \left(L' \frac{dt'}{dt} - L \right) dt = \int_a^b \psi_\mu [Z^\mu(\varepsilon) - \dot{x}^\mu T(\varepsilon)] \, dt$$
$$+ \, [Z^\mu(\varepsilon) p_\mu - T(\varepsilon) H]|_a^b. \tag{8.29}$$

For the functional to be invariant, by Noether's definition the integral on the left-hand side vanishes [Noether (1918)]. Thus there follows the integral form of the Noether identity,

$$\int_a^b \psi_\mu \left[Z^\mu(\varepsilon) - \dot{x}^\mu T(\varepsilon) \right] dt + [Z^\mu(\varepsilon) p_\mu - T(\varepsilon) H]|_a^b = 0. \tag{8.30}$$

This identity contains two important cases. They are Noether's first and second theorems.

(1) If $\int_a^b L \, dt$ is an extremal, then $\psi_\mu = 0$. The surviving contribution in the Noether identity is the boundary term:

$$[Z^\mu(\varepsilon) p_\mu - T(\varepsilon) H]|_a^b = 0. \tag{8.31}$$

Consider the special but important case when $\dot{\varepsilon} = 0$, so that $T(\varepsilon) = \varepsilon \tau$ and $Z^\mu(\varepsilon) = \varepsilon \zeta^\mu$. Eq. (8.31) becomes

$$\varepsilon [\zeta^\mu p_\mu - H\tau]|_a^b = 0. \tag{8.32}$$

Because it is infinitesimal but nonzero, the ε divides out, and because a and b are arbitrary, we recover Noether's first theorem,

$$\zeta^\mu p_\mu - H\tau = \text{const.} \tag{8.33}$$

(2) Without requiring $\psi_\mu = 0$, consider another possibility for the boundary conditions. Dismiss the boundary term by putting $[Z^\mu(\varepsilon) p_\mu - T(\varepsilon) H]|_a^b = 0$,[8] This leaves the Noether identity saying

$$\int_a^b [\psi_\mu Z^\mu(\varepsilon) - \psi_\mu \dot{x}^\mu T(\varepsilon)] \, dt = 0. \tag{8.34}$$

The generators $Z^\mu(\varepsilon)$ and $T(\varepsilon)$ take derivatives of their argument $\varepsilon(t)$. Because ε is arbitrary, we would like to factor it out, so the integral becomes something of the form $\int_a^b \varepsilon[\cdots] dt = 0$, which shifts to ε the role played

---

[8]Should the integral be over all space, such a conclusion follows for physically realistic fields that vanish at infinity.

by δq in our original analysis of variational problems. If that can be done, then by the fundamental lemma of the calculus of variations, the term in the square bracket may be set equal to zero. The result is Noether's second theorem. As the integral sits now, the ψ_μ standing to the left of $Z^\mu(\varepsilon)$ and $T(\varepsilon)$ gives us something to work with through integration by parts, thereby enabling the ε and the ψ_μ to swap places.

To carry out such a procedure, starting with $T(\varepsilon)$ we seek its "adjoint" operator $\tilde{T}$ such that[9] for any function $u(t)$,

$$uT(\varepsilon) = \varepsilon\tilde{T}(u) + \text{divergence terms.} \tag{8.35}$$

An illustration seems appropriate.

Consider a first-order $T(\varepsilon)$,

$$T(\varepsilon) = \tau\varepsilon + a\dot{\varepsilon} \tag{8.36}$$

where $\tau_0 \equiv \tau(t)$ and $\tau_1 \equiv a(t)$. For any $u(t)$ we may write

$$\begin{aligned} uT(\varepsilon) &= u[\tau\varepsilon + a\dot{\varepsilon}] \\ &= \varepsilon\tau u + a\left(\frac{d}{dt}(u\varepsilon) - \varepsilon\dot{u}\right) \\ &= \varepsilon[(\tau - \dot{a})u - a\dot{u}] + \frac{d}{dt}(au\varepsilon) \\ &\equiv \varepsilon\tilde{T}(u) + \frac{d}{dt}(au\varepsilon) \end{aligned} \tag{8.37}$$

where

$$\tilde{T}(u) = (\tau - \dot{a})u - a\dot{u}. \tag{8.38}$$

A similar calculation must be done for $u\tilde{Z}^\mu(\varepsilon)$. Now the identity is written in terms of adjoints and divergences, the divergences shifted to surface terms where they vanish, leaving for the Noether identity

$$\int_a^b \varepsilon[\tilde{Z}^\mu(\psi_\mu) - \tilde{T}(\dot{x}^\mu\psi_\mu)]dt = 0. \tag{8.39}$$

Since ε and the interval $[a, b]$ are arbitrary, by the fundamental lemma we obtain Noether's second theorem, the Noether identity in differential form,

$$\tilde{Z}^\mu(\psi_\mu) - \tilde{T}(\dot{x}^\mu\psi_\mu) = 0, \tag{8.40}$$

which holds without requiring ψ_μ to vanish.

---

[9]Students of quantum mechanics will recognize a similar move there: if A is an operator and ψ a wave function, the adjoint $A^\dagger$ of A may be defined as $\int \psi^*(A\psi)dx = \int (A^\dagger\psi^*)\psi dx$.

For instance, consider the case

$$uT(\varepsilon) = u[\varepsilon\tau_0 + \tau_1\dot\varepsilon]$$
$$uZ^\mu(\varepsilon) = u[\varepsilon\zeta_0^\mu + \zeta_1^\mu\dot\varepsilon].$$

(8.41)

The adjoint operators are

$$\varepsilon\tilde{T}(u) = \varepsilon\left[(\tau_0 - \dot\tau_1)u - \tau_1\dot u\right]$$

(8.42)

$$\varepsilon\tilde{Z}^\mu(u) = \varepsilon\left[(\zeta_0^\mu - \dot\zeta_1^\mu)u - \zeta_1^\mu\dot u\right].$$

(8.43)

Example (Local Gauge Invariance in the Electrodynamics of the Charged Scalar Field): The Lagrangian density for charged scalar particles coupled to the electromagnetic field is

$$\mathcal{L} = (D_\mu\varphi)^*(D^\mu\varphi) - m^2\varphi^*\varphi - \frac{1}{4\mu_o}F_{\mu\nu}F^{\mu\nu},$$

(8.44)

which determines the Euler-Lagrange operator for the electromagnetic field,

$$\psi^\nu \equiv \frac{\partial\mathcal{L}}{\partial A_\nu} - \partial_\mu\left(\frac{\partial\mathcal{L}}{\partial A_{\nu,\mu}}\right) = \partial_\mu F^{\mu\nu} - \mu_o J^\nu_{(\varphi)}.$$

(8.45)

Under a local gauge transformation $T(\varepsilon) = 0$, but the electromagnetic fields transform as

$$A'^\mu = A^\mu - \partial^\mu\varepsilon$$

(8.46)

which we now write as

$$A'^\mu = A^\mu + Z^\mu(\varepsilon)$$

(8.47)

where

$$Z^\mu = \zeta_0^\mu\varepsilon + \zeta_1^{\mu\nu}\partial_\nu\varepsilon + \zeta_2^{\mu\nu\rho}\partial_\nu\partial_\rho\varepsilon + \cdots.$$

(8.48)

Comparing the two expressions for A'^μ shows that $\zeta_1^{\mu\nu} = -\delta^{\mu\nu}$ and all other ζ-coefficients are zero. Noether's second theorem requires that $\tilde{Z}^\mu(\psi_\mu)$ vanish identically, where

$$\tilde{Z}^\mu(\psi_\mu) = (\zeta_0^\mu - \partial_\nu\zeta_1^{\mu\nu})\psi_\mu - \zeta_1^{\mu\nu}\partial_\nu\psi_\mu + \cdots$$
$$= \delta^{\mu\nu}\partial_\nu\psi_\mu$$
$$= \partial^\mu\psi_\mu$$
$$= \partial^\mu\partial^\rho F_{\rho\mu} - \mu_o\partial^\mu J_{\mu(\varphi)}.$$

(8.49)

Consider regions of spacetime for which no charged particles are nearby. Then $J_{\mu(\varphi)} = 0$, and Noether's second theorem requires

$$\partial^\mu\partial^\rho F_{\rho\mu} = 0.$$

(8.50)

Even if it had not been known already, Noether's second theorem *requires* $F_{\rho\mu} = -F_{\mu\rho}$ because $\partial^\mu \partial^\rho$ is symmetric under the exchange of μ and ρ. This is an example of a "dependency" of which Noether spoke with regard to her second theorem. Once the necessity of $\partial^\mu \partial^\rho F_{\mu\rho} = 0$ is realized to be independent of $J_{\mu(\varphi)}$, then even in the neighborhood of charged source particles, where $J_{\mu(\varphi)} \neq 0$, Noether's second theorem requires its divergence to vanish indentically, $\partial^\mu J_{\mu(\varphi)} = 0$. Note how all of this follows without requiring $\psi_\nu = 0$, and without needing to know the details of $J_{\mu(\varphi)}$ explicitly.[10]

The point is that: a Lagrangian (or, in generalizing, a Lagrangian density) determines the Euler-Lagrange operator ψ_μ because, by definition,

$$\psi_\mu \equiv \frac{\partial L}{\partial x^\mu} - \frac{d}{dt}\left(\frac{\partial L}{\partial \dot{x}^\mu}\right). \tag{8.51}$$

Noether's first theorem links conservation laws to the equation of motion: $\psi_\mu = 0$ is necessary for the conservation law to follow. The second theorem puts constraints on ψ_μ, and gives conservation laws from a definition of invariance, without ψ_μ having to be zero.

8.3 Parametric Invariance

Early in our study of functionals we saw how the integrand could be shifted to parametric form. Consider a functional defined in the xy plane,

$$\Gamma = \int_a^b L(x, y, y') \, dx \tag{8.52}$$

where $y = y(x)$ and $y' \equiv dy/dx$. A simple example trotted out whenever something important needs to be illustrated is found in the distance functional,

$$\begin{aligned} \Gamma &= \int_a^b \sqrt{dx^2 + dy^2} \\ &= \int_a^b \sqrt{1 + y'^2} \, dx. \end{aligned} \tag{8.53}$$

Let $y(x)$ be a particular curve in the xy plane, and switch over to a parametric representation of the curve, with some parameter t:

$$\begin{aligned} x &= x(t) \\ y &= y(t). \end{aligned} \tag{8.54}$$

---

[10]In other words, we do not have to work through $\partial^\nu J_{\nu(\varphi)}$ term by term to verify that it vanishes. This is true whether J_ν includes a covariant derivative or not.

This parameter could be arc length along the curve, a time, or in extensions to relativity, the proper time along a particle's world line. In terms of t, the arc length functional becomes

$$\Gamma = \int_a^b \sqrt{1 + y'^2} \, dx = \int_{t(a)}^{t(b)} \sqrt{1 + \left(\frac{\dot{y}}{\dot{x}}\right)^2} \, \dot{x} \, dt. \tag{8.55}$$

Two ways of writing the same functional now exist, with the Lagrangian $L(y') = \sqrt{1 + y'^2} \equiv L^{(x)}$ for the integration over x, and $L(\dot{x}, \dot{y}) = \dot{x}\sqrt{1 + (\frac{\dot{y}}{\dot{x}})^2} \equiv L^{(t)}$ for the integration over t. Same information, different representations.

Since it is possible to convert $L^{(x)}(x, y, y')dx$ into $L^{(t)}(x, y, \dot{x}, \dot{y})dt$ through $x \to x(t)$ and $y \to y(t)$, the question now arises: under what circumstances is the functional invariant under a transformation of the parameter t? Parametric invariance, which will turn out to be a special case of Noether's second theorem, is defined as follows.

Definition: Given the functional

$$\Gamma = \int_a^b L(t, x^\mu, \dot{x}^\mu) \, dt \tag{8.56}$$

where $\dot{x} \equiv dx/dt$, consider a transformation of the form

$$t' = \mathcal{T}(t)$$
$$x' = x(t') \tag{8.57}$$

where $\dot{\mathcal{T}} \equiv d\mathcal{T}/dt > 0$ for all t in the closed interval $[a, b]$. The functional is said to be "parameter invariant" if and only if[11]

$$\int_{t_1'}^{t_2'} L\left(t', x(t'), \frac{dx(t')}{dt'}\right) \, dt' = \int_{t_1}^{t_2} L\left(t, x(t), \frac{dx(t)}{dt}\right) \, dt \tag{8.58}$$

for all $[t_1, t_2] \subseteq [a, b]$. Notice that it is not necessary for the transformation to be infinitesimal.

In the original presentation of the functional, with the $L^{(x)}$ version of the Lagrangian,

$$\Gamma = \int_a^b L(x, y, y') \, dx \tag{8.59}$$

when $\delta\Gamma = 0$ one Euler-Lagrange equation results. It is

$$\frac{\partial L}{\partial y} - \frac{d}{dx}\frac{\partial L}{\partial y'} = 0. \tag{8.60}$$

---

[11]$x(t')$ means $x((\mathcal{T}(t))$ or, in composition notation, $x \circ \mathcal{T}^{-1}(t')$; see Logan (1977) 151.

In the parametric representation of the same functional, with the $L^{(t)}$ version of the Lagrangian,[12]

$$\Gamma = \int_{t(a)}^{t(b)} L(x, y, \dot{x}, \dot{y})\, dt \tag{8.61}$$

there are two Euler-Lagrange equations. They are

$$\frac{\partial L}{\partial x} - \frac{d}{dt}\frac{\partial L}{\partial \dot{x}} = 0 \tag{8.62}$$

and

$$\frac{\partial L}{\partial y} - \frac{d}{dt}\frac{\partial L}{\partial \dot{y}} = 0. \tag{8.63}$$

We expect a relation exists between the Euler-Lagrange operators

$$\psi_\mu \equiv \frac{\partial L}{\partial x^\mu} - \frac{d}{dt}\frac{\partial L}{\partial \dot{x}^\mu} \tag{8.64}$$

for $\dot{x}^\mu = \dot{x}$ and $\dot{y}$, because they came from $y' = dy/dx = \dot{y}/\dot{x}$. This is a form of parametric invariance, if we consider rescaling the velocities.

When rescaling the derivatives in the parametric version of the distance functional $L^{(t)}$ by letting $\dot{x} \to \lambda\dot{x}$ and $\dot{y} \to \lambda\dot{y}$, the Lagrangian itself gets linearly rescaled,

$$L^{(t)}(\lambda\dot{x}, \lambda\dot{y}) = \lambda L^{(t)}(\dot{x}, \dot{y}) \tag{8.65}$$

because

$$\int_{t(a)}^{t(b)} \sqrt{1 + \left(\frac{\lambda\dot{y}}{\lambda\dot{x}}\right)^2}\, \lambda\dot{x}\, dt = \int_{t(a)}^{t(b)} \lambda\sqrt{1 + \left(\frac{\dot{y}}{\dot{x}}\right)^2}\, \dot{x}\, dt. \tag{8.66}$$

This functional is also parameter invariant. Under $t' = \mathcal{T}(t)$ it follows that $dt' = \dot{\mathcal{T}} dt$, $\dot{y}' \equiv \frac{dy(t')}{dt'} = \frac{dy}{dt}\frac{dt}{dt'} = \dot{y}/\dot{\mathcal{T}}$, and similarly for $\dot{x}'$. Therefore,

$$\Gamma' = \int_{t'(a)}^{t'(b)} \sqrt{1 + \left(\frac{\dot{y}'}{\dot{x}'}\right)^2}\, \dot{x}'\, dt'$$

$$= \int_{t(a)}^{t(b)} \sqrt{1 + \left(\frac{\dot{y}/\dot{\mathcal{T}}}{\dot{x}/\dot{\mathcal{T}}}\right)^2}\, \frac{\dot{x}}{\dot{\mathcal{T}}}\, \dot{\mathcal{T}} dt \tag{8.67}$$

$$= \Gamma.$$

---

[12] $L^{(x)}$ and $L^{(t)}$ are the same Lagrangian insofar as the information packed into them is concerned. They describe the same system, but in different variables.

Not all functionals exhibit scaling linear in λ. For example, Hamilton's principle functional for a free particle in Newtonian mechanics is

$$\Gamma = \int_{t(a)}^{t(b)} \frac{1}{2} m \dot{x}^2 \, dt \tag{8.68}$$

and under the rescaling $\dot{x} \to \lambda \dot{x}$, the new Lagrangian L' rescales quadratically: $L' = \lambda^2 L$. This functional is not parameter invariant either: with $t' = \mathcal{T}(t)$,

$$\begin{aligned}
\Gamma' &= \int_{t'(a)}^{t'(b)} \frac{1}{2} m \left(\frac{dx'}{dt'} \right)^2 dt' \\
&= \int_{t(a)}^{t(b)} \frac{1}{2} m \left(\frac{\dot{x}}{\dot{\mathcal{T}}} \right)^2 \dot{\mathcal{T}} dt \neq \Gamma.
\end{aligned} \tag{8.69}$$

These examples suggest that linear scaling of generalized velocities and parametric invariance may go together. Scaling linearly or not, a well-known result from differential equations, Euler's theorem, says that if a function $\Psi(x^\mu)$ is homogeneous of degree k under rescaling, in other words if $\Psi(\lambda x^\mu) = \lambda^k \Psi(x^\mu)$,[13] then

$$x^\mu \frac{\partial \Psi}{\partial x^\mu} - k\Psi = 0. \tag{8.70}$$

Therefore if $L^{(t)}$ is homogeneous of degree 1 under a velocity rescaling, then by Euler's theorem,

$$\dot{x} \left(\frac{\partial L^{(t)}}{\partial \dot{x}} \right) + \dot{y} \left(\frac{\partial L^{(t)}}{\partial \dot{y}} \right) - L^{(t)} = 0. \tag{8.71}$$

Not coincidentally, this is equivalent to the Hamiltonian being identically zero. Notice that in the Lagrangians considered here, no explicit time dependence exists, so $\partial L / \partial t = 0$.

The following theorem can be readily proved: If $\partial L / \partial t = 0$ in the functional

$$\Gamma = \int_{t(a)}^{t(b)} L(x^\mu, \dot{x}^\mu) \, dt, \tag{8.72}$$

and if the Lagrangian is homogeneous of order 1 under velocity rescalings, then Γ is parameter invariant [Logan (1977)]. To prove this theorem, under the transformation the new functional takes the form

---

[13]Being homogeneous of degree k also means $\Psi(x, y) = x^k \Phi(y/x)$ for some function Φ. From this Euler's theorem follows. See ex. 8.6.

$$\Gamma' = \int_{t'(a)}^{t'(b)} L\left(x^\mu(\mathcal{T}(t)), \frac{dx^\mu(\mathcal{T}(t))}{d\mathcal{T}(t)}\right) d\mathcal{T}(t)$$

$$= \int_{t(a)}^{t(b)} L\left(x^\mu(t), \frac{\dot{x}^\mu}{\dot{\mathcal{T}}}\right) \dot{\mathcal{T}} dt. \tag{8.73}$$

With the rescalings $\dot{x} \to \lambda\dot{x}$ and $\dot{\mathcal{T}} \to \lambda\dot{\mathcal{T}}$, where λ is arbitrary, it follows that $\Gamma' - \Gamma = \int_a^b (L' - L)dt = \int_a^b (\lambda - 1)Ldt$, which vanishes by choosing $\lambda = 1$, resulting in Γ being parameter invariant.

The converse also holds: If Γ is parameter-invariant, then the Lagrangian satisfies $\partial L/\partial t = 0$ and $L(\lambda\dot{x}) = \lambda L(\dot{x})$. To prove this, invoke the invariance identity of chapter 4. That identity says that if the functional is invariant under the infinitesimal transformation $t' = t + \tau\varepsilon$ and $x' = x + \zeta\varepsilon$, then

$$\frac{\partial L}{\partial x^\mu}\zeta^\mu + \frac{\partial L}{\partial \dot{x}^\mu}\dot{\zeta}^\mu + \frac{\partial L}{\partial t}\tau - \left(\dot{x}^\mu \frac{\partial L}{\partial \dot{x}^\mu} - L\right)\dot{\tau} = 0. \tag{8.74}$$

If the functional is parameter invariant, then a Taylor expansion of $\mathcal{T}(t)$ about $t' = t$ gives

$$t' = t + (t' - t)\left(\frac{d\mathcal{T}}{dt'}\right)_{t'=t} + \frac{1}{2!}(t' - t)^2 \left(\frac{d^2\mathcal{T}}{dt'^2}\right)_{t'=t} + \cdots. \tag{8.75}$$

Denoting $\varepsilon \equiv t' - t$, the transformation $t' = \mathcal{T}(t), x' = x(t')$ becomes, to first order in ε, $t' = t + \varepsilon\tau$ where $\tau \equiv (d\mathcal{T}/dt)_0$ and $x' = x$. The condition $\dot{\mathcal{T}} > 0$ necessary for parameter invariance is readily satisfied because $dt'/dt = 1 + \varepsilon\dot{\tau}$ and ε can be made as small as one likes. With $\zeta = 0$, the invariance identity of eq. (8.74) reduces to

$$\frac{\partial L}{\partial t}\tau - \left(\dot{x}^\mu \frac{\partial L}{\partial \dot{x}^\mu} - L\right)\dot{\tau} = 0. \tag{8.76}$$

Since τ is arbitrary, the coefficients of both τ and $\dot{\tau}$ must vanish; therefore the necessary conditions for the functional to be parameter invariant are

$$\frac{\partial L}{\partial t} = 0 \tag{8.77}$$

and

$$\dot{x}^\mu \frac{\partial L}{\partial \dot{x}^\mu} - L = 0. \tag{8.78}$$

These are called the Zermelo conditions (1894) [Logan (1977)]. Thanks to the Euler-Lagrange equation $\dot{H} = -\partial L/\partial t$, parameter invariance is equivalent to the system's Hamiltonian being zero and constant.

Now we return to the question of relationships between the Euler-Lagrange operators ψ_x and ψ_y for the $L^{(t)}$ versions of the Lagrangian. Before switching to parametric form, the original functional

$$\delta \int_a^b L(x, y, y') \, dx = 0 \qquad (8.79)$$

produced one Euler-Lagrange equation,

$$\frac{\partial L}{\partial y} - \frac{d}{dx} \frac{\partial L}{\partial y'} = 0. \qquad (8.80)$$

But the parametric representation of the same functional,

$$\delta \int_{t(a)}^{t(b)} L(x, y, \dot{x}, \dot{y}) \, dt = 0 \qquad (8.81)$$

produces two Euler-Lagrange equations,

$$\frac{\partial L}{\partial x^\mu} - \frac{d}{dt} \frac{\partial L}{\partial \dot{x}^\mu} = 0 \qquad (8.82)$$

where $x^\mu = x$ or y. We anticipated that a relation should exist between the respective Euler-Lagrange operators

$$\psi_\mu \equiv \frac{\partial L}{\partial x^\mu} - \frac{d}{dt} \frac{\partial L}{\partial \dot{x}^\mu} \qquad (8.83)$$

whether or not $\psi_\mu = 0$, because $y' = dy/dx = \dot{y}/\dot{x}$ arises from within the two forms of the Lagrangian itself. Such a relation between the various ψ_μ follows at once if the functional is parameter invariant. The proof goes like this: Given that the $L^{(t)}$ version of the Lagrangian, $L = L(x^\mu, \dot{x}^\mu)$, is parameter invariant, then the Zermelo conditions hold. Let us differentiate one of them with respect to t, in particular, from eq. (8.78),

$$\frac{d}{dt} \left(\dot{x}^\mu \frac{\partial L}{\partial \dot{x}^\mu} \right) = \frac{dL}{dt}. \qquad (8.84)$$

By the other Zermelo equation, eq. (8.77), there will be no $\partial L/\partial t$ term when expanding dL/dt with the chain rule, so eq. (8.84) becomes

$$\ddot{x}^\mu \frac{\partial L}{\partial \dot{x}^\mu} + \dot{x}^\mu \frac{d}{dt} \frac{\partial L}{\partial \dot{x}^\mu} = \frac{\partial L}{\partial x^\mu} \dot{x}^\mu + \frac{\partial L}{\partial \dot{x}^\mu} \ddot{x}^\mu. \qquad (8.85)$$

Two terms cancel, leaving

$$\dot{x}^\mu \left(\frac{\partial L}{\partial x^\mu} - \frac{d}{dt} \frac{\partial L}{\partial \dot{x}^\mu} \right) = 0, \qquad (8.86)$$

or in the consise notation of the Euler-Lagrange operator,

$$\dot{x}^\mu \psi_\mu = 0. \tag{8.87}$$

This is the "Weierstrass representation" of parametric-invariant problems,[14] linking the various ψ_μ, whether or not the functional is an extremal.

The Weierstrass representation of parametric problems can now be seen as a special case of Noether's second theorem. In the notation of section 8.2, $Z^\mu = 0$ and the T there and the $\mathcal{T}$ here are the same, so Noether's identity reduces to $\dot{x}^\mu \psi_\mu T(\varepsilon) = 0$. Since ε and therefore $T(\varepsilon)$ are arbitrary,[15] it follows from Noether's second theorem that $\dot{x}^\mu \psi_\mu = 0$. Emmy Noether wrote,

The simplest example of Theorem II... is afforded by the Weierstrass parametric representation.... Another example is presented by the "general theory of relativity" of the physicists. [Noether (1918), Noether & Tavel (1971)]

Let us now return to the "general relativity of the physicists" and see what Emmy Noether will teach us about it.

8.4 Free Fall in a Gravitational Field

Noether's theorem grew out of Emmy Noether resolving an issue for David Hilbert and Felix Klein at the University of Göttingen, when they were corresponding with Albert Einstein in Berlin about Einstein's new general theory of relativity. In this section and the two to follow, let us review the main principles of general relativity that will be needed when applying Noether's second theorem to them.[16]

Based on the observation that all objects fall locally in a gravitational field at the same rate, the principle of equivalence for gravitational and inertial mass asserts that gravitation can be locally transformed away by going into a local free-fall frame. Because tensors transform the same as coordinate differentials, the content of an equation is preserved under coordinate transformations if it is written with tensors.

Special relativity and Newtonian relativity share a dependence on inertial reference frames. Between 1907 and 1915, Einstein expanded special

---

[14]Karl T. W. Weierstrass (1815–1897) is best known among calculus students for his foundational definitions of continuity, proof of the intermediate value theorem, and his work on uniform convergence.

[15]Notice, that, unlike $\psi_\mu Z^\mu(\varepsilon)$, where the ψ_μ and the Z^μ get mixed together in the sum, the $T(\varepsilon)$ factors out of the sum $\dot{x}^\mu \psi_\mu$.

[16]Since our goal is understanding Noether's theorem, much about general relativity itself and the tensor calculus must be left unsaid here. To fill in more details, consult general relativity or tensor calculus texts, such as Hobson et al. (2006), Misner et al. (1971), Neuenschwander (2015), Ohanian (1976), Schutz (2005), Taylor & Wheeler (2000), or Weinberg (1972).

relativity to general relativity by considering relative acceleration between frames. With the principle of equivalence, general relativity became a relativistic theory of gravitation. We write an equation for a law of physics in the no-gravity locally inertial reference frame, then carry out a coordinate transformation to the accelerated frame, which expresses that law of physics in the presence of gravitation.[17]

About 1912 Einstein realized the tools he needed were to be found in the non-Euclidean geometries and tensor calculus that had been developed by distinguished mathematicians Nikoli Lobachevski (1792–1856), Karl Friedrich Gauss (1777–1855), Janos Bolyai (1802–1860), Georg Friedrich Bernhard Riemann (1826–1866), Elwin Bruno Christoffel (1829–1900), Gregorio Ricci-Curbastro (1853–1925), Tullio Levi-Civita (1873–1941), and others.[18] For instance, Riemann generalized the non-Euclidean geometry of his predecessors by extending geometry on curved surfaces to geometry in spaces of n dimensions and variable curvature. Through this language, with its reliance on Riemann's curvature tensor, Einstein's theory leads to the notion of gravitation as the "curvature of spacetime."

To carry out the program, the equations of motion of a particle falling in a gravitational field, and the equations describing the field itself, had to be expanded beyond the Newtonian paradigm and made generally covariant. There are two sides to this problem, as Misner, Thorne, and Wheeler succintly summarize: "Space acts on matter, telling it how to move. In turn, matter reacts back on space, telling it how to curve" [Misner et al. (1971) 5]. Let's start with the equation that describes space telling matter how to move.

Imagine that an astronaut falling freely in a gravitational field $\mathbf{g}$ releases a camera of mass m. According to $\mathbf{F} = m\mathbf{a}$, and thanks to the equivalence of gravitational and inertial mass, as seen from the ground frame the camera and the astronaut fall together with the same acceleration, $\mathbf{a} = \mathbf{g}$. In terms of observers using ground frame coordinates x^μ who measure the gravitational potential Φ, where $\mathbf{g} = -\nabla\Phi$, the kth component of $\mathbf{a} = -\nabla\Phi$ may be written

$$\frac{d^2 x^k}{dt^2} - \partial^k \Phi = 0 \qquad (8.88)$$

for $k = 1, 2, 3$.

In the astronaut's frame, mapped with coordinates X^μ, the camera floats in front of the astronaut's face and does not accelerate. Extending

---

[17]This is a useful algorithm that usually works, but it is not a foundational principle. Exceptions exist—for example, it gives the wrong dynamics when the equations of a particle's spin without gravity are transformed to a frame with gravity; see Ohanian (1976) 259.

[18]In getting started down the road of tensors, Einstein collaboarted with his old friend Marcel Grossman (1878–1936). See Pais (1982) ch. 12.

Newtonian three-vectors to spacetime four-vectors, and Newtonian absolute time dt to the invariant proper time $d\tau$ of special relativity, the astronaut writes the "no accleration" condition for the floating camera as

$$\frac{d^2 X^\alpha}{d\tau^2} = 0. \tag{8.89}$$

Let's transform this expression for the camera's acceleration from the astronaut's Minkowskian frame to the ground frame coordinates, and see how the effects of gravitation enter through the change of reference frame. Each coordinate in the astronaut's frame is in general a function of all the ground-based coordinates,

$$X^\alpha = X^\alpha(x^\mu). \tag{8.90}$$

Therefore, by the chain rule,

$$\frac{dX^\alpha}{d\tau} = \frac{\partial X^\alpha}{\partial x^\mu} \frac{dx^\mu}{d\tau}, \tag{8.91}$$

and for the second derivative,

$$\frac{d}{d\tau}\left(\frac{dX^\alpha}{d\tau}\right) = \frac{\partial X^\alpha}{\partial x^\mu}\frac{d^2 x^\mu}{d\tau^2} + \frac{\partial^2 X^\alpha}{\partial x^\mu \partial x^\nu}\frac{dx^\mu}{d\tau}\frac{dx^\nu}{d\tau} = 0. \tag{8.92}$$

To make $d^2 x^\mu / d\tau^2$ stand alone—the camera's acceleration as measured in the ground frame—we need to peel off its $\partial X^\alpha / \partial x^\mu$ coefficient. This can be done by using

$$\frac{\partial x^\lambda}{\partial X^\alpha}\frac{\partial X^\alpha}{\partial x^\mu} = \delta^\lambda{}_\mu. \tag{8.93}$$

To include this in our calculation, we multiply eq. (8.92) by $\partial x^\lambda / \partial X^\alpha$, which gives as

$$\frac{\partial x^\lambda}{\partial X^\alpha}\frac{\partial X^\alpha}{\partial x^\mu}\frac{d^2 x^\mu}{d\tau^2} + \left(\frac{\partial x^\lambda}{\partial X^\alpha}\frac{\partial^2 X^\alpha}{\partial x^\mu \partial x^\nu}\right)\frac{dx^\mu}{d\tau}\frac{dx^\nu}{d\tau} = 0. \tag{8.94}$$

Recognizing the $\delta^\lambda{}_\mu$, we obtain

$$\frac{d^2 x^\lambda}{d\tau^2} + \Gamma^\lambda{}_{\mu\nu}\frac{dx^\mu}{d\tau}\frac{dx^\nu}{d\tau} = 0, \tag{8.95}$$

the "geodesic equation," which describes a particle falling freely through spacetime.[19] In it we encounter the "affine connection,"[20] which includes second derivatives of old coordinates with respect to the new ones:

---

[19]This equation, derived here in a physics context, also occurs in a geometry context when a particle follows the extremal path (or world line) between two points in a space (or two events in spacetime). In relativity, the elapsed proper time is maximized (not minimized) for reasons of physics, not for the calculus of variations.

[20]Also called Christoffel symbols, after Elwin Bruno Christoffel (1829–1900).

$$\Gamma^\lambda{}_{\mu\nu} \equiv \frac{\partial x^\lambda}{\partial X^\alpha} \frac{\partial^2 X^\alpha}{\partial x^\mu \partial x^\nu}. \tag{8.96}$$

Assuming continuity of the derivatives so that

$$\frac{\partial^2}{\partial x^\mu \partial x^\nu} = \frac{\partial^2}{\partial x^\nu \partial x^\mu}, \tag{8.97}$$

the affine connection is symmetric in its subscripted indices,

$$\Gamma^\lambda{}_{\mu\nu} = \Gamma^\lambda{}_{\nu\mu}. \tag{8.98}$$

As shown in tensor calculus and general relativity textbooks, [Neuen-schwander (2015) 107–109] within the x^μ system of coordinates the $\Gamma^\lambda{}_{\mu\nu}$ coefficients can be computed exclusively in terms of derivatives of that system's metric tensor (see appendices A and F), according to

$$\Gamma^\lambda{}_{\mu\nu} = \frac{1}{2} g^{\lambda\rho} [\partial_\mu g_{\nu\rho} + \partial_\nu g_{\mu\rho} - \partial_\rho g_{\mu\nu}]. \tag{8.99}$$

Now we are equipped to compare the Newtonian and relativistic descriptions of the camera's acceleration equations, as they are written relative to the ground frame. Upon comparing eqs. (8.88) and (8.95), we see the affine connection coefficient terms are analogous to the Newtonian gravitational field $\mathbf{g}$. Where in Newtonian gravitation we say that $g_k \neq 0$, in general relativity we say that $\Gamma^\lambda{}_{\mu\nu} \neq 0$. Continuing the analogy, because $\mathbf{g} = -\nabla\Phi$ and the affine connection coefficients include gradients of metric tensor components, the $g_{\mu\nu}$ evidently play the role of gravitational potentials. The $g_{\mu\nu}$ are, in general, functions of spacetime coordinates—they are gravitational fields.

Any equation whose content transcends the choice of this or that coordinate system must be expressible as a tensor equation in order to be generally covariant, because tensors transform the same as the coordinate displacements do. Since the new coordinates x'^μ are functions of the old coordinates x^ν, by the chain rule

$$dx'^\mu = \frac{\partial x'^\mu}{\partial x^\nu} \, dx^\nu. \tag{8.100}$$

A tensor of rank 2, for example, has two indices and transforms like the product of two coordinate displacements:

$$T'^{\mu\nu} = \frac{\partial x'^\mu}{\partial x^\rho} \frac{\partial x'^\nu}{\partial x^\sigma} \, T^{\rho\sigma}. \tag{8.101}$$

A scalar is a tensor of rank 0, a vector is a tensor of rank 1, and we have met rank-2 tensors including the metric, Faraday, and Maxwell tensors. Tensors of arbitrarily high rank can exist.

Most physics equations are differential equations, including the acceleration example just discussed. As vectors, the components of velocity and acceleration are supposed to transform the same way the coordinate differentials do, that is, as components of rank-1 tensors. Newtonian velocity $\mathbf{v} = d\mathbf{r}/dt$ is a vector because $\mathbf{r}$ and therefore $d\mathbf{r}$ are vectors; $\mathbf{v}$ is merely the result of vector addition to make $d\mathbf{r}$, and scalar multiplication with $1/dt$. But does the derivative of a vector with respect to a scalar always behave as a good little tensor of rank 1? To answer this question, we must investigate how $dx^\mu/d\tau$ behaves under a general spacetime transformation. Starting with

$$\frac{dx'^\mu}{d\tau} = \frac{d}{d\tau}\left(\frac{\partial x'^\mu}{\partial x^\nu}x^\nu\right), \tag{8.102}$$

by the product and chain rules for derivatives we find

$$\frac{dx'^\mu}{d\tau} = \frac{\partial x'^\mu}{\partial x^\nu}\frac{dx^\nu}{d\tau} + \frac{\partial^2 x'^\mu}{\partial x^\sigma \partial x^\nu}\frac{dx^\sigma}{d\tau}x^\nu. \tag{8.103}$$

The first term is expected if four-velocity is a rank-1 tensor. But if the second term is nonzero anywhere, the tensor transformation rule is spoiled. That does not happen in the orthogonal-Galilean transformations of Newtonian relativity, or in the Lorentz transformations of special relativity. But it happens in more general coordinate transformations.

A similar problem arises with the derivative of a vector component with respect to a coordinate, as encountered in divergences and curls. One finds[21]

$$(\partial_\lambda T^\alpha)' = \frac{\partial x'^\alpha}{\partial x^\rho}\frac{\partial x^\sigma}{\partial x'^\lambda}(\partial_\sigma T^\rho) + \frac{\partial^2 x'^\alpha}{\partial x^\rho \partial x^\sigma}\frac{\partial x^\sigma}{\partial x'^\lambda}T^\rho. \tag{8.104}$$

Here again the term with second derivatives prevents $\partial_\lambda T^\alpha$ from being a rank-2 tensor.

It looks like we may be in big trouble: If physics equations are to transcend the choice of this or that coordinate system, they must be written as tensors. Physics equations include derivatives, but the derivative of a tensor is not guaranteed to be another tensor!

However, under a coordinate transformation the nontensor parts of these derivatives resemble affine connection coefficients—which are not tensor components either! It is straighforward in principle, but tedious in practice, to show[22] they transform according to

---

[21]Neuenschwander (2015) 97–99.
[22]Neuenschwander (2015) 103–105.

$$\Gamma'^{\lambda}{}_{\mu\nu} = \frac{\partial x'^{\lambda}}{\partial x^{\rho}} \frac{\partial x^{\tau}}{\partial x'^{\mu}} \frac{\partial x^{\sigma}}{\partial x'^{\nu}} \Gamma^{\rho}{}_{\tau\sigma} + \frac{\partial x'^{\lambda}}{\partial x^{\rho}} \frac{\partial^2 x^{\rho}}{\partial x'^{\mu} \partial x'^{\nu}}. \tag{8.105}$$

The first term after the equals sign describes the transformation if the affine connection coefficients are components of a rank-3 tensor. The additional term, not identically zero in all spaces, prevents the affine connection from being a tensor. Significantly, the term spoiling the affine connection's third-order tensor character contains second derivatives similar to the terms that prevent the derivative of a tensor from being another tensor. To salvage tensor differentiation, we can extend the meaning of "derivative" by adding to the usual derivative a Γ-term, such that, in a coordinate transformation, the offending terms cancel out. This "$\partial + \Gamma$" redefinition of the derivative is called the "covariant derivative," denoted as $\nabla_{\mu}A$ or $A_{;\mu}$ (see appendix F). Here are some covaraint derivatives with their notations. They transform as respectable tensors.

1. Covariant derivative of a vector component A^{μ} or A_{μ} with respect to a coordinate:

$$\nabla_{\nu}A^{\mu} \equiv A^{\mu}{}_{;\nu} \equiv \partial_{\nu}A^{\mu} + \Gamma^{\mu}{}_{\rho\nu}A^{\rho}$$
$$\nabla_{\nu}A_{\mu} \equiv A_{\mu;\nu} \equiv \partial_{\nu}A_{\mu} - \Gamma^{\rho}{}_{\mu\nu}A_{\rho} \tag{8.106}$$

which transform as rank-2 tensors; for instance,

$$(\nabla_{\nu}A^{\mu})' = \frac{\partial x^{\rho}}{\partial x'^{\nu}} \frac{\partial x'^{\mu}}{\partial x^{\sigma}} (\nabla_{\rho}A^{\sigma}). \tag{8.107}$$

2. Covariant divergence of a vector:

$$\nabla_{\mu}A^{\mu} \equiv A^{\mu}{}_{;\mu} \equiv \partial_{\mu}A^{\mu} + \Gamma^{\mu}{}_{\mu\nu}A^{\nu}. \tag{8.108}$$

By noting that

$$\Gamma^{\mu}{}_{\mu\lambda} = \frac{1}{2}g^{\mu\rho}(\partial_{\lambda}g_{\rho\mu}) \tag{8.109}$$

the divergence of a vector can also be written

$$\nabla_{\mu}A^{\mu} = \frac{1}{\sqrt{||g||}}\partial_{\mu}(\sqrt{||g||}A^{\mu}). \tag{8.110}$$

3. Covariant derivatives of rank-2 tensors:

$$\nabla_{\rho}T^{\mu\nu} \equiv T^{\mu\nu}{}_{;\rho} \equiv \partial_{\rho}T^{\mu\nu} + \Gamma^{\mu}{}_{\sigma\rho}T^{\sigma\nu} + \Gamma^{\nu}{}_{\sigma\rho}T^{\mu\sigma} \tag{8.111}$$

$$\nabla_{\rho}T_{\mu\nu} \equiv T_{\mu\nu;\rho} \equiv \partial_{\rho}T_{\mu\nu} - \Gamma^{\sigma}{}_{\mu\rho}T_{\sigma\nu} - \Gamma^{\sigma}{}_{\nu\rho}T_{\mu\sigma} \tag{8.112}$$

$$\nabla_{\rho}T^{\mu}{}_{\nu} \equiv T^{\mu}{}_{\nu;\rho} \equiv \partial_{\rho}T^{\mu}{}_{\nu} + \Gamma^{\mu}{}_{\sigma\rho}T^{\sigma}{}_{\nu} - \Gamma^{\sigma}{}_{\nu\rho}T^{\mu}{}_{\sigma}. \tag{8.113}$$

4. Covariant divergence of a rank-2 tensor:

$$\nabla_\mu T^{\mu\nu} \equiv \partial_\mu T^{\mu\nu} + \Gamma^\mu{}_{\mu\lambda} T^{\lambda\nu} + \Gamma^\nu{}_{\mu\lambda} T^{\mu\lambda}$$

$$= \frac{1}{\sqrt{||g||}} \partial_\mu (\sqrt{||g||} T^{\mu\nu}) + \Gamma^\nu{}_{\mu\lambda} T^{\mu\lambda}. \tag{8.114}$$

Notably, the covariant derivative of the metric tensor vanishes identically:

$$\nabla_\rho g_{\mu\nu} = 0. \tag{8.115}$$

One typically thinks of a vector field as something differentiable throughout all space, such as an electomagnetic field or the pressure field in a fluid. But when a particle follows a trajectory through the space, some vectors (or other tensors) may exist only along that trajectory. For example, a single particle's momentum vector exists only along the particle's trajectory. Locations on the trajectory can be parameterized by a scalar, such as proper time. The covariant derivative of a vector components A^μ with respect to a scalar τ is a relative of the covariant derivative, called the "intrinsic derivative":[23]

$$\frac{DA^\mu}{D\tau} \equiv \frac{dA^\mu}{d\tau} + \Gamma^\mu{}_{\rho\sigma} \frac{dx^\rho}{d\tau} A^\sigma \tag{8.117}$$

which transforms as a rank-1 tensor.

Let's revisit the camera that floats in the astronaut's reference frame with zero acceleration, but falls with nonzero acceleration relative to the ground-based frame. Return to eq. (8.95), written in terms of the four-velocity $u^\mu = dx^\mu / d\tau$,

$$\frac{du^\lambda}{d\tau} + \Gamma^\lambda{}_{\mu\nu} u^\mu u^\nu = 0. \tag{8.118}$$

We now recognize this as the intrinsic version of the covariant derivative,

$$\frac{Du^\lambda}{D\tau} = 0, \tag{8.119}$$

the equation of a geodesic in spacetime (as confirmed in section 8.5). This says that the rate of change of a vector that moves along a curve with

---

[23]If A^μ is defined throughout a region of space (or spacetime) and not merely along a specific curve—for example when taking the intrinsic derivative along a streamline within a fluid that fills the space—the intrinsic and covariant derivatives are related by [Hobson et al. (2006) 72]

$$\frac{DA^\mu}{D\tau} = (\nabla_\nu A^\mu) \frac{dx^\nu}{d\tau}. \tag{8.116}$$

the particle, such as the particle's velocity or momentum, does not change if observed from a reference frame that is locally inertial at $x^\mu(\tau)$. By the principle of equivalence, that is always possible for freely falling particles. In the astronaut's locally inertial coordinate system, the camera does not accelerate. But from the ground observer's perspective, in the global coordinate system, the falling camera follows a geodesic through spacetime.

One can describe the distinction between the intrinsic derivative and ordinary derivative in terms of something being conserved on the one hand, and something being exchanged on the other. If $du^\mu/d\tau = 0$, the four-velocity component u^μ is conserved. But if $Du^\mu/D\tau = 0$, the four-velocity component is not conserved. $Du^\mu/D\tau = 0$ means $du^\mu/d\tau = -\Gamma^\mu{}_{\nu\sigma}u^\nu u^\sigma$, which describes acceleration in a gravitational field.

An algorithm now exists for doing physics in the presence of gravitation.[24] Begin by writing the relevant equations in a locally inertial free-fall frame that has no gravity, where spacetime is "flat." To transform those equations to a frame that "sees" gravity, replace the usual derivative ∂_μ with the covariant derivative ∇_μ, and replace the local Minkowskian coordinates and their metric tensor $\eta_{\mu\nu}$ with the coordinates and metric tensor $g_{\mu\nu}$ of the curvilinear system.

We have seen that, given the metric tensor components $g_{\mu\nu}$, eq. (8.118) describes how gravity tells a particle to move. In section 8.5 we turn to the inverse problem: Given a source of gravity, such as a ponderable mass or any other form of energy, what gravitational potentials $g_{\mu\nu}$ are produced?

8.5   The Gravitational Field Equations

The other side of gravitation theory—matter telling space how to curve—predicts the field components $g_{\mu\nu}$ in terms of gravitational sources. Einstein published his opus on general relativity, "The Foundations of the General Theory of Relativity," in 1916 [Einstein (1916), Kennedy (2012)]. In 1922 he expanded that great paper into a book, *The Meaning of Relativity*, which has enjoyed multiple reprintings [Einstein (1974)]. Einstein's point of departure from Newtonian gravitation theory comes through Poisson's equation, in which a mass density ρ produces the gravitational potential Φ according to

$$\nabla^2\Phi = 4\pi G\rho \qquad (8.120)$$

where G denotes Newton's gravitational constant. Einstein wrote in *Meaning* that "Poisson's equation...of the Newtonian theory must serve as a model"[Einstein (1974) 82]. Creativity extends what already exists into something new.

---

[24]Recall that this is a procedure that usually works, not a principle guaranteed to work.

The result of ex. 8.5 shows that, in the Newtonian limit, the time-time component of the metric tensor is approximately

$$g_{00} \approx 1 + 2\Phi, \qquad (8.121)$$

in terms of which Poisson's equation, eq. (8.120) takes the form

$$\nabla^2 g_{00} = 8\pi G \rho. \qquad (8.122)$$

One way[25] to extend this to a generally relativistic paradigm is to recognize g_{00} as a special case of $g_{\mu\nu}$. Covariant generalizations in spacetime for the Laplacian and for the mass density must also be constructed.

Guided by his deep intuition, Einstein put his gravitational field equations together in such a heuristic way. Since special relativity showed mass to be a form of energy, the covariant generalization of mass density should a tensor with components $T_{\mu\nu}$ describing the densities of energy and momentum carried by particles and radiation (which Einstein lumped together in the term "matter," a practice we follow here for simplicity). These include the electrodynamic energy-momentum tensor (the Maxwell tensor) of section 7.3, and, for another example, the local stress tensor for a continuous fluid

$$T_{\mu\nu} = g_{\mu\nu} P + (\rho + P)\frac{dx^\mu}{d\tau}\frac{dx^\nu}{d\tau}, \qquad (8.123)$$

where P and ρ are, respectively, the fluid's pressure and energy density.[26] Since the energy and momentum of matter and radiation are supposed to be locally conserved, an equation of continuity would, we expect, hold for the relevant stress tensor, $\partial_\mu T^{\mu\nu} = 0$. However, when dealing with graviation it's not that simple.

Any form of energy—stars and planets, galaxies and black holes, electromagnetic radiation, a swarm of neutrinos, anything with energy—sets up a gravitational field. The gravitational field, itself, also has energy. In Newtonian mechanics, the gravitational field's energy density is $-|\mathbf{g}|^2/8\pi G$. The energy of a mass density ρ interacting with a gravitational potential Φ is $\rho\Phi$; thus in introductory physics the energy density of gravitation and its interaction with matter is

$$\eta_g = -\frac{|\mathbf{g}|^2}{8\pi G} + \rho\Phi. \qquad (8.124)$$

---

[25]For an alternative approach that starts with the weak-field approximation and builds on it, see Ohanian (1976) ch. 3.

[26]P becomes P/c^2 in conventional units. Equations of state exist that relate P and ρ. Some of them take the form (in conventional units) $P = w\rho c^2$, where w is a dimensionless constant. For instance, with static dust $w = 0$, for electromagnetic radiation $w = \frac{1}{3}$, and for the enigmatic "dark energy" of cosmology $w = -1$.

Because it is part of matter's interactions, a term such as $\rho\Phi$ is included in $T^{\mu\nu}$, but $-|\mathbf{g}|^2/8\pi G$ is not included.

Thus whenever matter and the gravitational field interact, neither the energy of matter or the gravitational field are separately conserved. For general covaraince we have to replace $\partial_\mu T^{\mu\nu} = 0$ with the covariant divergence,

$$\nabla_\mu T^{\mu\nu} = 0. \tag{8.125}$$

This is not the usual equation of continuity; or to say it another way, this is an equation of continuity in the context of a covariant divergence. The difference can be seen by splitting out the usual divergence from the affine connection terms, whence eq. (8.125) says

$$\partial_\mu T^{\mu\nu} = -\Gamma^\mu{}_{\mu\sigma} T^{\sigma\nu} - \Gamma^\nu{}_{\sigma\mu} T^{\mu\sigma}. \tag{8.126}$$

The energy of matter is not conserved because matter and the gravitational field *exchange* energy. Einstein emphasized this point after writing down the covariant divergence being zero: "It must be remembered that besides the energy density of matter there must also be given an energy density of the gravitational field, so there can be no talk of principles of conservation of energy and momentum for matter alone"[Einstein (1974) 83].

We have seen this sort of thing before. As mentioned in chapter 7, the charged scalar field coupled to the electromagnetic field admits from Noether's first theorem the vanishing divergence,

$$\partial_\mu J^\mu_{(\varphi)} = 0. \tag{8.127}$$

Whatever is being conserved, it is not just electric charge, whose current density is $j^\mu_{(\varphi)}$, because eq. (8.127) says

$$\partial_\mu j^\mu_{(\varphi)} = -2e^2 \partial_\mu(\varphi^* A^\mu \varphi) \tag{8.128}$$

and the term $\partial_\mu(\varphi^* A^\mu \varphi)$ does not vanish in an arbitrary gauge. Similarly, Poynting's theorem,

$$\nabla \cdot \mathbf{S} + \frac{\partial \eta}{\partial t} = -\mathbf{j} \cdot \mathbf{E}, \tag{8.129}$$

becomes a proper equation of continuity only when $\mathbf{j} \cdot \mathbf{E} = 0$, which typically happens only when the current of charged particles vanishes. Covariantly, Poynting's theorem may be written

$$\partial_\nu U^{\mu\nu} = \mu_o(\partial^\rho A^\mu) j_\rho, \tag{8.130}$$

where the electromagnetic energy-momentum tensor components $U^{\mu\nu}$ are given by eq. (7.82). The $(\partial^\rho A^\mu) j_\rho$ term does not vanish in an arbitrary gauge. Matter and the field exchange energy and momentum. We could invent a special-case "covariant derivative $\mathcal{D}_\nu$" and rewrite this result as

$$\partial_\nu U^{\mu\nu} - \mu_o(\partial^\rho A^\mu)j_\rho \equiv \mathcal{D}_\nu U^{\mu\nu} = 0, \qquad (8.131)$$

to make a nonzero divergence look like a zero divergence. But the two divergences have different meanings. We must "remember that there can be no talk of principles of conservation of energy and momentum for matter alone."

Einstein built into general relativity the conservation of energy coupled to matter and gravitation, but the way he did it originally depended on choosing the right reference frame, or to say it another way, the conservation law was gauge dependent, as we shall see. What then of general covariance?

Jumping across the equal sign in modifying Poisson's equation, we need another rank-2 tensor, whose job will be to generalize the Laplacian's role when operating on the $g_{\mu\nu}$. Seeking a theory no more complicated than necessary, Einstein postulated that the required tensor be linear in second derivatives of the metric tensor, analogous to the Laplacian's linearity in the second derivatives. Thanks to mathematicans who developed tensor calculus, it was known that only one tensor could meet that criterion, the Riemann curvature tensor:[27]

$$R^\alpha{}_{\beta\mu\nu} = \partial_\mu\Gamma^\alpha{}_{\beta\nu} - \partial_\nu\Gamma^\alpha{}_{\beta\mu} + \Gamma^\sigma{}_{\beta\nu}\Gamma^\alpha{}_{\sigma\mu} - \Gamma^\sigma{}_{\beta\mu}\Gamma^\alpha{}_{\sigma\nu}. \qquad (8.132)$$

Since $T_{\mu\nu}$ is a rank-2 tensor, we need to contract the rank-4 Riemann tensor and thereby introduce the symmetric rank-2 Ricci tensor, with components $R_{\beta\mu}$:[28]

$$R_{\beta\mu} \equiv R^\alpha{}_{\beta\mu\alpha} = R_{\mu\beta}. \qquad (8.133)$$

In his work leading up to the final form of general relativity, Einstein assumed that the allowed coordinate transformations had to be unimodular, that is, the Jacobian equals ± 1. In that case the determinant of the metric tensor becomes an invariant,[29] and the Ricci tensor reduces to two sets of terms, $R_{\mu\nu} = \tilde{R}_{\mu\nu} + S_{\mu\nu}$, where

$$\tilde{R}_{\mu\nu} = \partial_\rho\Gamma^\rho{}_{\mu\nu} + \Gamma^\sigma{}_{\mu\rho}\Gamma^\rho{}_{\nu\sigma} \qquad (8.134)$$

---

[27]Georg Friedrich Bernhard Riemann (1826–1866) is known for differential geometry, the Riemann sum, the Riemann hypothesis, the zeta function, and Riemann surfaces, to name a few of his other contributions to mathematics.

[28]Gregorio Ricci-Curbastro (1853–1925) and his former student Tullio Levi-Civita (1873–1941) are credited with inventing tensor calculus, in their famous 1900 paper "Méthodes de calcul différentiel absolu et leurs applications" (*Mathematische Annalen* **54** [125–201]), studied by Einstein when he was developing general relativity.

[29]As seen in appendix E, the Jacobian J and the determinants of the metric tensors of the original and transformed coordinate systems are related by $J = \sqrt{|g|/|g'|}$.

and

$$S_{\mu\nu} = \frac{\partial^2(\ln\sqrt{||g||})}{\partial x^\mu \partial x^\nu} - \Gamma^\rho_{\ \mu\nu}\frac{\partial(\ln\sqrt{||g||})}{\partial x^\rho}. \qquad (8.135)$$

Einstein originally set $|g| = 1$, a restriction with the advantage of making $S_{\mu\nu} = 0$. Considering first a local region of spacetime devoid of matter, the local values of nonzero affine connection coefficients are due to matter elsewhere, and we will temporarily denote such affine connection coefficients with tildes. Thus for a tentative field equation Einstein set [Kennedy (2012) 194]

$$\tilde{R}_{\mu\nu} = \partial_\rho\tilde{\Gamma}^\rho_{\ \mu\nu} + \tilde{\Gamma}^\sigma_{\ \mu\rho}\tilde{\Gamma}^\rho_{\ \nu\sigma} = 0. \qquad (8.136)$$

With the metric tensor components as dependent variables, for this to be the Euler-Lagrange equation

$$\frac{\partial\mathcal{L}}{\partial g^{\mu\nu}} = \partial^\rho\left(\frac{\partial\mathcal{L}}{\partial g^{\mu\nu,\rho}}\right), \qquad (8.137)$$

(where $g^{\mu\nu,\rho} \equiv \partial^\rho g^{\mu\nu}$) requires the Lagrangian density

$$\mathcal{L} = g^{\mu\nu}\tilde{\Gamma}^\sigma_{\ \mu\rho}\tilde{\Gamma}^\rho_{\ \nu\sigma} = \mathcal{L}(g^{\alpha\beta}, g^{\alpha\beta,\gamma}). \qquad (8.138)$$

A Lagrangian density $\mathcal{L}$ determines a Hamiltonian tensor,

$$H^\rho_{\ \sigma} = (\partial^\rho g^{\mu\nu})\left(\frac{\partial\mathcal{L}}{\partial g^{\mu\nu,\sigma}}\right) - \delta^\rho_{\ \sigma}\mathcal{L}, \qquad (8.139)$$

which in this context is proportional to the energy-momentum tensor $t^\rho_{\ \sigma}$ of the pure gravitational field. For convenience when introducing the gravitational constant G into the field equations, where it plays the role of a coupling constant, the definition of the energy-momentum tensor of pure gravitation includes the factor $\kappa \equiv 8\pi G$ carried over from the modified Poisson's equation,[30] so that

$$t^\rho_{\ \sigma} \equiv -\frac{1}{2\kappa}H^\rho_{\ \sigma}. \qquad (8.140)$$

After some manipulation [Kennedy (2012) 195] this can be written

$$\kappa t^\rho_{\ \sigma} = \frac{1}{2}\delta^\rho_{\ \sigma}g^{\mu\nu}\tilde{\Gamma}^\lambda_{\ \mu\beta}\tilde{\Gamma}^\beta_{\ \nu\lambda} - g^{\mu\nu}\tilde{\Gamma}^\rho_{\ \mu\beta}\tilde{\Gamma}^\beta_{\ \nu\sigma}. \qquad (8.141)$$

This energy-momentum tensor satisfies the equation of continuity,

$$\partial_\rho t^\rho_{\ \sigma} = 0, \qquad (8.142)$$

provided $|g| = 1$ and no matter exists locally, so that $T^{\mu\nu} = 0$.

---

[30]In SI units $\kappa \equiv 8\pi G/c^4$.

After multiplying eq. (8.136) with $g^{\nu\sigma}$ and writing some of the affine connection coefficients in terms of $t^\rho{}_\sigma$ (using the $g\tilde{\Gamma}\tilde{\Gamma}$ common to both), Einstein obtained an alternate form for his field equation,

$$\partial_\alpha(g^{\nu\sigma}\tilde{\Gamma}^\alpha{}_{\mu\nu}) = -\kappa\left(t^\sigma{}_\mu - \frac{1}{2}\delta^\sigma{}_\mu t\right) \tag{8.143}$$

where $t \equiv t^\lambda{}_\lambda$. This equation suggests a way to incorporate $T^\mu{}_\nu$ when it is allowed to be nonzero: $\tilde{\Gamma}^\lambda{}_{\mu\nu}$ becomes a more general $\Gamma^\lambda{}_{\mu\nu}$, and eq. (8.143) gets replaced with

$$\partial_\alpha(g^{\nu\sigma}\Gamma^\alpha{}_{\mu\nu}) = -\kappa\left[(t^\sigma{}_\mu + T^\sigma{}_\mu) - \frac{1}{2}\delta^\sigma{}_\mu(t + T)\right] \tag{8.144}$$

where $T \equiv T^\lambda{}_\lambda$. Working backward to the original version of the Ricci tensor,[31] Einstein obtained

$$\partial_\alpha\Gamma^\alpha{}_{\mu\nu} + \Gamma^\alpha{}_{\mu\beta}\Gamma^\beta{}_{\nu\alpha} = -\kappa\left(T_{\mu\nu} - \frac{1}{2}g_{\mu\nu}T\right) \tag{8.145}$$

and $|g|$ remains restricted to 1. The energy-momentum tensor of the gravitational field itself is embedded in the new Γs, which in eq. (8.145) are different from the $\tilde{\Gamma}$s for pure gravitation. In other words, when all the $T_{\mu\nu} = 0$, then $\Gamma^\lambda{}_{\mu\nu}$ reverts back to $\tilde{\Gamma}^\lambda{}_{\mu\nu}$, and thus the $t^{\mu\nu}$ are built into the left-hand side of eq. (8.145).

Taking the divergence of eq. (8.144), Einstein was able to show that

$$\partial_\sigma(t^\sigma{}_\mu + T^\sigma{}_\mu) = 0, \tag{8.146}$$

a proper equation of continuity for the matter + gravitational field system [Kennedy (2012) 246–248]. However, this result still depends on $|g| = 1$, and is thus frame-dependent, or to say it another way, not generally covariant. The covariant derivative $\nabla_\mu(t^{\mu\nu} + T^{\mu\nu}) = 0$ reduces to $\partial_\mu(t^{\mu\nu} + T^{\mu\nu}) = 0$ only when $|g| = 1$. Therefore, whether or not the proper divergence of $t^{\mu\nu} + T^{\mu\nu}$ vanishes depends on the choice of coordinates.[32]

Sometime between the summer and November 25 of 1915 when Einstein publicly unveiled the final version of his field equations, he decided that $|g| = 1$ was "no equation of principle but rather an important guide to the choice of convenient coordinate systems" [Pais (1982) 256]. During that time he derived the form of the field equations as we know them today.

---

[31]Of course now the Lagrangian density must include a contribution from matter, in addition to pure gravitation.

[32]For more discussion on this point see Landau & Lifshitz (1962) 341–342, and Byers (1999) 14.

The field equations are typically introduced in textbooks by postulating a covariant extension of Poisson's equation as follows. Upon generalizing the mass density to a rank-2 energy-momentum tensor, the Laplacian side of Poisson's equation gets generalized to a superposition of the rank-2 tensors derivable from the Riemann tensor: the Ricci tensor $R_{\mu\nu}$ along with $g_{\mu\nu}R^\lambda{}_\lambda$, where $R \equiv R^\lambda{}_\lambda$ is called the curvature scalar. Thus one begins with the template

$$AR_{\mu\nu} + Bg_{\mu\nu}R = \kappa T_{\mu\nu} \qquad (8.147)$$

where A and B are constants to be determined.[33] This is a set of 16 equations, but only 10 are independent because the Ricci, metric, and stress tensors are symmetric. It remains to impose three constraints to determine A and B as follows.

1. Local conservation of energy and momentum in the sense of a vanishing covariant divergence of the stress tensor, $\nabla_\mu T^{\mu\nu} = 0$.

2. The Bianchi identity that holds for covariant derivatives of the Riemann curvature tensor.[34] Its most common version demonstrates a cyclic symmetry in the curvature tensor's covariant derivatives [Weinberg (1972) 146–147]:

$$R_{\lambda\mu\nu\rho;\sigma} + R_{\lambda\mu\rho\sigma;\nu} + R_{\lambda\mu\sigma\nu;\rho} = 0. \qquad (8.148)$$

Another version expresses the identity in terms of the Ricci tensor and curvature scalar:

$$R^{\mu\nu}{}_{;\nu} = \frac{1}{2}g^{\mu\nu}R_{;\nu}. \qquad (8.149)$$

3. The Newtonian limit. General relativistic field equations must reduce to the classical Poisson equation in the low-speed, weak-field limit.

In the customary pedagogical presentation, we take the covariant divergence of eq. (8.147), set $\nabla_\mu T^{\mu\nu} = 0$, and recall that $\nabla_\mu g^{\mu\nu} = 0$, which leaves

$$AR^{\mu\nu}{}_{;\mu} + Bg^{\mu\nu}R_{;\mu} = 0. \qquad (8.150)$$

The Bianchi identity of eq. (8.149) requires that $B = -A/2$. The requirement that the field equations also reduce to the Newtonian static, weak-field limit puts $A = -1$. [Neuenschwander (2015) 150–151]. Therefore, Einstein's field equations are

$$R_{\mu\nu} - \frac{1}{2}g_{\mu\nu}R = -\kappa T_{\mu\nu}. \qquad (8.151)$$

---

[33]General covariance also allows a term $\Lambda g_{\mu\nu}$, where Λ is a constant unrelated to the Riemann curvature tensor. Einstein set $\Lambda = 0$ initially, perhaps because he was committed to seeing gravitation exclusively in terms of the curvature of spacetime. He would resurrect Λ in 1917, in his paper on the closed universe cosmology [Einstein et al. (1952)], where Λ became known as the cosmological constant, but that's another story.

[34]The Bianchi identity was reportedly first discovered by Aurel Voss in 1880, then rediscovered independently by Ricci in 1889 and again in 1902 by Felix Klein's former student, Luigi Bianchi (1856–1928) [Pais (1982) 256, 276].

However, in the autumn of 1915, neither Einstein or Hilbert were aware of the Bianchi identities. This makes Noether's involvement in general relativity even more significant, because her second theorem applied to gravitation *derives* the Bianchi identity, just as her second theorem applied to electrodynamics derives the antisymmetry of the Faraday tensor (eq. [8.50]). The Bianchi identities and the antisymmetry of the Faraday tensor are essential to the development of the gravitational and electrodynamic field equations, respectively. Noether's second theorem produces these identities, or "dependencies" as she called them, independent of other methods.

Back to Einstein in the fall of 1915. To arrive at his final form of the field equations without the Bianchi identity, he proceeded [Einstein (1916), Pais (1982) 256, Kennedy (2012) ch. 5] to generalize the left-hand side of eq. (8.145) to the more general Ricci tensor $R_{\mu\nu}$, consider that the factor of $\frac{1}{2}$ may not be the most general, and replace it with a parameter α to be determined; after all, the factor of $\frac{1}{2}$ was inspired by matter-free spacetime. So he started with

$$R^{\mu\nu} = -\kappa(T^{\mu\nu} - \alpha g^{\mu\nu}T). \tag{8.152}$$

Multiply this by $g^{\mu\nu}$, note the appearance of the Ricci scalar $R \equiv g_{\mu\nu}R^{\mu\nu}$ and note also that $g_{\mu\nu}g^{\mu\nu} = 4$. This gives $T = -R/[\kappa(1-4\alpha)]$. Eq. (8.152) may now be written

$$R^{\mu\nu} + \frac{\alpha}{1-4\alpha}g^{\mu\nu}R = -\kappa T^{\mu\nu}. \tag{8.153}$$

Einstein determined the constant α by imposing the vanishing of the covariant derivative of the matter energy tensor, $T^{\mu\nu}{}_{;\nu} = 0$. Thus

$$R^{\mu\nu}{}_{;\nu} + \frac{\alpha}{1-4\alpha}g^{\mu\nu}R_{;\nu} = 0, \tag{8.154}$$

using $\nabla_\nu g^{\mu\nu} = 0$. Unaware of the Bianchi identity, Einstein fell back on choosing coordinates for which $|g| = 1$. His friend and biographer Pais wrote, "See if there is a solution for α. One finds $\alpha = \frac{1}{2}$. Einstein's choice of coordinates is of course admissible, but it is an unnecessary restriction" [Pais (1982) 256].

Einstein dervied his field equations by working around the lack of the Bianchi identity. Noether's second theorem derives the Bianchi identity as a "dependency" on the field equation's Lagrangian density, and as a byproduct offers clarity to the issue of energy conservation in general relativity. But before we can apply Noether's second theorem to general relativity, we need the theory's functionals.

8.6 The Functionals of General Relativity

In this section we examine general relativity's two main differential equations in terms of their functionals. Let's begin with the simpler case—the free-fall equation where space tells matter how to move [Misner et al. (1971) 5]. This equation was already known in differential geometry and tensor calculus as the "equation of the geodesic." Because the proper time between two events is an invariant between all frames (inertial or not), we start from

$$d\tau^2 = g_{\mu\nu}dx^\mu dx^\nu. \tag{8.155}$$

The equations of motion will follow by making an extremum the functional

$$\Gamma = \int_a^b d\tau$$

$$= \int_a^b \sqrt{g_{\mu\nu}dx^\mu dx^\nu}$$

which can be converted into parametric form,

$$\Gamma = \int_{\tau(a)}^{\tau(b)} \sqrt{g_{\mu\nu}u^\mu u^\nu} \, d\tau \tag{8.156}$$

where $u^\mu \equiv dx^\mu/d\tau$ with τ the proper time. This Lagrangian is a function of the coordinates (because the metric tensor components are) and their first derivatives,[35]

$$L(x^\rho, u^\mu, g_{\mu\nu}) = \sqrt{g_{\mu\nu}(x^\rho)u^\mu u^\nu}. \tag{8.157}$$

The Euler-Lagrange equation

$$\frac{\partial L}{\partial x^\rho} = \frac{d}{d\tau}\frac{\partial L}{\partial u^\rho} \tag{8.158}$$

applied to this Lagrangian density produces the free-fall equation, eq. (8.118).[36] Now we turn to the field equation, where matter tells space how to curve [Misner et al. (1971) 5].

After Albert Einstein visited Göttingen in the summer of 1915, a race was on between Hilbert and Einstein to complete general relativity. Einstein got there first, but Hilbert's great contribution to the theory was

---

[35]Notice that this Lagrangian, a *function* of spacetime coordinates and four-velocities, is *numerically* equal to 1.

[36]The intrinsic derivative did not need to be used in the Lagrangian of eq. (8.156); see Q 8.b.

articulating the final form of the Lagrangian density whose Euler-Lagrange equations are Einstein's field equations. In the absence of matter, within an over-all constant the Hilbert-Einstein functional that does the job is

$$\Gamma = \int_{\mathcal{R}} R\sqrt{||g||}\ d^4x \qquad (8.159)$$

where $\mathcal{R}$ is a domain of spacetime, $||g||$ the absolute value of the metric tensor's determinant, d^4x the four-dimensional volume element of spacetime,[37] and R the curvature scalar. The fields $g_{\mu\nu}$ are the dependent variables, and the spacetime components are the independent variables. Since $R \sim \Gamma\partial\Gamma + \Gamma\Gamma$ and $\Gamma \sim g(\partial g)$, it follows that $R \sim g(\partial g)^3 + g^2(\partial g)^2 + g^2(\partial g)(\partial^2 g)$: the curvature scalar depends on spacetime coordinates through the metric tensor, its first and its second derivatives; it is nonlinear in $g_{\mu\nu}$ and its first derivatives, but linear in the second derivative. Therefore the Euler-Lagrangian equation coming from the Hilbert-Einstein Lagrangian is complicated in derivatives of $g_{\mu\nu}$ (compare to eq. [3.111]):

$$\frac{\partial \mathcal{L}}{\partial(g_{\mu\nu})} = \partial_\rho\left(\frac{\partial \mathcal{L}}{\partial(g_{\mu\nu,\rho})}\right) - \partial_\rho\partial_\sigma\left(\frac{\partial \mathcal{L}}{\partial(g_{\mu\nu,\rho,\sigma})}\right). \qquad (8.160)$$

This Euler-Lagrange equation yields (after much work[38]) Einstein's matter-free gravitational field equation,[39]

$$R_{\mu\nu} - \frac{1}{2}g_{\mu\nu}R = 0. \qquad (8.161)$$

Now to introduce matter into the neighborhood of the event where the field equations are applied. For the program to be generally covariant when matter interacts with gravity, derivatives with respect to spacetime coordinates in the Lagrangian density of matter must be covariant derivatives. We can put the $\Gamma^\lambda{}_{\mu\nu}$ terms explicitly into the Lagrangian density of matter and work out the Euler-Lagrange equation by explicitly differentiating the affine connection pieces. Alternatively, it might be more efficient to re-derive the Euler-Lagrange equations for matter with the covariant derivatives already built in.

For example, suppose that before interacting with the gravitational field the Lagrangian density of matter is that of the uncharged scalar field,[40] $\mathcal{L}^{(\varphi)}(x^\mu, \varphi, \varphi_{,\mu})$ where $\varphi_{,\mu} = \partial_\mu\varphi$.

---

[37] $\sqrt{||g||}d^4x$ is an invariant volume element; see appendix E.

[38] Hobson, Efstathiou, and Lasenby adopt another approach, saying "the task of evaluating each term in the above equation involves a formidable amount of algebra, albeit straightforward" [Hobson et al. (2006) 539].

[39] This equation for the field equations of pure gravitation is analogous to the second-order sourceless electrodynamic field equations that result when the Lagrangian density is only $-(1/4\mu_o)F_{\mu\nu}F^{\mu\nu}$ for the pure electromagnetic field.

[40] See ex. 8.7 for charged matter coupled to electromagnetism and gravity.

When the matter interacts with gravitation, to be generally covariant the partial derivatives of matter fields are replaced with covariant derivatives, $\partial_\mu \varphi \to \nabla_\mu \varphi$ or, in the comma-to-semicolon notation, $\varphi_{,\mu} \to \varphi_{;\mu}$. Dropping the (φ) superscript on its Lagrangian density[41] the generally covariant functional for matter is

$$\Gamma = \int_{\mathcal{R}} \mathcal{L}(x^\mu, \varphi, \varphi_{;\mu}) \sqrt{\|g\|} d^4 x. \qquad (8.162)$$

Consider a variation on the field φ

$$\varphi' = \varphi + \varepsilon \zeta \qquad (8.163)$$

where ζ vanishes on the boundary of the region $\mathcal{R}$. The functional is now a function of ε, and required to be an extremum:

$$0 = \left[\frac{d\Gamma(\varepsilon)}{d\varepsilon} \right]_0 = \int_{\mathcal{R}} \left[\left(\frac{\partial \mathcal{L}}{\partial \varphi} \right) \zeta + \left(\frac{\partial \mathcal{L}}{\partial \varphi_{;\mu}} \right) \zeta_{;\mu} \right] \sqrt{\|g\|} d^4 x. \qquad (8.164)$$

It is a straightforward exercise (ex. 8.3) to show that the covariant derivative respects the usual product rule for derivatives,

$$\nabla_\rho (AB) = (\nabla_\rho A)B + A(\nabla_\rho B). \qquad (8.165)$$

With this, and by defining the canonical momentum density with the covariant derivative,

$$p^\mu \equiv \frac{\partial \mathcal{L}}{\partial \varphi_{;\mu}}, \qquad (8.166)$$

$\delta\Gamma = 0$ gives

$$0 = \int_{\mathcal{R}} \left[\frac{\partial \mathcal{L}}{\partial \varphi} - \nabla_\mu p^\mu \right] \zeta \sqrt{\|g\|} d^4 x + \int_{\mathcal{R}} \nabla_\mu (p^\mu \zeta) \sqrt{\|g\|} d^4 x. \qquad (8.167)$$

The divergence theorem works for the covariant derivative (see appendix F) [Hobson et al. (2006) 532], which casts the covariant divergence term onto a surface integral, where ζ vanishes. The integrand of the remaining integral must vanish, presenting the Euler-Lagrange equation in terms of covariant derivatives:

$$\frac{\partial \mathcal{L}}{\partial \varphi} = \nabla_\mu p^\mu. \qquad (8.168)$$

Let $\mathcal{L}^{(m)}$ denote the Lagrangian density of matter (including radiation) in whatever form it may take. The Hilbert-Einstein functional, including matter coupled to gravitation, becomes

$$\Gamma = \int_{\mathcal{R}} \left(\mathcal{L}^{(m)} - \frac{1}{2\kappa} R \right) \sqrt{\|g\|} \, d^4 x \qquad (8.169)$$

---

[41] The integration measure $\sqrt{\|g\|} d^4 x$ is invariant; see appendix F.

where $\kappa = 8\pi G$. Requiring $\delta\Gamma = 0$, the Euler-Lagrange equation (8.168) yields Einstein's field equations, including matter sources,

$$R_{\mu\nu} - \frac{1}{2}g_{\mu\nu}R = -\kappa T_{\mu\nu}. \tag{8.170}$$

Einstein, Hilbert, and Klein must have been uneasy with the covariant divergence $\nabla_\mu T^{\mu\nu} = 0$ standing in for the conservation of energy, since the covariant divergence being zero is not an equation of continuity in Hilbert's "proper" sense. Evidently it was the closest the theory could come to local energy conservation, since matter and the gravitational field exchange energy.

The problem becomes readily apparent in the weak-field approximation, where matter produces a small perturbation on the spacetime of an originally Minkowskian region. This business involves infinitesimal transformations of the spacetime coordinates, which through the metric tensor carry over into local gauge transformations of the gravitational field. Let's look at gauge transformations in general relativity and apply Noether's second theorem to them.

8.7 Gauge Transformations on Spacetime

In gauge theories of elementary particle physics, spacetime coordinates are left alone and matter fields are gauged. In general relativity, where the fields are components of the metric tensor, a locally variable transformation of spacetime coordinates is equivalent, through their induced effect on the metric tensor, to a local gauge transformation.

In flat spacetime with no gravitation, the metric tensor components are the $\eta_{\mu\nu}$ of Minkowskian spacetime, whose elements expressed in Cartesian coordinates are all 0 or ± 1. In the presence of modest gravitational sources that produce weak fields, the metric tensor $g_{\mu\nu}$ acquires a perturbation added to the Minkowskian metric:

$$g_{\mu\nu} = \eta_{\mu\nu} + h_{\mu\nu}, \tag{8.171}$$

where for the nonzero components, $|h_{\mu\nu}| \ll |\eta_{\mu\nu}|$. In the transformation from coordinates x^μ to x'^μ, each new coordinate is a function of all the old coordinates. Therefore the transformation from the old to the new system is given in general by the chain rule for coordinate differentials,

$$dx'^\mu = \frac{\partial x'^\mu}{\partial x^\nu} dx^\nu. \tag{8.172}$$

As a tensor with two indices, the metric tensor transforms the same as the product of two coordinate differentials:

$$g'_{\mu\nu} = \frac{\partial x^\rho}{\partial x'^\mu} \frac{\partial x^\sigma}{\partial x'^\nu} g_{\rho\sigma}$$

$$\equiv \Lambda^\rho{}_\mu \Lambda^\sigma{}_\nu g_{\rho\sigma}. \tag{8.173}$$

If you complete ex. 8.1, you will have shown that if each $\Lambda^\alpha{}_\beta = $ const., as occurs in a Lorentz transformation from one inertial frame to another, then $h'_{\mu\nu} = \Lambda^\rho{}_\mu \Lambda^\sigma{}_\nu h_{\rho\sigma}$. In such cases, $h_{\mu\nu}$ is a tensor introduced into the flat spacetime, similar to how the Faraday tensor $F_{\mu\nu}$ was, in special relativity, introduced into flat spacetime without changing it.

However, $h_{\mu\nu}$ does not transform as a tensor if the $\Lambda^\rho{}_\mu$ vary locally, when $\Lambda^\rho{}_\mu = \Lambda^\rho{}_\mu(x^\nu)$.[42] Consider

$$x'^\mu = x^\mu + \varepsilon^\mu(x^\nu). \tag{8.174}$$

Under these circumstances (see ex. 8.1),

$$\Lambda^\mu{}_\nu \equiv \frac{\partial x^\mu}{\partial x'^\nu} = \delta^\mu{}_\nu - \partial_\nu \varepsilon^\mu. \tag{8.175}$$

Since $g'_\mu = \eta'_{\mu\nu} + h'_{\mu\nu}$ and the $\eta_{\mu\nu}$ components are all 0 or ± 1 in both reference frames, it follows that

$$h'_{\mu\nu} = h_{\mu\nu} - (\partial_\nu \varepsilon_\mu + \partial_\mu \varepsilon_\nu). \tag{8.176}$$

Here $h_{\mu\nu}$ does not transform as a second-rank tensor. This result may, however, be interpreted as a gauge transformation instead. As a gauge transformation on the metric tensor field, it is in effect a gauge transformation on spacetime itself.

To first order in the $h_{\mu\nu}$, and with $h \equiv h^\lambda{}_\lambda$, after a lot of hard work gets expended writing the affine connection coefficients, the Ricci tensor, and the curvature scalar to first order in $h_{\mu\nu}$, the field equations become

$$\partial_\lambda \partial^\lambda h^{\mu\nu} - 2\partial_\lambda(\partial^\nu h^{\mu\lambda} + \partial^\mu h^{\nu\lambda}) + \partial^\mu \partial^\nu h - \eta^{\mu\nu} \partial_\lambda \partial^\lambda h$$
$$+ \eta^{\mu\nu} \partial_\lambda \partial_\sigma h^{\lambda\sigma} = -\kappa T^{\mu\nu}. \tag{8.177}$$

This is invariant under the gauge transformation

$$h'_{\mu\nu} = h_{\mu\nu} - \frac{1}{2}(\partial^\mu \chi^\nu + \partial^\nu \chi^\mu) \tag{8.178}$$

---

[42]Despite their indices, the $\Lambda^\mu{}_\nu$ are not tensors themselves. They don't transform; they *are* the transformation.

for an arbitrary function $\chi = \chi(x^\nu)$, and is equivalent to the coordinate transformation that induces $h_{\mu\nu} \to h'_{\mu\nu}$ described above, if we identify $\varepsilon^\nu = \frac{1}{2}\chi^\nu$. Analogous to the differential operator $R_{\mu\nu} - \frac{1}{2}g_{\mu\nu}R$ that appears on the left-hand side of Einstein's field equations, the corresponding quantity for the perturbative gravitational field is

$$\varphi_{\mu\nu} \equiv h_{\mu\nu} - \tfrac{1}{2}\eta_{\mu\nu}h. \tag{8.179}$$

If we choose the "Hilbert gauge"

$$\partial_\mu \varphi^{\mu\nu} = 0, \tag{8.180}$$

the perturbative field equation takes a simple-appearing form:

$$\partial_\lambda \partial^\lambda \varphi^{\mu\nu} = -\kappa T^{\mu\nu}. \tag{8.181}$$

When we take its divergence, we find the right-hand side yields a proper equation of continuity, $\partial_\mu T^{\mu\nu} = 0$, thanks to the explicit appearance of the Hilbert gauge on the left-hand side, for which by eq. (8.180)

$$\partial_\mu(\partial_\lambda \partial^\lambda \varphi^{\mu\nu}) = \partial_\lambda \partial^\lambda(\partial_\mu \varphi^{\mu\nu}) = 0. \tag{8.182}$$

The problem is, this equation of continuity for the coupled matter-gravitational system has become gauge dependent. In the context of general relativity this means reference-frame dependent, which conflicts with the requirement of general covariance. That is why Hilbert asserted energy conservation to be an "improper" conservation law in general relativity.

8.8 Noether's Resolution of an Enigma in General Relativity

Klein observed [in 1918], "As one will see, in the following presentation [of the conservation laws] I really do not any longer need to calculate but only to make use of the most elementary formulae of the calculus of variations."
It was the year of the Noether theorem.

In November 1915, neither Hilbert nor Einstein was aware of this royal road to the conservation laws. —Abraham Pais, *Subtle Is the Lord* (1982), 274.

Noether's second theorem offers a powerful tool for addressing the relation between Einstein's field equations and energy conservation. We have seen how the coordinate transformation $x'^\mu = x^\mu + \varepsilon^\mu(x^\nu)$ leads to a local gauge transformation:

$$h'^{\mu\nu} = h^{\mu\nu} - (\partial^\mu \varepsilon^\nu + \partial^\nu \varepsilon^\mu). \tag{8.183}$$

In the formalism of Noether's second theorem, this transformation of the dependent variables may be expressed as

$$h'^{\mu\nu} = h^{\mu\nu} + Z^{\mu\nu}(\varepsilon) \qquad (8.184)$$

where by definition, and for now using usual partial derivatives,

$$Z^{\mu\nu}(\varepsilon) = \zeta_0^{\mu\nu}\varepsilon + \zeta_1^{\mu\nu}{}_{\rho\sigma}\,\partial^\rho\varepsilon^\sigma + \dots. \qquad (8.185)$$

Comparison with eq. (8.183) yields

$$\zeta_1^{\mu\nu}{}_{\rho\sigma} = -(g^\mu{}_\rho\,g^\nu{}_\sigma + g^\nu{}_\rho\,g^\mu{}_\sigma) \qquad (8.186)$$

with all other ζ-coefficients equal to zero. The functional of interest is the Hilbert-Einstein action plus a Lagrangian density $\mathcal{L}^{(m)}$ for matter (including radiation), in which ordinary derivatives have been replaced by covariant derivatives, thereby coupling matter and radiation to gravity:

$$\Gamma = \int_{\mathcal{R}} \left(\mathcal{L}^{(m)} - \frac{1}{2\kappa}R\right) \sqrt{||g||}\, d^4x. \qquad (8.187)$$

This Lagrangian density determines the Euler-Lagrange operator $\psi_{\mu\nu}$:

$$\psi_{\mu\nu} = R_{\mu\nu} - \frac{1}{2}g_{\mu\nu}R + \kappa T^{\mu\nu}. \qquad (8.188)$$

Here we do *not* require the functional to be an extremal, so $\psi_{\mu\nu}$ is not set to zero. Our interest turns to invariance for its own sake, regardless of the equation of motion. For Γ to be invariant under the transformation of eq. (8.184), by Noether's second theorem we set

$$\psi_{\mu\nu}Z^{\mu\nu}(\varepsilon) = 0. \qquad (8.189)$$

To isolate the ε we must write $\psi_{\mu\nu}Z^{\mu\nu}(\varepsilon)$ as its adjoint[43]

$$\varepsilon \tilde{Z}^{\mu\nu\cdots}(\psi_{\mu\nu}) + \text{divergence terms}, \qquad (8.190)$$

this still being within the integral. The divergence terms can be put onto the surface of the integrated region, where they vanish. Then, because ε is arbitrary, by the fundamental lemma of the variational calculus, we may set

$$\tilde{Z}^{\mu\nu\cdots}(\psi_{\mu\nu}) = 0, \qquad (8.191)$$

---

[43]The dots in $\tilde{Z}^{\mu\nu\cdots}$ are put there because we do not know in advance how many indices will be needed; it depends on which ζ-coefficients and which derivatives of ε survive.

even if $\psi_{\mu\nu} \neq 0$, and see what happens. Before we can find out what happens, we need to construct $\tilde{Z}^{\mu\nu\cdots}$. We proceed as follows, remembering now to swap the usual partial derivative for the covariant derivative. To first order in ε and its derivatives we start with

$$\psi_{\mu\nu} Z^{\mu\nu}(\varepsilon) = \psi_{\mu\nu} \zeta_1^{\mu\nu}{}_{\rho\sigma}(\nabla^\rho \varepsilon^\sigma) \tag{8.192}$$

and use the product rule for derivatives on $\psi_{\mu\nu} \zeta_1^{\mu\nu}{}_{\rho\sigma} \varepsilon^\sigma$:

$$\nabla^\rho(\psi_{\mu\nu} \zeta_1^{\mu\nu}{}_{\rho\sigma} \varepsilon^\sigma) = \varepsilon^\sigma \zeta_1^{\mu\nu}{}_{\rho\sigma}(\nabla^\rho \psi_{\mu\nu}) + \psi_{\mu\nu} \varepsilon^\sigma(\nabla^\rho \zeta_1^{\mu\nu}{}_{\rho\sigma})$$
$$+ \psi_{\mu\nu} \zeta_1^{\mu\nu}{}_{\rho\sigma}(\nabla^\rho \varepsilon^\sigma). \tag{8.193}$$

Recalling eq. (8.186) and the fact that covariant derivatives of the metric tensor vanish, the second term in eq. (8.193) also vanishes. Therefore eq. (8.192) becomes

$$\psi_{\mu\nu} Z^{\mu\nu}(\varepsilon) = -\varepsilon^\sigma \zeta_1^{\mu\nu}{}_{\rho\sigma}(\nabla^\rho \psi_{\mu\nu}) + \nabla^\rho(\psi_{\mu\nu} \zeta_1^{\mu\nu}{}_{\rho\sigma} \varepsilon^\sigma). \tag{8.194}$$

Recalling eq. (8.186) again, this becomes

$$\psi_{\mu\nu} Z^{\mu\nu}(\varepsilon) = \varepsilon^\sigma(g^\mu{}_\rho \, g^\nu{}_\sigma + g^\nu{}_\rho \, g^\mu{}_\sigma)(\nabla^\rho \psi_{\mu\nu}) + \nabla^\rho(\cdots)$$
$$= 2\varepsilon^\sigma(\nabla^\mu \psi_{\mu\sigma}) + \nabla^\rho(\cdots) \tag{8.195}$$
$$\equiv \varepsilon^\sigma \tilde{Z}^\mu(\psi_{\mu\sigma}) + \nabla^\rho(\cdots)$$

where

$$\tilde{Z}^\mu(\psi_{\mu\sigma}) = 2 \, \nabla^\mu \psi_{\mu\sigma}. \tag{8.196}$$

The divergence terms are disposed of by shifting them to the surface integral. For Γ to be invariant under the gauge transformation, because ε is arbitrary Noether's second theorem says $\tilde{Z}^\mu(\psi_{\mu\nu}) = 0$ for $\nu = 0, 1, 2, 3$, which becomes

$$\nabla^\mu \left[R_{\mu\nu} - \frac{1}{2} g_{\mu\nu} R + \kappa T_{\mu\nu} \right] = 0. \tag{8.197}$$

In source-free regions of spacetime, devoid of matter and radiation, $T_{\mu\nu}$ vanishes, and Noether's second theorem produces the Bianchi identity

$$\nabla^\mu R_{\mu\nu} - \frac{1}{2} g_{\mu\nu} \nabla^\mu R = 0. \tag{8.198}$$

Recall that Einstein and Hilbert were unaware of the Bianchi identities when they were developing the field equations of general relativity. Einstein had to try another approach that required an unnecessary restriction on $|g|$, even though he still derived the correct field equations. But

Noether's second theorem shows that the Hilbert-Einstein Lagrangian density *requires* the Bianchi identities. We saw the analogous "dependency" emerge in electrodynamics, when Noether's second theorem, applied to the Lagrangian density of the electromagnetic field, required the Faraday tensor to be antisymmetric.

With the Bianchi identity in hand, allowing a nonzero $T_{\mu\nu}$ in eq. (8.197) gives at once the "royal road" to the covariant conservation law,

$$\nabla^{\mu} T_{\mu\nu} = 0 \qquad\qquad (8.199)$$

independent of gauge, and independent of the detailed structure of the energy-momentum tensor.

According to Noether's second theorem, there is no escaping the appearance of the covariant divergence in general relativity. Conservation laws in general relativity are necessarily improper; matter and gravitation are inherently coupled. One can only speak of proper conservation of energy for the *entire* system by enclosing, with a surface at infinity, all the matter (including radiation) and gravitational fields. Out there, away from all matter and fields, spacetime asymptotically becomes Minkowskian and the covariant divergence goes over to a proper divergence. Only in that sense does general relativity contain a proper energy conservation law.

In her 1918 paper, Noether explained that, from the perspective of group theory, the apparent paradox that Einstein, Hilbert, and Klein encountered was to be expected. In the group of transformations, proper conservation laws come from a countable set of generators that apply globally. Such a set of transformations is merely a subgroup of an infinite group of transformations. The infinite group includes generators that vary from one spacetime event to another, and therefore an uncountably infinite number of generators because spacetime is a continuum. In other words, proper conservation laws came from the countable subset of global generators, but improper conservation laws—which are intrinsic to general relativity—come from a continuous local gauging of spacetime itself. If you want general covariance, you must accept covariant derivatives and, with them, improper conservation laws. Noether summarized,

From the foregoing, finally, we also obtain a proof of a Hilbertian assertion about the connection of the failure of laws of conservation of energy proper with "general relativity." ...

Hilbert enunciates his assertion to the effect that the failure of proper laws of conservation of energy is a characteristic feature of the "general theory of relativity." In order for this assertion to hold good literally, therefore, the term "general relativity" should be taken in a broader sense than is usual, and extended to the foregoing groups depending on n arbitrary functions. [Noether (1918), Noether & Tavel (1971)]

Questions for Reflection and Discussion

Q.8.a. Discuss the similarities and differences between the covariant derivative of general relativity and the covariant derivative of electrodynamics. What are the similarities, and what are the differences, between the two?

Q.8.b In the derivation of the free-fall equation, also called the equation of the geodesic, why not include in the Lagrangian the intrinsic derivative,

$$L = \sqrt{g_{\mu\nu} \frac{Dx^\mu}{D\tau} \frac{Dx^\nu}{D\tau}} \qquad (8.200)$$

instead of

$$L = \sqrt{g_{\mu\nu} \frac{dx^\mu}{d\tau} \frac{dx^\nu}{d\tau}} \qquad (8.201)$$

as was done?

Q8.c. When writing the Euler-Lagrange equation for the gravitational field (eq. 8.160), why did we not replace partial derivatives ∂_μ with covariant derivatives ∇_μ? When the electromagnetic field interacts with the gravitational field, should the partial derviatives in the Faraday tensor, $\partial_\mu A_\nu - \partial_\nu A_\mu$, be replaced with the covariant derivatives of Riemannian geometry, $\nabla_\mu A_\nu - \nabla_\nu A_\mu$?

Exercises

8.1. a. Recall the weak-field approximation,

$$g_{\mu\nu} = \eta_{\mu\nu} + h_{\mu\nu} \qquad (8.202)$$

where $\eta_{\mu\nu}$ is the metric tensor of Minkowskian spacetime, and $|h_{\mu\nu}| << |\eta_{\mu\nu}|$ for nonzero components. Show that if each $\Lambda^\alpha{}_\beta$ is a constant, as occurs in a Lorentz transformation from one inertial coordinate system to another, then $h'_{\mu\nu} = \Lambda^\rho{}_\mu \Lambda^\sigma{}_\nu h_{\rho\sigma}$. In this case $h_{\mu\nu}$ is a tensor introduced into the flat spacetime of special relativity.

b. Consider $h_{\mu\nu}$ under an infinitesimal transformation that varies from one event in spacetime to another,

$$x'^\mu = x^\mu + \varepsilon^\mu(x^\nu). \qquad (8.203)$$

Show that, under these circumstances,

$$\frac{\partial x^\mu}{\partial x'^\nu} = \delta^\mu{}_\nu - \partial_\nu \varepsilon^\mu \qquad (8.204)$$

to first order in ε and its derivatives.

c. Now show that, to first order in the small quantities,

$$h'_{\mu\nu} = h_{\mu\nu} - (\partial_\nu \varepsilon_\mu + \partial_\mu \varepsilon_\nu). \tag{8.205}$$

Although $h_{\mu\nu}$ transforms as a second-rank tensor under global Lorentz transformations, it does not transform as a second-rank tensor under locally variable transformations. The transformation of eq. (8.205) may, however, be considered a gauge transformation.

d. List analogies and differences between the procedure just outlined and the gauge transformation of electrodynamics, $A'^\mu = A^\mu - \partial^\mu \varepsilon$.

8.2. Consider a classical mechanics problem in one spatial dimension, where the motion of a particle of mass m is governed by the action $\Gamma = \int_a^b L(t, x, \dot{x})\, dt$.

a. In the context of Noether's first theorem, under the transformation $t' = t + \varepsilon\tau$ and $x' = x + \varepsilon\zeta$, allow for the possibility that the parameter ε is a function of time, $\varepsilon = \varepsilon(t)$, and re-derive the invariance identity. In particular, denoting total derivatives with respect to t with overdots, show that

$$\frac{\partial L}{\partial x}\zeta + p\dot{\zeta} + \frac{\partial L}{\partial t}\tau - H\dot{\tau} + \alpha p + \gamma\left(\frac{\partial L}{\partial x}\zeta + \frac{\partial L}{\partial t}\tau + L\dot{\tau}\right) = 0, \tag{8.206}$$

where p is the momentum conjugate to x, H the Hamiltonian, and

$$\alpha = \dot{\varepsilon}(\tau\dot{\zeta} - \dot{\tau}\zeta), \tag{8.207}$$

$$\gamma \equiv \dot{\varepsilon}(2\tau + \dot{\varepsilon}\tau^2) \approx 2\dot{\varepsilon}\tau. \tag{8.208}$$

If the functional is also made an extremal, show that

$$\frac{d}{dt}(p\zeta - H\tau) + \gamma(\dot{p}\zeta + L\dot{\tau} - \dot{H}\tau) + \alpha p = 0. \tag{8.209}$$

b. Now treat $t' = t + \varepsilon\tau$ and $x' = x + \varepsilon\zeta$ in the context of Noether's second theorem. Write $T(\varepsilon)$ and $Z(\varepsilon)$ and find their adjoints $\tilde{T}$ and $\tilde{Z}$. Show that

$$(\zeta - \dot{x}\tau)\psi = 0, \tag{8.210}$$

where ψ is the Euler-Lagrange operator. If $\psi \neq 0$, what sense can be made of the other option, $\zeta - \dot{x}\tau = 0$?

8.3. Show that the covariant derivative respects the usual rule for differentiating a product, for instance

$$\nabla_\rho(T^\mu R^\nu) = (\nabla_\rho T^\mu)R^\nu + T^\mu(\nabla_\rho R^\nu). \tag{8.211}$$

8.4. Consider the Lagrangian of a particle in free fall, $L = \sqrt{g_{\mu\nu}u^\mu u^\nu}$, where $u^\mu = dx^\mu/d\tau$. Note that L is numerically equal to 1 but is still a function of x^μ and u^μ.

a. Show that the canonical momenta are

$$p_\rho \equiv \frac{\partial L}{\partial u^\rho} = \frac{u_\rho}{L} = \frac{g_{\rho\sigma}u^\sigma}{L}. \tag{8.212}$$

b. Show that the Euler-Lagrange equation

$$\frac{\partial L}{\partial x^\rho} = \frac{d}{d\tau}\frac{\partial L}{\partial u^\rho} \tag{8.213}$$

gives the equation of the geodesic, eq. (8.118). Why is it not necessary to use the intrinsic derivative, that is, to write the Euler-Lagrange equation as

$$\frac{\partial L}{\partial x^\rho} = \frac{D}{D\tau}\frac{\partial L}{\partial u^\rho}? \tag{8.214}$$

8.5. In the slow-speed, weak-field limit, where $d\tau \approx dt$ so that $u^0 \approx 1$ and $u^k \approx v^k \equiv dx^k/dt$ for $k = 1, 2, 3$, show that the equation of the freely falling particle,

$$\frac{d^2 x^\lambda}{d\tau^2} + \Gamma^\lambda{}_{\mu\nu}u^\mu u^\nu = 0 \tag{8.215}$$

yields in the slow-speed, weak-field Newtonian limit

$$g_{00} \approx 1 + 2\Phi \tag{8.216}$$

where Φ is the Newtonian gravitational potential.[44]

8.6. Prove Euler's theorem: If $\Psi(\lambda x, \lambda y) = \lambda^k \Psi(x, y)$ then

$$x\frac{\partial \Psi}{\partial x} + y\frac{\partial \Psi}{\partial y} = k\Psi. \tag{8.217}$$

8.7. Go back to eq. (8.169) and for $\mathcal{L}^{(m)}$ put the Lagrangian density of charged scalar particles coupled to the electromagnetic field (recall eq. [7.99]). What is the Euler-Lagrange equation for the charged particles? For the electromagnetic field? What energy-momentum tensor $T_{\mu\nu}$ appears in Einstein's field equations, eq. (8.151)?

---

[44]For hints see Neuenschwander (2015) 101–103 or Weinberg (1972) 77–79.

Part IV

TRANS-NOETHER INVARIANCE

Chapter 9

Invariance in Phase Space

The advantages of the Hamiltonian formulation lie not in its use as a calculational tool, but rather in the deeper insight it affords into the formal structure of mechanics. The equal status accorded to coordinates and momenta as independent variables encouarges a greater freedom in selecting the physical quantities to be designated as "coordinates" and "momenta."... These more abstract formulations are primarily of interest to us today because of their essential role in constructing the more modern theories of matter.
—Herbert Goldstein, *Classical Mechanics*, 1950, 237

9.1 Phase Space

The Lagrangian depends on the coordinates and coordinate velocities; the Hamiltonian depends on the coordinates and their conjugate momenta. A coordinate q and its conjugate momentum p form a pair of coordinates in "phase space," where the Hamiltonian takes center stage.

In the first section of this chapter, phase space is introduced. Noether's theorem is expressed in its language in the second section. Through the Hamilton-Jacobi formalism introduced in the third section comes a transformation generator—the action itself!—that stands deep within classical mechanics but points suggestively towards quantum mechanics. We have been considering transformations on the variables that appear *in* the action integral, but in quantum theory the action itself requires something to operate *on*—the quantum state—as is described in chapter 10.

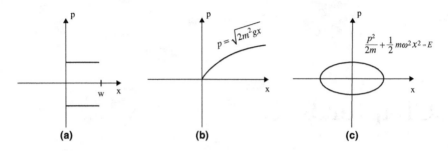

Figure 9.1: *Phase space trajectories of (a) a particle making elastic colli-sions with a box of width w, where $U = 0$ between the walls; (b) a particle falling freely with $U = -mgx$ (vertical downward taken as positive); and (c) a simple harmonic oscillator for which $U = \frac{1}{2}m\omega^2 x^2$.*

Newton's second law says force changes momentum. In one spatial dimension,

$$F = \frac{dp}{dt}$$

$$= \frac{dp}{dx}\frac{dx}{dt} \tag{9.1}$$

$$= \frac{dp}{dx}\frac{p}{m}.$$

If the force can be written as the negative gradient of a potential energy function $U(x)$, this becomes

$$-\frac{dU}{dx} = \frac{p}{m}\frac{dp}{dx}, \tag{9.2}$$

which integrates to

$$\frac{p^2}{2m} + U = E, \tag{9.3}$$

with integration constant E, the mechanical energy.

At any given moment, the state of the particle is given parametrically by $x(t)$ and $p(t)$. As time advances, a plot of the particle's coordinates through (x, p) space—the phase space—sweeps out a trajectory. From eq. (9.3), the momentum can be written as a function of x,

$$p = p(x) = \pm\sqrt{2m[E - U(x)]}. \tag{9.4}$$

Sample phase space trajectories are sketched in Figures 9.1(a)–(c).

The equation of motion (eq. [9.1]) for a particle can be written as one second-order differential equation,

$$-\frac{\partial U}{\partial x} = m\ddot{x}. \qquad (9.5)$$

Alternatively, it may be written as a pair of first-order equations,

$$\dot{p} = -\frac{\partial U}{\partial x} \qquad (9.6)$$

and $\dot{x} = p/m$, or in light of eq. (9.3),

$$\dot{x} = \frac{\partial E}{\partial p}. \qquad (9.7)$$

The program of extending eqs. (9.6) and (9.7) to generalized coordinates, in terms of their canonical momenta and the Hamiltonian, is called Hamiltonian dynamics. Here the Hamiltonian, rather than the Lagrangian, occupies center stage.

Our interest within these pages is Noether's theorem. We may wonder: If one wants to work within phase space by starting with the Hamiltonian in lieu of the Lagrangian, how is Noether's theorem affected? Let's develop Hamiltonian dynamics in the next section, then in the subsequent section use its language to approach Noether's theorem.

9.2 Hamilton's Principle in Phase Space

Consider a functional of a single-integral problem:

$$\Gamma = \int_a^b L(t, q^\mu, \dot{q}^\mu) \, dt. \qquad (9.8)$$

The Hamiltonian is related to the Lagrangian by a Legendre transformation,

$$H(t, q^\mu, p_\mu) = p_\mu \dot{q}^\mu - L(t, q^\mu, \dot{q}^\mu), \qquad (9.9)$$

where $p_\mu \equiv \partial L/\partial \dot{q}^\mu$. When there are N independent generalized coordinates or degrees of freedom, there are $2N$ independent phase space coordinates, the q^μ and p_μ. In phase space they are considered independent variables that depend parametrically on t. A specific set of coordinates $(q^\mu(t), p_\mu(t))$ defines an instantaneous vector in the $2N$-dimensional phase space. As time advances, the tip of that state vector sweeps out a trajectory in phase space.

In terms of the Hamiltonian, the functional of eq. (9.8) becomes

$$\Gamma = \int_a^b (p_\mu \dot{q}^\mu - H)\, dt. \tag{9.10}$$

To determine the equation of motion, Γ must be an extremum. As in chapter 3, consider varied functions, which in phase space take the form

$$q'^\mu = q^\mu + \varepsilon \zeta^\mu \tag{9.11}$$

and

$$p'_\mu = p_\mu + \varepsilon \chi_\mu, \tag{9.12}$$

where both ζ^μ and χ_μ vanish at $t = a$ and $t = b$. As before, we make the functional an extremal by setting $[d\Gamma/d\varepsilon]_0 = 0$, and with the same steps as in chapter 3, come to[1]

$$0 = \int_a^b \left(\dot{q}^\mu \chi_\mu + p_\mu \dot{\zeta}^\mu - \frac{\partial H}{\partial q^\mu} \zeta^\mu - \frac{\partial H}{\partial p_\mu} \chi_\mu \right) dt. \tag{9.13}$$

Integration by parts with $d(p_\mu \zeta^\mu)/dt$ turns the integral into

$$0 = \int_a^b \left[\left(\dot{q}^\mu - \frac{\partial H}{\partial p_\mu} \right) \chi_\mu - \left(\dot{p}_\mu + \frac{\partial H}{\partial q^\mu} \right) \zeta^\mu \right] dt + (p_\mu \zeta^\mu)|_a^b. \tag{9.14}$$

Because ζ^μ vanishes at the endpoints, the integrand in the square brackets must vanish whatever $\chi_\mu(t)$ and $\zeta^\mu(t)$ happen to be, leading to *Hamilton's equations*:

$$\dot{p}_\mu = -\frac{\partial H}{\partial q^\mu}, \tag{9.15}$$

$$\dot{q}^\mu = \frac{\partial H}{\partial p_\mu}. \tag{9.16}$$

These coupled equations are generalized coordinate versions of eqs. (9.6) and (9.7).

Hamilton's equations describe the evolution in time of the phase space coordinates, in terms of the phase-space gradient of the Hamiltonian. How

---

[1] You may have noticed the absence of a $\dot{\chi}_\mu$ term. Its absence is due to the lack of $\dot{p}_\mu$ in $L = p_\mu \dot{q}^\mu - H$.

does the Hamiltonian itself change in time? Deploy the chain rule on $H = H(t, q^\mu, p_\mu)$ to write

$$\frac{dH}{dt} = \frac{\partial H}{\partial t} + \frac{\partial H}{\partial q^\mu}\dot{q}^\mu + \frac{\partial H}{\partial p_\mu}\dot{p}_\mu. \qquad (9.17)$$

By virtue of Hamilton's equations, the second and third terms cancel, so the total derivative of H with respect to time depends only on the Hamiltonian's *explicit* time dependence. Furthermore, we recall the version of the Euler-Lagrange equation which says that $-\partial L/\partial t = \dot{H}$; therefore, we can write the three-way relation that holds whenever the functional is an extremal:

$$\frac{dH}{dt} = \frac{\partial H}{\partial t} = -\frac{\partial L}{\partial t}. \qquad (9.18)$$

If either H or L does not depend on time explicitly, then the Hamiltonian is conserved.

9.3 Noether's Theorem and Hamilton's Equations

Our original development of Noether's theorem proceeded through the Lagrangian. Having swapped out the Lagrangian for the Hamiltonian in the functional, do we get the same Noether's theorem result? We expect that we should, but let's confirm it.

The definition of invariance required $L'(dt'/dt) - L \sim \varepsilon^s, s > 1$. If we write L as $p_\mu\dot{q}^\mu - H$, this definition requires

$$\left(p'_\mu\frac{dq'^\mu}{dt'} - H(t', q'^\nu, p'_\nu)\right)\frac{dt'}{dt} - \left(p_\mu\frac{dq^\mu}{dt} - H(t, q^\nu, p_\nu)\right) \sim \varepsilon^s \qquad (9.19)$$

where $q'^\mu = q'^\mu(t')$ and $p'_\mu = p'_\mu(t')$ and $s > 1$.

Now consider the same kind of infinitesimal transformation as discussed before, including a generator χ_μ for the transformation of p_μ:

$$t' = t + \varepsilon\tau, \qquad (9.20)$$

$$q'^\mu = q^\mu + \varepsilon\zeta^\mu, \qquad (9.21)$$

$$p'_\mu = p_\mu + \varepsilon\chi_\mu. \qquad (9.22)$$

Does the inclusion of the generator χ_μ add something new compared to our previous discussions based on the Lagrangian? No, it does not, because of the relation between $\dot{q}^\mu$ and p_μ, through $p_\mu \equiv \partial L/\partial \dot{q}^\mu$. Because of this relation, χ_μ may be written in terms of ζ^μ and τ. For instance, in one spatial dimension with $q = x$ and $p = m\dot{x}$,

$$p' = m\frac{dx'}{dt'}$$

$$= m\frac{d(x + \varepsilon\zeta)}{d(t + \varepsilon\tau)} \tag{9.23}$$

$$= m(\dot{x} + \varepsilon\dot{\zeta})(1 + \varepsilon\dot{\tau})^{-1},$$

which gives, to first order in ε,

$$\chi = m(\dot{\zeta} - \dot{x}\dot{\tau}). \tag{9.24}$$

Returning to the question of invariance, differentiate eq. (9.19) with respect to ε, which requires differentiating $H'(t', q'^\nu, p'^\nu)$ by the chain rule. Then use the transformation equations. You will be rewarded with this version of the invariance identity,

$$-\frac{\partial H}{\partial t}\tau - H\dot{\tau} + \left(\dot{q}^\mu - \frac{\partial H}{\partial p_\mu}\right)\chi_\mu + p_\mu\dot{\zeta}^\mu - \frac{\partial H}{\partial q^\mu}\zeta^\mu = 0. \tag{9.25}$$

When Hamilton's equations are invoked—in other words, when the functional is also made an extremal—χ_μ drops out. What's left reduces to the familiar Noether conservation law,

$$p_\mu\zeta^\mu - H\tau = \text{const.} \tag{9.26}$$

Reassuringly, we affirm that Noether's theorem does not depend on whether we use $\Gamma = \int_a^b L\,dt$ with its generalized coordinates and their velocities, or $\Gamma = \int_a^b (p_\mu\dot{q}^\mu - H)dt$ with its generalized coordinates and their conjugate momenta. In the former, the equation of motion is the second-order Euler-Lagrange equation; the latter leads to the pair of coupled first-order Hamilton's equations. The Noether conservation laws do not depend on the choice of this partitioning.

9.4   Hamilton-Jacobi Theory

Hamilton-Jacobi theory offers a cross-cultural experience in the world of functional invariance, and points suggestively from classical mechanics to quantum mechanics.

Consider a system described by a Hamiltonian $H(t, q, p)$ (suppressing indices on phase space coordinates). Hamilton's equations govern the system's evolution,

$$\dot{q} = \frac{\partial H}{\partial p} \tag{9.27}$$

$$\dot{p} = -\frac{\partial H}{\partial q}. \tag{9.28}$$

Suppose we make a transformation from the original phase space variables (q, p) to new phase space variables (q', p'). A new Hamiltonian H' will result. In the new phase space, Hamilton's equations are

$$\dot{q}' = \frac{\partial H'}{\partial p'} \tag{9.29}$$

$$\dot{p}' = -\frac{\partial H'}{\partial q'}. \tag{9.30}$$

A change of phase space variables that respects the covariance of Hamilton's equations—in other words that leaves the form of the equations unaltered—is called a "canonical transformation." Changes of variable are made to transform a problem we don't know how to solve into one we already know how to solve. The simplest situation I can think of where Hamilton's equations are easily solved occurs when the new system's phase space coordinates and the new Hamiltonian are all constants,

$$q' = \text{const.} \equiv q_o, \tag{9.31}$$

$$p' = \text{const.} \equiv p_o \tag{9.32}$$

$$H' = \text{const.} \equiv H_o. \tag{9.33}$$

If such a new phase space can be found, the problem will be immediately solved by specifying initial conditions. Then all we have to do is reverse the transformation from (q', p') back to (q, p) to display in triumph the solution in our original phase space coordinate system. Hamilton-Jacobi theory aims to develop a strategy for carrying out this program.

Let's reflect on this strategy for a moment. In all our previous work, we took a given Lagrangian, and conservation laws *followed* from the equations of motion and the invariance identity. In Hamilton-Jacobi theory this procedure is inverted. Working in phase space with Hamilton's equations, which hold provided the functional is an extremal, a conservation law is *imposed* by requiring that the new phase space coordinates, and therefore the new Hamiltonian, are all constants. Then (as explained below) we seek the transformation that makes this possible. Thus Hamilton-Jacobi theory is analogous to finding and solving the Killing equations as was done within the Lagrangian paradigm.

Go back to the original action functional written in terms of the Lagrangian (we switch over to the Hamiltonian shortly). Consider a functional

$$\Gamma = \int_a^b L(t, q, \dot{q})dt \qquad (9.34)$$

and a transformation

$$t' = t, \qquad (9.35)$$

$$q' = q + \varepsilon\zeta \qquad (9.36)$$

(the transformation from p to p' arises when we swap the Lagrangian for the Hamiltonian). Because $dt' = dt$, the defintion of invariance becomes

$$\int_a^b (L' - L)dt \sim \varepsilon^s, \quad s > 1. \qquad (9.37)$$

For the purposes of Hamilton-Jacobi theory, the definition of invariance is tightened up to read[2]

$$\int_a^b (L - L')dt = 0. \qquad (9.38)$$

A way to make this happen without L' and L having to be identical requires that they differ by an exact differerential, so that

$$(L - L')dt = dS, \qquad (9.39)$$

where, as a function of t, $S(b) = S(a)$, which guarantees that the integral in eq. (9.38) vanishes. In the new phase space where q' and p' are constants, it follows that L' will also be a constant, so the integrated version of eq. (9.39) says

$$S = \int Ldt + \text{const.} \qquad (9.40)$$

As an indefinite integral of the Lagrangian, S is a function of time and the phase space variables.

Now we switch to the Hamiltonian. In Hamiltonian dynamics, q and p evolve independently as functions of time. Therefore, eq. (9.39) requires that

$$(\dot{q}p - H) - (\dot{q}'p' - H') = \frac{dS}{dt}. \qquad (9.41)$$

We are considering here a transformation $q \to q'$ and $p \to p'$. It could be written in the usual style

$$q' = q + \varepsilon\zeta \qquad (9.42)$$

$$p' = p + \varepsilon\chi, \qquad (9.43)$$

---

[2]I write $L - L'$ instead of $L' - L$ so the signs in what follows will match traditional literature on this subject.

but expressions for the generators ζ and χ will not concern us here. Their job has been taken over by the generating function S.

Besides being a function of time, S must also be a function of one of the original phase space coordinates and one of the new phase space coordinates, to bridge the old and new systems. Four possible combinations exist:

$$(q, q'), \quad (q, p'), \quad (p, q'), \quad (p, p'). \tag{9.44}$$

Suppose we choose $S = S(t, q, p')$. Return to eq. (9.41), expand dS/dt with the chain rule, and regroup the terms, to obtain

$$\dot{q}\left(\frac{\partial S}{\partial q} - p\right) + \left(H - H' + \frac{\partial S}{\partial t}\right) + \left(\dot{p}'\frac{\partial S}{\partial p'} + p'\dot{q}'\right) = 0, \tag{9.45}$$

which must hold for all times and all phase space coordinates. This will be guaranteed to happen if the grouped terms vanish separately, which gives the set of conditions

$$p = \frac{\partial S}{\partial q}, \tag{9.46}$$

$$H + \frac{\partial S}{\partial t} = H', \tag{9.47}$$

and

$$p'\dot{q}' + \dot{p}'\frac{\partial S}{\partial p'} = 0. \tag{9.48}$$

In the transformed phase space, we require that $q' = \text{const.} \equiv q_o$ and $p' = \text{const.} \equiv p_o$, which means $H' = \text{const.}$ Without loss of generality we may set $H' = 0$. Then Hamilton's equations collapse to $0 = 0$ in the primed frame. With $H' = 0$, eq. (9.47) relates the original known Hamiltonian H to the unknown $S(t, q, p')$ by the Hamilton-Jacobi equation:

$$H(t, q, p) = -\frac{\partial S}{\partial t}. \tag{9.49}$$

Our task is to solve the Hamilton-Jacobi equation for this S that enforces $H' = 0$, by using the original Hamiltonian H, which is a known function of the original phase space coordinates q and p. Once we have found S, we can transform back to the original coordinates, via eqs. (9.46) and (9.48), and declare the problem solved.

For example, in nonrelativistic mechanics problems, H typically has the form (with x as q)

$$H(t, x, p) = \frac{p^2}{2m} + U(t, x). \tag{9.50}$$

With this Hamiltonian and eq. (9.46), the Hamilton-Jacobi equation becomes

$$\frac{1}{2m}\left(\frac{\partial S}{\partial x}\right)^2 + U(t,x) = -\frac{\partial S}{\partial t}, \qquad (9.51)$$

to be solved once a potential energy function has been specified.

Because the Hamiltonian numerically equals the energy E, the Hamilton-Jacobi equation also tells us that

$$E = -\frac{\partial S}{\partial t}. \qquad (9.52)$$

This can be integrated at once because E is a number, not a function. Since S is a function of q and $p' = p_o$, the integration "constant" that arises in evaluating $\int (\partial S/\partial t)dt = -\int E dt$ will be some function of q and p_o, so that

$$S(t,q,p') = -Et + W(q,p_o). \qquad (9.53)$$

Placing this into eq. (9.51), if U has no explicit time dependence this procedure yields the time-independent Hamilton-Jacobi equation,

$$\frac{1}{2m}\left(\frac{\partial W}{\partial x}\right)^2 + U(x) = E \qquad (9.54)$$

to be solved for W.

Our objective, remember, is to find the evolution of q and p in the original phase space by transforming back to it from the second phase space in which q' and p' are constants. How do we carry that out when S has been found by solving the Hamilton-Jacobi equation? Once S is known, we have $p(t)$ at once from eq. (9.46). To find $q(t)$ takes a bit more work. If we impose

$$q' = \frac{\partial S}{\partial p'}, \qquad (9.55)$$

then eq. (9.48) reduces to $q'p' = $ const., a solution consistent with our assumption that q' and p' are separately constant. So to find $q(t)$ in the original phase space, we evaluate the derivative suggested in eq. (9.55), and after that set q' and p' equal to constants. That turns eq. (9.55) into an algebraic equation to solve for q, because we started with $S = S(t,q,p')$.

Let's work through a simple example, the case of a free particle, for which $U = 0$. Let the particle move to the right with energy E and momentum $p' = p_o = \sqrt{2mE}$. Eq. (9.54) becomes

$$\frac{1}{2m}\left(\frac{\partial W}{\partial x}\right)^2 = E \qquad (9.56)$$

with solution $W = p_o x + \delta$, with δ an integration constant. From eq. (9.53) we have S:

$$S(t, x, p_o) = p_o x - Et + \delta$$

$$= p_o x - \left(\frac{p_o^2}{2m}\right) t + \delta. \tag{9.57}$$

From eq. (9.46) it follows that $p = \partial S / \partial x = p_o$. To obtain $x(t)$, we invoke eq. (9.55),

$$x' = \frac{\partial S}{\partial p_o}$$

$$= x - \frac{p_o}{m} t \tag{9.58}$$

and obtain the familiar result for free-particle motion,

$$x = x_o + \frac{p_o}{m} t, \tag{9.59}$$

where x' has been set equal to a constant x_o.

Of course, solving the free-particle problem with Hamilton-Jacobi formalism is like slicing bread with a chainsaw. It is cumbersome, messy, and brings far more horsepower to the job than needed. The existence of less convoluted methods, such as good old $\mathbf{F} = m\mathbf{a}$, or the Euler-Lagrange or Hamilton's equations, means that the Hamilton-Jacobi equation is of little use for grinding out answers to practical problems. Rather, Hamilton-Jacobi theory is celebrated for suggesting a deeper way of thinking about the foundations of mechanics. In particular, even from deep within classical mechanics the Hamilton-Jacobi formalism foreshadows a union of the complementary paradigms in physics, the particle and wave models that are so essential to quantum mechanics. We can see it as follows.

Stepping outside our free particle example, we have been working with the generator $S = S(t, x, p')$ where $p' = \text{const.}$ The change in S under an infinitesimal change in x and t would be

$$dS = \frac{\partial S}{\partial x} dx + \frac{\partial S}{\partial t} dt. \tag{9.60}$$

But we have also seen that, in terms of the original phase space coordinates, $\partial S / \partial x = p$ and $\partial S / \partial t = -H$, so eq. (9.60) may be written $dS = p \, dx - H \, dt$, and upon integrating,

$$S = \int p \, dx - \int H \, dt, \tag{9.61}$$

which offers a consistency check on our doings, since this brings us back to eq. (9.40). But notice something else. Returning to the free particle example, if p and H are constants (dropping the o subscripts), these indefinite integrations yield

$$S = px - Ht + \delta(t, x), \qquad (9.62)$$

which looks strikingly like the phase of a sinusoidal harmonic wave function,

$$kx - \omega t + \delta, \qquad (9.63)$$

where k denotes the wave number (2π divided by the wavelength), ω the angular frequency (2π divided by the temporal period), and δ a phase shift, which could depend on t and x. Had classical physics encountered a daring postulate that would associate a free particle's momentum p with the wave number k of a harmonic wave, and that would associate the Hamiltonian H with the harmonic wave's angular frequency ω; or had someone been sufficiently nonconformist to suggest that frequency and energy are two names for the same reality, hypotheses equivalent to those of Max Planck (1858–1947), Albert Einstein (1870–1955), and Louis de Broglie (1892–1987) could have been made half a century earlier![3] Indeed, striking resemblances of the Hamilton-Jacobi equation to the Schrödinger equation will be noticed. So will crucial differences.

Such analogies are surely more than coincidences. Richard Feynman's path integral formulation of quantum mechanics sees the evolution of the quantum mechanical wave function in terms of $S = \int L \, dt$: when something happens to the system, the new quantum wave function ψ' is produced from the original one ψ by a unitary transformation of the form

$$\psi' = \exp\left(\frac{i}{\hbar} \int L \, dt\right) \psi. \qquad (9.64)$$

Such transformations occupy our attention in the next chapter.

Questions for Reflection and Discussion

Q9.a. a. Compare the Hamilton-Jacobi statements $p = \partial S / \partial q$ and $H = -\partial S / \partial t$ with the momentum and Hamiltonian operators of quantum theory.

---

[3]The Planck-Einstein and de Broglie postulates are statements about a correspondence between *free* particles and *harmonic* (sinusoidal) waves. Planck-Einstein (1900, 1905): Corresponding to a harmonic wave of angular frequency ω there exists a free particle of energy E, where $E = \hbar\omega$. De Broglie (1924): Corresponding to motion of a free particle of momentum p there exists a harmonic wave of wavenumber k, where $p = \hbar k$. In both statements, $\hbar$ is the reduced Planck's constant, $h/2\pi$.

b. Make a list of similarities, and another list of differences, between the Hamilton-Jacobi equation and the Schrödinger equation. In making your list, consider both the time-dependent and time-independent versions of both equations. For example, what are the consequences for each equation regarding whether or not $\partial U/\partial t = 0$?

Q9.b. Suppose Newtonian physicists before 1900 had come up with a way to associate a unique wavelength to a free particle's momentum, and a unique frequency to the particle's energy. Could the Hamilton-Jacobi equation have led to a version of quantum mechanics?

Q9.c. Can one make an analogy between finding a phase space in which the coordinates are all constants, and using the principle of covariance in relativity to solve a problem in a "rest frame" then transforming back to the original frame?

Q9.d. a. The development of the Hamilton-Jacobi equation uses both a definition of invariance and the equations of motion, Hamilton's equations. Review the similarities and differences, in motivation and approach, between Noether's theorem and Hamilton-Jacobi theory.
b. Discuss the similarities and differences between the conservation laws as they are revealed in Noether's theorem formalism, and their role in Hamilton-Jacobi formalism. To be specific, consider the conservation of mechanical energy and linear momentum. Do these conservation principles emerge, or are they imposed?

Q9.e. Can Hamilton-Jacobi formalism be extended to special relativity?

Q9.f. Comment on the ubiquitous appearance, in Hamilton-Jacobi theory, in the invariance identity, and in the Noether conservation laws, of the structure $pX - HT$ where X is a space quantity and T is a time quantity. How does this combination of variables show up in the physics of waves? How does it show up in special relativity?

Q9.g. Are Hamilton's equations and the Euler-Lagrange equation equivalent? Consider replacing H with $p_\mu \dot{q}^\mu - L$ in Hamilton's equations, and replacing L with $p_\mu \dot{q}^\mu - H$ in the Euler-Lagrange equation.

Q9.h. A puzzled student asks, "Which is more fundamental, the Lagrangian or the Hamiltonian?" Answer the student's question, making sure to include the following considerations:

a. In the spacetime of special relativity, the Lagrangian density is normally a Lorentz scalar (or scalar density in some field theories). In contrast, a particle's energy, and thus the Hamiltonian, is the time component of a four-vector, and the canonical momenta are the space components of that same four-vector.

b. Formally, the functional is defined in terms of the Lagrangian, and the momenta and Hamiltonian are derived from the Lagrangian. Could the roles of what is defined and what is derived be reversed?

Q9.i. In order to solve Hamilton's equations, do you have to combine the pair after all into a second-order equation? Try it with a simple problem, such as the one-dimensional simple harmonic oscillator.

Q9.j. Can the Hamiltonian be considered a momentum canonically conjugate to time? See the paper by Donald Kobe on this topic [Kobe (1993)].

Q9.k. Find out whatever you can about "contact transformations" as discussed in some mechanics books (e.g., Finkelstein [1973]).

Q9.l. Under what circumstances could Hamilton-Jacobi theory and parametric invariance (section 8.3) be the same?

Exercises

9.1. a. The Hamiltonian for the simple harmonic oscillator is

$$H(x,p) = \frac{p^2}{2m} + \frac{1}{2}m\omega^2 x^2. \qquad (9.65)$$

Write Hamilton's equations. Lo, they are coupled. How do you go about uncoupling them? In so doing, what equation results?

b. Generalize to a particle of mass m moving in one spatial dimension, under the influence of the potential energy $U(x)$. Describe a strategy for finding the solution $x(t)$ when starting from Hamilton's equations.

9.2. Compare eq. (9.64) to the Wentzel-Kramers-Brillouin (WKB) approximation. The WKB approximation, also called the semiclassical approximation, is useful for deriving approximate solutions to the Schrödinger equation when the potential energy changes very little over one de Broglie wavelength. Start with the time-independent Schrödinger equation

$$-\frac{\hbar^2}{2m}\frac{d^2\psi(x)}{dx^2} + U(x)\psi(x) = E\psi(x). \qquad (9.66)$$

Since $\psi(x)$ is a complex number, factor it into an amplitude and phase according to

$$\psi(x) = A(x)e^{iS(x)/\hbar}. \qquad (9.67)$$

Typically, phases vary rapidly with distance compared to amplitudes. The rapid oscillation of the second derivative of $\psi(x)$ gets shifted over to the phase $S(x)$, leaving the second derivative of slowly varying $A(x)$ as the coefficient of $\hbar^2$. Since $\hbar$ is small, this quadratic term in $\hbar$ is neglected. Show $S(x)$ to be the classical action

$$S(x) = \pm\sqrt{2m}\int \sqrt{E - U(x)}\; dx \qquad (9.68)$$

and

$$\frac{d(A^2 p)}{dx} = 0. \qquad (9.69)$$

Demonstrate eq. (9.69) to be a statement of probability conservation. Write the approximate wave function $\psi(x)$ in terms of a boundary value $\psi(x_o)$. Why is this approximation valid only in regions whose width is much smaller than a de Broglie wavelength (see e.g., Saxon [1968] 174–177)?

9.3. The Poisson bracket of A and B, denoted $[A, B]$, is defined as

$$[A, B] \equiv \frac{\partial A}{\partial q^\mu}\frac{\partial B}{\partial p_\mu} - \frac{\partial B}{\partial q^\mu}\frac{\partial A}{\partial p_\mu}, \qquad (9.70)$$

where the repeated indices are summed.
a. Calculate the Poisson brackets $[x^\mu, p_\nu]$, $[x^\mu, H]$, $[p_\nu, H]$.
b. For any function $f(x^\mu, p_\nu)$, derive an expression for df/dt in terms of Poisson brackets.
c. Show the Poisson bracket to be invariant under a canonical transformation. The covariance of Hamilton's equations and the invariance of the Poisson brackets in going from (x^μ, p_μ) to (x'^μ, p'_μ) are equivalent definitions of a canonical transformation.

9.4. This exercise is included to offer a feel for phase space as a visual tool, independent of Hamilton's equations. The method of isoclines offers a graphical approach to the solution of first-order differential equations, working with the first derivative as a slope [Rainville & Bedient (1974) 17; Ritger & Rose (1968) 97, 314]. Let's develop such a method, via some rather generic equations for $\dot{x}$ and $\dot{p}$ (see Figure 9.2)

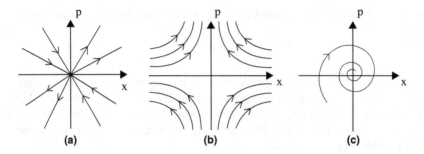

Figure 9.2: *A case with (a)* $\delta = 0$, *(b)* $\delta > 0$, *and (c)* $\delta < 0$.

Assume that both $\dot{x}$ and $\dot{p}$ vanish at the phase space origin. A Taylor series about the origin has the leading terms

$$\dot{x} = Rx + Sp, \tag{9.71}$$

$$\dot{p} = Ax + Bp \tag{9.72}$$

where R, S, A, and B are constants. Suppose the time dependence of x and p are parameterized by $x(t) = \alpha e^{\lambda t}$ and $p(t) = \beta e^{\lambda t}$. Then the pair of equations for $\dot{x}$ and $\dot{p}$ becomes

$$(R - \lambda)\alpha + S\beta = 0, \tag{9.73}$$

$$A\alpha + (B - \lambda)\beta = 0. \tag{9.74}$$

a. Show that a nontrivial solution exists for this pair of equations only if

$$\lambda = \tfrac{1}{2}(R + B) \pm \tfrac{1}{2}\sqrt{\delta}, \tag{9.75}$$

where $\delta \equiv (R - B)^2 + 4AS$. Consider the three possibilities: $\delta = 0, \delta > 0$, and $\delta < 0$, to see what can be learned.

b. $\delta = 0$: One way for this to happen has $R = B \neq 0$, and $A = S = 0$. Show that the coupled equations become $\dot{x} = Bx$ and $\dot{p} = Bp$, leading to similar exponential solutions, $x \sim p \sim e^{Bt}$. Show that, near the origin, the phase space trajectory of the system's possible evolutions form a family of straight lines passing through the origin (see Figure 9.2 [a]). If $B > 0$ the system is unstable, with all the states repelled from the phase space origin. On the other hand, if $B < 0$ the system is stable and all phase space trajectories converge to the origin as the system finds itself attracted to it. When the phase space trajectories converge onto or diverge away from a fixed point in phase space, that point is called a "node."

c. $\delta > 0$: The simplest way here is to have $A = S = 0$, and $R \neq B$. Show that the coupled equations now become $\dot{p} = Bp$ and $\dot{x} = Rx$ with solutions

$p = \beta e^{Bt}$ and $x = \alpha e^{Rt}$. Show also that, when t is eliminated between these equations, a power law trajectory in phase space results:

$$x = \alpha \left(\frac{p}{\beta}\right)^{R/B}. \tag{9.76}$$

Show that if R and B have the same sign then the origin is a node; if R and B carry opposite signs then the node is a "saddle point," whereby in the evolution of the system the phase space trajectory swerves away from the origin (Figure 9.2 [b]).

d. If $\delta < 0$, for instance when $R = B$ and $S = -A$, then $\dot{p} = Bp + Ax$ and $\dot{x} = Bx - Ap$. Try a solution of the form $x(t) = v(t)e^{Bt}$ and $p(t) = u(t)e^{Bt}$. Show that $v = -\alpha \sin(At)$, $u = \alpha \cos(At)$, and therefore,

$$x^2 + p^2 = \alpha^2 e^{2Bt}. \tag{9.77}$$

Demonstrate that this phase space trajectory spirals inward if $B < 0$, spirals outward if $B > 0$, and forms an ellipse if $B = 0$ (Figure 9.2.[c]).

e. Test this program with the force $-kx$ where $k > 0$, and interpret the locus in phase space.

f. In chaos theory, a nonlinear system's phase space trajectory exhibits extreme sensitivity to initial conditions. Sometimes these trajectories converge around a point called an "attractor." Read about and coment on the role of attractors in chaos theory, as in Gleick (1987).

g. What Hamiltonian would give the same results in each of parts (b), (c) and (d)?

9.5. Solve the Hamilton-Jacobi equation for
a. a particle in a uniform gravitational field, $U(x) = mgx$;
b. the simple harmonic oscillator, $U(x) = \frac{1}{2}m\omega^2 x^2$.

9.6. Develop the Hamilton-Jacobi equation approach for at least one of the other three options for the argument of S (recall the four possibilities mentioned in eq. [9.44]).

9.7. Hamilton-Jacobi theory can be used to find periods of motion under a broad set of conditions by introducing "action-angle" (or "action and angle") variables [Goldstein (1965) 288; Finkelstein (1973) 284]. In Hamilton-Jacobi theory as described in section 9.4, we considered a transformation $(q, p) \to (q', p')$ where $p' \equiv p_0 = \text{const.}$, with W a function of q and p', and where $p = \partial W/\partial q$, $q' = \partial W/\partial p'$, and H numerically equals E. Consider periodic motion. Define the "action variable" J according to

$$J \equiv \oint p\, dq = \oint \frac{\partial W}{\partial q}\, dq \tag{9.78}$$

with integration over one period of the motion. Since q gets integrated out, J is a function only of p_o, so that $J = J(p_o)$, which can be inverted to write $p_o = p_o(J)$. Let us now identify a new p' as J, and introduce a new coordinate θ, the "angle variable," as the q' conjugate to J. Thus the Hamilton-Jacobi expression $q' = \partial W/\partial p'$ becomes $\theta = \partial W/\partial J$. The original Hamiltonian, for which $H = E$, may be rewritten in terms of J and θ. Thus by Hamilton's equations, for these new phase space coordinates,

$$\dot{\theta} = \frac{\partial H}{\partial J} \tag{9.79}$$

which is a constant because $J = J(p_o)$ and p_o is constant. Denoting the constant as ω and integrating eq. (9.79) over time gives $\theta = \omega t + \text{const.}$ Thus the frequency of a periodic motion can be found by calculating J, rewriting H in terms of J, then evaluating $\partial H/\partial J = \omega$.

a. Demonstrate the viability of this idea with the elementary example of a particle of mass m moving with uniform speed in a circle of radius R. Recall that the particle's moment of inertia about the circle's center is mR^2.

b. Apply this procedure to the simple harmonic oscillator, for which

$$H(x, p) = \frac{p^2}{2m} + \frac{1}{2}kx^2. \tag{9.80}$$

Let the amplitude be A, evaluate J and show that, when $H = E$ is written in terms of J, we recover the frequency $1/T = (1/2\pi)\sqrt{k/m}$.

9.8. Consider a wave equation for light moving through a medium with refractive index n. The wave's speed is $v = c/n$ and the wave equation becomes

$$\frac{\partial^2 \varphi}{\partial x^2} = \frac{n^2}{c^2} \frac{\partial^2 \varphi}{\partial t^2}, \tag{9.81}$$

which features a harmonic solution,

$$\begin{aligned}
\varphi(t, x) &= A e^{i(kx - \omega t + \delta)} \\
&= A e^{ik_o(nx - ct) + i\delta},
\end{aligned} \tag{9.82}$$

where k_o is the wave number in vacuum. Suppose the refractive index is a function of x, so we may write $nx = N(x)$. The amplitude may also vary with position, so we parameterize it as $A(x) \sim e^{i\alpha(x)}$.

a. Show that, in the short-wavelength limit, where the refractive index varies little over one wavelength, and thus the dominant terms are those quadratic in k_o, the wave equation becomes the "eikonal equation,"

$$\left(\frac{dN}{dx}\right)^2 = n^2. \tag{9.83}$$

b. Compare the eikonal equation to the time-independent Hamilton-Jacobi equation (9.54) and draw an analogy between the potential energy and the refractive index. See also ex. 3.18.

c. Discuss how Hamilton-Jacobi theory supports the notion that Newtonian mechanics is the short-wavelength limit of quantum mechanics.[4]

---

[4]For applications of the eikonal approximation to classical mechanics see Finkelstein (1973), and Perl (1974) for quantum mechanics.

Chapter 10

The Action as
a Generator

It is a most beautiful and awe-inspiring fact that all the fundamental laws of Classical Physics can be understood in terms of one mathematical construct called the Action. It yields the classical equations of motion, and analysis of its invariances leads to quantities conserved in the course of the classical motion. In addition, as Dirac and Feynman have shown, the Action acquires its full importance in Quantum Physics.
—Pierre Ramond, *Field Theory*, 1981

Let us now take a dynamical system, having a classical analogue and let us take the ξ's to be the coordinates q. Put $\langle q'_t | q'' \rangle = e^{iS}$ [Dirac's eq. 52] and so define the function S on the variables $q'_t, q'' \ldots$.

The solution of the Hamilton-Jacobi equations ... is the action functional of classical mechanics for the time interval t_o to t, i.e., it is the time integral of the Lagrangian L,

$$S = \int_{t_o}^{t} L(t') \, dt'. \tag{10.1}$$

Thus the S defined by (52) is the quantum analogue of the classical action function and equals it in the limit $\hbar \to 0$.
—Paul Dirac, *The Principles of Quantum Mechanics*, 1947

10.1 Conservation of Probability and Continuous Transformations

The state of a system is described in quantum mechanics by a state function ψ, or state vector $|\psi\rangle$ or its matrix representation. Consider an operator U that transforms $|\psi\rangle$ into a new state vector $|\psi'\rangle$ according to

$$|\psi\rangle \rightarrow |\psi'\rangle = U|\psi\rangle. \tag{10.2}$$

Noether's theorem applies to symmetries parameterized by a variable ε (or set of variables ε^k) that can be varied continuously from zero. But discrete transformations are also crucial to physics. Examples include:

- Charge conjugation, where particles are replaced with their antiparticles;

- Parity transformations, which reflect the coordinate axes through the spatial origin, so the position vector $\mathbf{r}$ gets inverted to $-\mathbf{r}$;

- Time reversal, which interchanges initial and final states;

- Exchanging identical particles, by swapping their positions and spin orientations;

- Isospin exchanges. For example, the proton and neutron can be thought of as two states of the nucleon, an isospin doublet with $I_z = +\frac{1}{2}$ for the neutron and $I_z = -\frac{1}{2}$ for the proton. To exchange the isospin quantum numbers is to turn protons into neutrons and neutrons into protons.

No continuum of intermediate charge states exists that carries charge continuously from $-e$ for an electron to $+e$ for a positron. In a time reversal, one cannot go continuously from t to $-t$. How then does one define invariance for discrete symmetries? For example, if ψ is a function of spatial coordinate x then the parity operator $\mathcal{P}$ gives $\psi'(x) = \mathcal{P}\psi(x) = \psi(-x)$. But no infinitesimal ε exists between $+x$ and $-x$ for nonzero x. So far as I know, no "Noether's theorem for discrete symmetries" exists that offers a unified treatment to all discrete symmetries, with the same generality and élan as the Noether theorem we know for continuously parameterized symmetries.

Be that as it may, whether a transformation happens to be discrete or continuous, let us return to the fundamental concept that an operator U produces a new state vector $|\psi'\rangle$ from an old one according to $|\psi'\rangle = U|\psi\rangle$. In matrix representation, by the rules of matrix algebra the corresponding row vectors are

$$\langle\psi'| = \langle\psi|U^\dagger, \tag{10.3}$$

where $U^\dagger$ denotes the matrix made by forming the transpose and complex conjugate, or adjoint, of U. The conservation of probability requires $\langle\psi|\psi\rangle = 1$. Invariance of length squared of the state vector means

$$\langle\psi'|\psi'\rangle = \langle\psi|\psi\rangle \tag{10.4}$$

which requires U to be unitary,

$$U^\dagger U = 1, \tag{10.5}$$

where 1 denotes the unit matrix. Any unitary operator can be written as a complex exponential of a Hermitian operator,

$$U = e^{i\Lambda} \tag{10.6}$$

where Λ is Hermitian, that is,[1]

$$\Lambda^\dagger = \Lambda. \tag{10.7}$$

If Λ depends on some continuous real parameter ε, then for infinitesimal $\varepsilon, U = e^{i\Lambda(\varepsilon)}$ becomes

$$e^{i\Lambda(\varepsilon)} = 1 + i\varepsilon\frac{d\Lambda(0)}{d\varepsilon} + \dots, \tag{10.8}$$

and thus

$$U|\psi\rangle = |\psi'\rangle = \psi + i\varepsilon\lambda|\psi\rangle + \dots \tag{10.9}$$

where $\lambda \equiv d\Lambda(0)/d\varepsilon$.[2] The generator of the field transformation is $\zeta = i\lambda|\psi\rangle$.[3] Our task in this section is to find the generators for various transformations in the evolution of the quantum state $|\psi\rangle \to |\psi'\rangle$.

Let's go back to an important result of the Hamilton-Jacobi equation. Recall eq. (9.40), repeated here:

$$S = \int L\,dt + \text{const.}, \tag{10.10}$$

where S denotes the action that generates a canonical transformation, the indefinite integral of the Lagrangian, whereas the functional Γ that started this business is a definite integral,

$$\Gamma = \int_a^b L\,dt. \tag{10.11}$$

---

[1] Hermitian operators have real eigenvalues. Therefore, the act of making a measurement of a system gets represented in the mathematics as a Hermitian operator operating on the state vector, which (if not an eigenstate of the Hermitian operator itself) may be expressed as a superposition of the operator's eigenstates. The transformation of one state into another proceeds through a unitary operator.

[2] This λ is generic, not to be confused with the $\frac{1}{2}\lambda_k$ of SU(N). The latter is a special case of the former.

[3] Should more than one parameter participate in the transformation, put indices on ε's and the generators, and sum over them.

When we put a variable upper limit time T on the integral, the action as a function of time evolves from its initial value $S(0)$ to a later value $S(T)$ according to

$$S(T) = S(0) + \int_0^T L\,dt \qquad (10.12)$$

which looks like a transformation of S, with $\int_0^T L\,ds$ playing the role of a generator, or in other notation,

$$S' = S + \int_0^\varepsilon L\,dt. \qquad (10.13)$$

Applied to quantum mechanics, whose transformations proceed by unitary operators, this offers motivation for a generic transformation that changes a quantum wave function ψ according to

$$\psi' = e^{iS/\hbar}\psi \qquad (10.14)$$

where S has been divided by Planck's constant $\hbar$, the unit of action, to make dimensionless the argument of the exponential. Such was the idea of Paul Dirac, extended by Richard Feynman, that a generic continuous transformation of a quantum state, $\psi \to \psi'$, arises from

$$\psi' = \exp\left(\frac{i}{\hbar}\int L\,dt\right)\psi. \qquad (10.15)$$

To consider spacetime translations, express the Lagrangian in terms of the Hamiltonian, $L = p_\mu \dot{q}^\mu - H$. Then eq. (10.15) becomes (suppressing indices on phase space coordinates)

$$\psi' = \exp\left(\frac{i}{\hbar}\int (p\dot{q} - H)\,dt\right)\psi$$
$$= \exp\left(\frac{i}{\hbar}\int (p\,dq - H\,dt)\right)\psi. \qquad (10.16)$$

So far we have indicated no limits on the integral, so the action S may or may not be an extremum, and there is no reason, even if it were to be minimized, that it would be small. To produce the infinitesimal transformation, where $\psi' - \psi$ is to be small, consider $dS = p\,dq - H\,dt$, which peels off the integral, so that

$$\psi' = \left[1 + \frac{i}{\hbar}(p\,dq - H\,dt)\right]\psi + \dots$$
$$= \psi + \frac{i}{\hbar}(p\psi)dq - \frac{i}{\hbar}(H\psi)dt + \dots \qquad (10.17)$$

Here dq and dt are playing the role of the infinitesimal parameters ε we have been discussing throughout this book. Comparing this to

$$\psi' = \psi + \varepsilon^k \zeta_k + \dots \tag{10.18}$$

we can identify generators of spacetime transformations for a quantum state ψ:

$$\zeta_q = \frac{i}{\hbar} p\psi \tag{10.19}$$

and[4]

$$\zeta_t = -\frac{i}{\hbar} H\psi. \tag{10.20}$$

We say that momentum is the generator that translates the wave function in space, and the Hamiltonian is the generator that translates the wave function in time.

Denoting $\psi' - \psi = d\psi$, we see that, to first order in an infinitesimal spacetime transformation,

$$d\psi = \frac{i}{\hbar} p\psi \, dq - \frac{i}{\hbar} H\psi \, dt. \tag{10.21}$$

On the other hand, because $\psi = \psi(t, q)$, by the chain rule we may also write $d\psi$ as

$$d\psi = \frac{\partial \psi}{\partial q} dq + \frac{\partial \psi}{\partial t} dt. \tag{10.22}$$

Comparing the differentials between eqs. (10.21) and (10.22) leads to significant conclusions.

Time Translation: Equating the coefficients of dt between eqs. (10.21) and (10.22), we obtain the Schrödinger equation

$$H\psi = -\frac{\hbar}{i} \frac{\partial \psi}{\partial t}. \tag{10.23}$$

Space Translation: Equating the coefficients of dq between eqs. (10.21) and (10.22), should q be a length we find the quantum prescription of the linear momentum operator,

$$p\psi = \frac{\hbar}{i} \frac{\partial \psi}{\partial q}. \tag{10.24}$$

Putting these results together, the classical mechanics Hamiltonian for a particle of mass m moving in response to a potential energy $U(x)$ is typically

$$H = \frac{p^2}{2m} + U(x). \tag{10.25}$$

---

[4]We used ζ_t rather than τ here because the question is not about the change in t itself, but the change in the quantum state under $\psi(t) \to \psi(t + dt)$.

When this H is inserted into eq. (10.23), and eq. (10.24) is used for p, the familiar time-dependent Schrödinger equation for the wave function $\psi(t, x)$ results:

$$-\frac{\hbar^2}{2m}\frac{\partial^2\psi}{\partial x^2} + U\psi = -\frac{\hbar}{i}\frac{\partial\psi}{\partial t}. \tag{10.26}$$

Rotation: Let the generalized coordinate be an angle, $dq = d\varphi$, where $d\varphi$ describes an infinitesimal rotation about, say, the z-axis. The momentum conjugate to φ will be a component of angular momentum $p_\varphi = \ell_z$, in this role the generator of rotations. Now $p = \frac{\hbar}{i}\frac{\partial}{\partial q}$ becomes

$$\ell_z\psi = \frac{\hbar}{i}\frac{\partial\psi}{\partial\varphi}. \tag{10.27}$$

In our original uses of the action that led to Noether's theorem, the action functional was not operating *on* anything else; rather, we did things *to it*. When required to be an extremal, it gave the equations of motion. When invariant and extremal, it give the Noether conservation laws. Quantum theory finds an additional role for the action functional, by giving it something to operate on: the quantum state ψ or state vector $|\psi\rangle$. To fully develop this role of the action in quantum mechanics requires the Feynman path integral formulation of the theory. In classical mechanics, the particle follows only the extremal path. In quantum theory, *all* possible paths are summed over in the Feynman path integral. The ones closest to the classical path, where S is a minimum, dominate $e^{iS/\hbar}$ because $\hbar$ is small. In other words, in the limit as $\hbar \to 0$, the classical trajectory is recovered.

In classical mechanics, Hamilton's principle could seem a bit teleological—how could the particle "know" the extremal path it was "supposed to" follow, seemingly in advance? Particles move according to local, not global instructions; their equations of motion are differential equations, even though the functional they come from is a definite integral, a global description. But with action as a generator, we see how, according to quantum theory, all paths are sampled. The classical trajectory merely gets most heavily weighted.

10.2 The Poetry of Nature

The useful combinations are precisely the most beautiful.
—Henri Poincaré, *Science and Method*, 1908

It seems that if one is working from the point of view of getting beauty into one's equations, and if one has really a sound insight, one is on a sure line of progress.—Paul Dirac, *Scientific American*, May 1963

Physics is the art of creating, testing, and improving a network of concepts in terms of which the physical universe becomes comprehensible. Physics is guided by evidence and logic, but inspired by aesthetics. Because we can imagine more hypotheses than we could ever test, we have to rule out some in advance. The criteria for making those decisions include value judgments, such as simplicity and unity. We seek not only a solution, but an *elegant* solution.

Creativity, Jacob Bronowski wrote, is the fusion of ideas:

> The discoveries of science, the works of art are explorations . . . of hidden likeness. The discoverer or the artist presents in them two aspects of nature and fuses them into one . . . and it is the same act in original science and original art. But it is not therefore the monopoly of the man who wrote the poem or who made the discovery. . . . The poem or the discovery exists in two moments of vision: the moment of appreciation as much as that of creation. [Bronowski (1956) 19].

Seeking unity beneath surface appearances has been a recurring and fruitful strategy throughout the history of physics. Hero of Alexandria and Pierre Fermat unified the principles of geometrical optics; Isaac Newton united Earth and sky with universal gravitation; James Joule established the mechanical equivalence of heat and work; Michael Faraday and James Maxwell joined electricity and magnetism into electromagnetism; Albert Einstein merged electromagnetism and light with space and time in special relativity, and fused them with matter and gravitation in general relativity; Richard Feynman, Julian Schwinger, Sin-itiro Tomonaga, and Freeman Dyson made electrodynamics consistent with quantum theory; Steven Weinberg, Adbus Salam, and Sheldon Glashow blended electrodynamics and the weak nuclear force into the electroweak interaction. The ongoing quest for field theories that unify all the fundamental interactions include superstring theories that link bosons to fermions.

When I was an undergraduate physics major, the Society of Physics Students distributed a lapel pin that held a deep truth. It said, "Physics is the poetry of nature." It was the beauty of connections between concepts that drew this student into physics. They are the bonding agents that hold the structure together so beautifully.

Amid the rich and sometimes bewildering dynamics, our understanding of the world hangs on those factors that do not change, thanks to underlying symmetries, expressed as invariances and conservation laws. Standing tall amid it all, we find Emmy Noether and her elegant, wonderful theorem.

Figure 10.1: *Emmy Noether (1882–1935)*. (Photo courtesy Bryn Mawr College)

Questions for Reflection and Discussion

Q10.a. Make a list of differences and a list of similarities between the Poisson bracket of ex. 9.3 and the commutator of quantum mechanics, $(A, B) \equiv AB - BA$. In particular, suggest analogies between Poisson brackets and the commutators, especially with regard to canonically conjugate variables. In the Poisson bracket of classical mechanics, we insert A and B *into* the brackets; the commutator as used in quantum mechanics needs a state vector or function on which to operate. This echoes the point that, in quantum mechanics, the action needs something on which to operate.

Q10.b. Find out whatever you can about CP violation, where C means charge conjugation and P means parity. What does CP violation say about cosmic history and the lack of symmetry in the abundance of matter over antimatter?

Q10.c. Find out whatever you can about the CPT theorem, where C stands for charge conjugation, P for parity, and T for time reversal.

Q10.d. In developing the operator calculus from the action S, why not work with the Lagrangian itself and be content with

$$d\psi = \frac{i}{\hbar} \, L\psi \, dt \, ? \tag{10.28}$$

Q10.e. Notice in eq. (10.21) how the factor $p_\mu dq^\mu - H dt$ resembles a scalar product in spacetime, such as the invariant spacetime interval or the energy-momentum relation. Show how the arguments leading to eq. (10.21) carry through in special relativity, and thus how eqs. (10.23), (10.24), and (10.27) can be interpreted beyond Newtonian Hamiltonians.

Q10.f. Consider the expectation value of some operator Ω, $\langle\psi|\Omega|\psi\rangle$. Under a transformation, some authors write $\Omega' = U^\dagger\Omega U$, whereas we wrote $|\psi'\rangle = U|\psi\rangle$.
a. Are these two ways of describing the transformation equivalent, and if so, why?
b. Find out whatever you can about the "Schrödinger picture" and the "Heisenberg picture" of quantum mechanics.

Q10.g. In developing quantum mechanics, Paul Dirac postulated that when a particle propagates from a position described by quantum state $|q\rangle$ at time $t = 0$ to a position described by the state $|q'\rangle$ at time $t = T$, the amplitude for that transition carries a phase shift proportional to the action:

$$\langle q'|q\rangle \sim \exp\left(\frac{i}{\hbar}\int_0^T Ldt\right). \tag{10.29}$$

Although his legendary insight here was as solid as elsewhere, Dirac made this postulate using assumptions that he did not rigorously justify at the time. Richard Feynman later made the supporting arguments rigorous. The Feynman path integral works like this: Break up each possible trajectory into a connect-the-dots trajectory between N points, and evaluate the integral with a formidable measure

$$\langle q'|q\rangle = \lim_{N\to\infty} A^N \int\left(\prod_{i=1}^{N-1} dq_i\right)\exp\left(\frac{i}{\hbar}\int_0^T L(q,\dot{q})dt\right), \tag{10.30}$$

where A denotes a normalization factor. In a field theory, the classical action is thereby extended to an all-spacetime integral over a Lagrangian density. In principle, one can apply the Feynman path integral to any system (although the list of path integrals that can be evaluated analytically is short) [Feynman & Hibbs (1965)]. The quantum mechanical amplitude that carries a particle from $|q\rangle$ at $t = 0$ to $|q'\rangle$ at $t = T$ requires us to sum over all possible paths, and each path gets weighted by the exponential of $i/\hbar$ times the classical action S. Since $\hbar$ is macroscopically tiny, $e^{iS/\hbar} = \cos(S/\hbar) + i\sin(S/\hbar)$ produces very rapid oscillations unless S is very small. So the smallest value of S dominates the exponential. And of course the classical trajectory is the one for which S is a minimum!

So in classical mechanics, the particle follows only one path, the extremal. In quantum theory, the particle follows all paths (visualize a spreading wave function or a diffusion of probability); the extremal path is merely the one that contributes the greatest weight (see ex. 10.4). Find out whatever you can about the Feynman path integral [Dirac, (1947); Feynman & Hibbs (1965); Ramond (1981) and references therein].

Exercises

10.1 Calculate the following commutators for a nonrelativistic particle moving in one dimension x under the influence of a potential energy $U(x)$.
a. (x, p)
b. (p, H)
c. (x, H)
d. Let $\boldsymbol{\ell} = \mathbf{r} \times \mathbf{p}$ denote angular momentum. Show that the commutators of the components of angular momentum, $(\ell_x, \ell_y), (\ell_y, \ell_z), (\ell_z, \ell_x)$, together are equivalent to the weird relation $\boldsymbol{\ell} \times \boldsymbol{\ell} = i\hbar\boldsymbol{\ell}$.
e. For each component ℓ_k, show that $(\ell^2, \ell_k) = 0$.

10.2. Show that $d\langle p \rangle/dt = -\langle \partial U/\partial x \rangle$. This important result, called Ehrenfest's theorem, illustrates the correspondence principle requirement, that Newtonian mechanics be contained within quantum mechanics as a special case.

10.3. Suppose the potential energy is given an imaginary part, $U = U_0 - iZ$, where U_0 and Z are real.
a. Show that, according to the Schrödinger equation, probability is not conserved, namely $d\langle \psi | \psi \rangle/dt \neq 0$.
b. Let $P \equiv \langle \psi | \psi \rangle$, suppose $Z = $ const., and show that $P(t) = P(0)e^{-2Z/\hbar}$. How might this relate to radioactive decay?

10.4. Let us end with a problem similar to those in chapter 2, but here seen in the context of the connection between classical and quantum physics. Recall once again the simple harmonic oscillator, for which

$$L = \tfrac{1}{2}m\dot{x}^2 - \tfrac{1}{2}m\omega^2 x^2. \tag{10.31}$$

Consider three paths through t-x space, as the particle moves from $x = 0$ at $t = 0$ to $x = A$ at $t = T/4$, where T denotes the period and A the amplitude:
a. $x(t) = A\sin(\omega t)$;
b. $x(t) = 4At/T$;
c. $x(t) = 16At^2/T^2$.
Compare the values they give for the action functional

$$\frac{i}{\hbar} \int_0^{T/4} L \, dt. \tag{10.32}$$

Which of these three actions will dominate in the exponential in the Feynman path integral? Recall that $|\langle q'|q \rangle|^2$ is the probability that the particle makes the trip from $x = 0$ to $x = A$ in the allotted time.

The time is gone the song is over,
thought I'd something more to say

—Pink Floyd, "Time," *Dark Side of*
the Moon, lyrics by Roger Waters, EMI Records, 1973

Appendixes

Appendix A

Scalars, Vectors, and Tensors

Coordinate systems, or reference frames, are not part of nature. They are maps introduced by us. Whenever we project a problem onto a set of coordinates, we must distinguish features intrinsic to the physical system from artifacts that arise from the coordinate system. Whenever we shift from one coordinate map to another, we must understand what stays the same and what changes and how. The principles that tell us how to convert quantities and relationships between coordinate systems are theories of relativity.

Theories of relativity are founded on those quantities that are postulated to be invariant among a well-defined set of coordinate systems. For example, Newtonian-Galilean relativity postulates that, in all inertial frames, length and time intervals are separately invariant. Consequently, it follows that the speed of light depends on the relative motion between the light source and the observer. The special theory of relativity, in contrast, postulates the speed of light to be invariant among all inertial reference frames. Consequently, length and time intervals are not invariant.

When a quantity is not invariant between two coordinate systems, we need to know how it changes when we change reference frames. Consider a familiar introductory physics example, to illustrate the issues involved: a block of mass m sliding down an inclined plane. The dominant forces are those of gravity $m\mathbf{g}$ and contact forces: friction $\mathbf{f}$ tangent to the surface, and the normal force $\mathbf{N}$. A coordinate system can be oriented any number of ways, including the xy and $x'y'$ axes shown in Figure A.1.

The components of any vector are frame dependent. For instance, in the $x'y'$ system of Figure A.1, $\mathbf{F}' = m\mathbf{a}'$ becomes

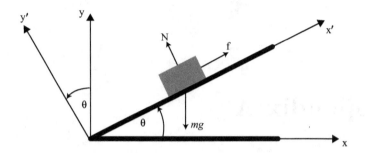

Figure A.1: *The two coordinate axes xy and x'y'.*

$$x' \text{ component: } f - mg\sin\theta = ma_{x'} \tag{A.1}$$

$$y'\text{component: } N - mg\cos\theta = 0. \tag{A.2}$$

But when $\mathbf{F} = m\mathbf{a}$ is projected onto the xy axes, the components look different:

$$x \text{ component: } f\cos\theta - N\sin\theta = ma_x, \tag{A.3}$$

$$y \text{ component: } N\cos\theta + f\sin\theta - mg = ma_y. \tag{A.4}$$

The theory of Newtonian relativity says that time intervals are invariant, and displacements in the x and y directions map to their x' and y' counterparts according to an orthogonal transformation, so that

$$dt' = dt, \tag{A.5}$$

$$dx' = dx\cos\theta + dy\sin\theta, \tag{A.6}$$

$$dy' = -dx\sin\theta + dy\cos\theta. \tag{A.7}$$

You can verify that the last two equations guarantee the Newtonian assumption of invariant length by checking that

$$(dx')^2 + (dy')^2 = (dx)^2 + (dy)^2. \tag{A.8}$$

Upon differentiating the spatial coordinate transformations with respect to time, and because time is assumed in Newtonian relativity to be invariant, we readily find the relativity of the velocity and acceleration components:

$$v_{x'} = v_x\cos\theta + v_y\sin\theta, \tag{A.9}$$

$$v_{y'} = -v_x\sin\theta + v_y\cos\theta, \tag{A.10}$$

and

$$a_{x'} = a_x \cos\theta + a_y \sin\theta, \qquad\qquad (A.11)$$

$$a_{y'} = -a_x \sin\theta + a_y \cos\theta. \qquad\qquad (A.12)$$

You can confirm that these transformations convert Newton's second law for the block on the inclined plane from the unprimed version to the primed version. The vectors $\mathbf{F}$ and $\mathbf{a}$ have a relationship independent of coordinates. Although $a_{x'} \neq a_x$ and $F_{x'} \neq F_x$, the relation between force and acceleration is the same in both frames: $F_x = ma_x$ and $F_{x'} = ma_{x'}$. The relation $F_k = ma_k$ is said to be *covariant* between the frames. For this to happen, the vectors had to transform the same way as the coordinate displacements. This, indeed, serves as the formal definition of a vector.

Does the kinetic energy depend on the choice of the coordinate system? Let $K = \frac{1}{2}m\mathbf{v} \cdot \mathbf{v}$ denote the kinetic energy as measured in the xy coordinate system, and let $K' = \frac{1}{2}m\mathbf{v}' \cdot \mathbf{v}'$ as measured in the $x'y'$ system. Because kinetic energy depends only in the invariant mass and the square of $\mathbf{v}$, it is invariant between the two systems, as you can verify by explicit calculation. Any quantity that is precisely invariant under a coordinate transformation is called a "scalar." The kinetic energy is a scalar in Newtonian relativity, but contributes to the time component of a spacetime four-vector in special relativity.

Scalars have no coordinate indices. Vector components carry one index, to label the dimension of space to which it belongs. Scalars are rank-0 and vectors are rank-1 tensors, respectively. Rank-2 tensors are also important to the topics of this book.

Within a given coordinate system, a rank-2 tensor can be represented as a square matrix. Vectors can be thought of as column or row matrices. For the purposes of calculation, you can substitute the word "matrix" for "vector" and "tensor" and worry no more about it. But a tensor is not merely an array of numbers. It has to respect precise transformation properties when you change coordinate systems. If you want to know more about tensors, why they can be represented with matrices but are not synonomous with them, please read on.

Let's start with the convention of labeling coordinates with superscripts. The set $\{x^\mu\}$ denotes the coordinates in an N-dimensional space, where μ ranges over all N coordinates. For example, in three-dimensional Euclidean space, we can map the space with rectangular coordinates $(x^1, x^2, x^2) = (x, y, z)$; or we may use spherical coordinates, $(x^1, x^2, x^2) = (r, \theta, \varphi)$.

Consider now a transformation from one coordinate system to another, two different *maps* of the same space. Under the transformation, a scalar, by definition, is invariant: λ is a scalar if and only if

$$\lambda' = \lambda. \qquad\qquad (A.13)$$

In contrast, each new coordinate x'^μ will in general be a function of *all* the original coordinates:

$$x'^\mu = x'^\mu(x^\nu). \tag{A.14}$$

Therefore coordinate displacements transform, by virtue of the chain rule for differentials, according to

$$dx'^\mu = \frac{\partial x'^\mu}{\partial x^\nu} dx^\nu, \tag{A.15}$$

where repeated indices are summed (e.g., 1, 2, 3 for space; 0, 1, 2, 3 for spacetime; see below). The differential dx^ν defines a component of a displacement vector.[1] Thanks to vector addition and scalar multiplication, from the displacement vector come other vectors such as velocity, momentum, acceleration, force, electric and magnetic fields, the Newtonian gravitational field, Poynting's vector, and so on. For the purposes of theories of relativity, vectors are formally defined as a set of components $\{A^\mu\}$, one for each dimension of the space, that transform the same as the coordinate differentials:

$$A'^\mu = \frac{\partial x'^\mu}{\partial x^\nu} A^\nu. \tag{A.16}$$

A tensor of rank 2 transforms like the product of two coordinate displacements. For example, consider the inertia tensor, with components

$$I^{ij} = \int (\delta^{ij} r^2 - x^i x^j) dm = I^{ji} \tag{A.17}$$

where $\delta^{ij} = 1$ if $i = j$ and 0 if $i \neq j$ (the Kronecker delta) and where $r^2 = x^2 + y^x + z^2$. The inertia tensor relates one component of angular momentum ℓ^i to all the components of the angular velocity ω, through $\ell^i = I^{ij}\omega_j$.[2] Carrying products of two vector components, an inertia tensor component transforms as

$$I'^{ij} = \frac{\partial x'^i}{\partial x^k} \frac{\partial x'^j}{\partial x^n} I^{kn}. \tag{A.18}$$

The symmetric metric tensor $g_{\mu\nu} = g_{\nu\mu}$ forms an especially important second-rank tensor. It relates coordinate displacements to distances. The square of the distance ds between two infinitesimally nearby points, related by coordinates x^μ and $x^\mu + dx^\mu$, is given by

$$(ds)^2 = g_{\mu\nu} dx^\mu dx^\nu. \tag{A.19}$$

---

[1] A coordinate such as x is, itself, a displacement from the origin, because x describes the displacement $x - 0$. The transformation is given in terms of differential displacements, to allow the possibility of non-Euclidean geometries.

[2] The distinction between upper and lower indices is explained below.

For instance, in Euclidean space in rectangular coordinates, $(ds)^2$ is the familiar Pythagorean expression (the parentheses, as in $(ds)^2$, are usually omitted when the context is clear),

$$ds^2 = dx^2 + dy^2 + dz^2, \tag{A.20}$$

so the metric tensor components are the elements of the identity matrix, the Kronecker delta $\delta_{\mu\nu}$. In the same space mapped with spherical coordinates,

$$ds^2 = dr^2 + r^2 d\theta^2 + r^2 \sin^2 \theta d\varphi^2, \tag{A.21}$$

the nonzero metric tensor components are $g_{rr} = 1, g_{\theta\theta} = r^2, g_{\varphi\varphi} = r^2 \sin^2 \theta$. The length squared is a scalar, as one can show explicitly from eqs. (A.8) and (A.13).

In the special theory of relativity (appendix B), events in spacetime are mapped with four coordinates $\{x^\mu\}$, such as $(x^0, x^1, x^2, x^3) = (t, x, y, z)$ or (t, r, θ, φ). The invariant "length" is the proper time $d\tau$ between nearby events, the *spacetime interval*. In Cartesian coordinates it reads

$$d\tau^2 = g_{\mu\nu} dx^\mu dx^\nu = dt^2 - dx^2 - dy^2 - dz^2, \tag{A.22}$$

so that $g_{00} = 1$ and $g_{11} = g_{22} = g_{33} = -1$, and all other $g_{\mu\nu}$ vanish.[3]

Because of the opposite signs in the spacetime interval, we cannot write eq. (A.22) as the square of a four-dimensional spacetime vector unless we either modify the definition of scalar product or introduce two kinds of vectors. We choose the latter.[4] Therefore, corresponding to the vector with components A^μ we define a "dual" vector with components denoted A_μ, according to the sum

$$A_\mu \equiv g_{\mu\nu} A^\nu. \tag{A.23}$$

In Euclidean spaces, $g_{\mu\nu} = \delta_{\mu\nu}$, so the upper and lower index distinction is redundant. In contrast, in the spacetime of special relativity, $A_0 = A^0$, but $A_k = -A^k$ for $k = 1, 2, 3$.

Conversely, given a vector with lower indices, its dual with upper indices can be constructed by first defining the multiplicative inverse to the metric tensor. In terms of matrices, $g^{-1}g = gg^{-1} = 1$ (1 means the unit matrix), so the components $g^{\mu\nu}$ of the inverse matrix are defined according to

$$g^{\mu\lambda} g_{\lambda\nu} = \delta^\mu{}_\nu \tag{A.24}$$

---

[3]Please note that in eq. (A.22) we are measuring time in meters, where dt denotes the quantity $c\,dT$, where dT is a time increment measured in seconds and c is the speed of light in vacuum. Thus 1 meter of time $= \frac{1}{3} \times 10^{-8}$ s. If c is retained, then the spacetime interval appears as $c^2 d\tau^2 = c^2 dt^2 - (dx^2 + dy^2 + dz^2)$.

[4]Arithmetic faced an analogous issue when subtraction was invented: One could think of subtraction as another operation distinct from addition, or define subtraction as addition by introducing negative numbers—for every number x there exists a number $-x$ such that $x + (-x) = 0$, which defines $-x$. Therefore subtraction became a part of addition: $y - x$ means $y + (-x)$.

where $\delta^{\mu}{}_{\nu}$ is again the Kronecker delta.[5] Then

$$A^{\mu} = g^{\mu\nu} A_{\nu}. \qquad (A.25)$$

Besides defining vectors with lower indices through the ones with upper indices and the metric tensor, a prototype for a vector with "naturally" lower indices may be found in the gradient of a scalar function,

$$\frac{\partial \lambda}{\partial x^{\mu}} \equiv \partial_{\mu} \lambda. \qquad (A.26)$$

The upper index on the x^{μ} in the denominator of the gradient suggests the lower index of ∂_{μ}. To evaluate this gradient component with respect to some other coordinate system, from the chain rule we have

$$\begin{aligned}
\partial_{\mu'} \lambda &= \frac{\partial \lambda}{\partial x^{\rho}} \frac{\partial x^{\rho}}{\partial x'^{\mu}} \\
&= \frac{\partial x^{\rho}}{\partial x'^{\mu}} \partial_{\rho} \lambda.
\end{aligned} \qquad (A.27)$$

This motivates the transformation rule for any vector component with lower indices. They transform in the same way as the gradient:

$$A_{\mu'} \equiv A'_{\mu} = \frac{\partial x^{\rho}}{\partial x'^{\mu}} A_{\rho}. \qquad (A.28)$$

These transformation rules generalize to tensors of higher rank. One can also construct transformations of tensors that carry mixed upper and lower indices.

The indices labeling a tensor of rank 2 can be represented in a square matrix. In matrix notation, the left index, whether up or down, denotes the row, and the right index denotes the column where that tensor component appears in the matrix, like this: $\mathcal{H}_{\text{row}}{}^{\text{column}}, \mathcal{H}^{\text{row}}{}_{\text{column}}, \mathcal{H}_{\text{row column}}$, or $\mathcal{H}^{\text{row column}}$. Notice that the superscripts never stand directly over a subscript; these subscripts and superscripts can be raised and lowered like toggle switches.

---

[5]The upper/lower locations in the Kronecker delta make no difference, since each Kronecker delta symbol is either 0 or 1. So the indices are located to be consistent with the positions of other indices in an equation.

Appendix B

Special Relativity

A theory of relativity tells us how physical quantities and relationships are affected by changing the reference frame. Noether's theorem concerns itself with quantities that are invariant under a change of reference frame.

The unaccelerated inertial reference frames play a central role in Newtonian physics and in special relativity. A postulate of the special theory of relativity asserts the invariance among all inertial frames of the speed of light c in vacuum.[1] Coordinate changes between them are described by the Lorentz transformation. These statements are equivalent to the invariance of the spacetime interval. What is the "spacetime interval"? For observers in two inertial reference frames recording data for the same pair of events, the time between the events is not invariant ($dt \neq dt'$), nor is the distance between them invariant ($ds \neq ds'$). But the following combination of the time and space intervals *is* invariant. The spacetime interval, numerically equal to the value of proper time $d\tau$ squared (times c^2):

$$c^2 dt^2 - ds^2 = c^2 d\tau^2 = c^2 dt'^2 - ds'^2. \qquad (B.1)$$

In terms of a reference frame employing xyz spatial coordinates, the spacetime interval becomes[2] the "Minkowskian spacetime" of special relativity,

$$d\tau^2 = dt^2 - dx^2 - dy^2 - dz^2. \qquad (B.2)$$

For any other reference frame described by generalized coordinates x^μ, with the convention of summing over repeated indices, the invariant spacetime interval can be written generically as

$$d\tau^2 = g_{\mu\nu} dx^\mu dx^\nu. \qquad (B.3)$$

---

[1]The symbol c for the speed of light in vacuum comes from the Latin *celeritas*, for "swift."

[2]Henceforth the c will be absorbed into the time, so the symbols dt or $d\tau$ will be measured in meters.

The set of quantities $g_{\mu\nu}$ are components of the metric tensor in a specific reference frame (see appendix A).

From the spacetime displacement four-vector

$$\{dx^\mu\} = (dt, dx, dy, dz)$$
$$= (dt, d\mathbf{r}),$$

(B.4)

the components u^μ of the so-called four-velocity are constructed by rescaling the coordinate displacements with $1/d\tau$. Within a given frame a component of the velocity four-vector is

$$u^\mu \equiv \frac{dx^\mu}{d\tau} = \left(\frac{dt}{d\tau}, \frac{d\mathbf{r}}{d\tau} \right)$$

$$= \frac{dt}{d\tau} \left(1, \frac{d\mathbf{r}}{dt} \right)$$

(B.5)

$$= \gamma(1, \mathbf{v})$$

and from the invariant spacetime interval we obtain

$$\left(\frac{d\tau}{dt} \right)^2 = 1 - \left(\frac{d\mathbf{r}}{dt} \right)^2$$

$$= 1 - v^2$$

(B.6)

and where $\mathbf{v} = d\mathbf{r}/dt$ is a particle's velocity in the frame. From this it follows that[3]

$$\frac{dt}{d\tau} = \frac{1}{\sqrt{1 - v^2}} \equiv \gamma.$$

(B.7)

The four-momentum is defined analogous to Newtonian "$m\mathbf{v}$" momentum, but with four-velocity u^μ replacing the three-velocity $\mathbf{v}$,

$$p^\mu = mu^\mu$$
$$= m\gamma(1, \mathbf{v}).$$

(B.8)

The spatial components of four-momentum form the relativistic generalization of the Newtonian momentum, $\mathbf{p} = m\gamma\mathbf{v}$, which of course reduces to the Newtonian $m\mathbf{v}$ when $v \ll 1$. The interpretation of $p^0 = m\gamma$ takes a bit more thought.[4] It becomes clear by expanding γ in powers of v^2:

$$p^0 = m + \tfrac{1}{2}mv^2 + O(v^4) + \ldots$$

(B.9)

---

[3] In conventional units, replace v with v/c in γ, and the u^μ become $\gamma(c, \mathbf{v})$.
[4] In conventional units, $p^0 = mc^2\gamma$.

which contains the mass (more famously, mc^2 in conventional units), plus velocity-dependent terms whose leading member is the Newtonian kinetic energy. Since there are no interaction terms in this series expansion of p^0, we interpret p^0 as the energy E of a free particle, mass plus kinetic energy: $p^0 = m + K = E = m\gamma$. Therefore the energy-momentum four-vector takes the form

$$\{p^\mu\} = m\gamma(1, \mathbf{v})$$
$$= (E, \mathbf{p}) \tag{B.10}$$
$$= (m + K, \mathbf{p})$$

where
$$K = m(\gamma - 1) \tag{B.11}$$

is the kinetic energy.

Lowering indices with the metric tensor yields the dual energy-momentum four-vector, with components $p_\mu = g_{\mu\nu}p^\nu$:

$$(p_0, p_2, p_2, p_3) = (E, -\mathbf{p}). \tag{B.12}$$

Notice that, with summing over repeated indices, the magnitude squared of the energy-momentum four-vector is the square of the particle's mass m, an invariant Lorentz scalar:[5]

$$p_\mu p^\mu = E^2 - \mathbf{p} \cdot \mathbf{p} = m^2. \tag{B.13}$$

Newton's second law, generalized to special relativity, says that

$$f^\mu = \frac{dp^\mu}{d\tau}, \tag{B.14}$$

where f^μ is a component of the "four-force." Its spatial components are the three-vector force $\mathbf{F}$ components made relativistic, and the time component f^0 is the power produced or consumed by the particle's interactions with its surroundings.

In most of the textbooks that introduce special relativity, the simplest Lorentz transformation, which gets to the essence of this business, imagines a coasting rocket frame moving with uniform velocity $\mathbf{v}_r$ in the $+x$-direction relative to the laboratory frame [Taylor & Wheeler (1966)]. The x and x' axes are parallel, as are yy' and zz'. Furthermore, it is assumed that, at the event where the two frame's origins instantaneously coincide, each frame's arrays of clocks, previously synchronized separately within each frame to read the time of events at each clock's location, record $t = 0$ and $t' = 0$.

---

[5]In conventional units, $E^2 - (pc)^2 = (mc^2)^2$.

Under these simplifying circumstances, the Lorentz transformation from the laboratory frame coordinates (t, x, y, z) to the coasting rocket frame coordinates (t', x', y', z') is the "Lorentz boost"[6]

$$t' = \gamma_r(t - v_r x), \tag{B.15}$$

$$x' = \gamma_r(x - v_r t), \tag{B.16}$$

$$y' = y, \tag{B.17}$$

$$z' = z, \tag{B.18}$$

where

$$\gamma_r \equiv \frac{1}{\sqrt{1 - v_r^2}}. \tag{B.19}$$

Because the Newtonian limit is recovered in the limit $c \to \infty$, the Galilean transformation of Newtonian mechanics is subsumed in the Lorentz transformation as a special case. An especially elegant way to parameterize the Lorentz transformation arises by introducing the *rapidity* ε through $v_r \equiv \tanh \varepsilon_r$.[7] For the transformation of eqs. (B.15) through (B.19), the appearance of the Lorentz transformation equations are analogous to those for a rotation of axes:

$$\begin{pmatrix} t' \\ x' \\ y' \\ z' \end{pmatrix} = \begin{pmatrix} \cosh \varepsilon_r & -\sinh \varepsilon_r & 0 & 0 \\ -\sinh \varepsilon_r & \cosh \varepsilon_r & 0 & 0 \\ 0 & 0 & 1 & 0 \\ 0 & 0 & 0 & 1 \end{pmatrix} \begin{pmatrix} t \\ x \\ y \\ z \end{pmatrix}. \tag{B.20}$$

To first order in ε (dropping the r subscript), the infinitesimal form of eq. (B.20) can be written

$$x'^\mu \approx x^\mu + \varepsilon \omega^\mu{}_\nu x^\nu, \tag{B.21}$$

where the constant matrix $\{\omega^\mu{}_\nu\}$ is given by

$$\{\omega^\mu{}_\nu\} = \begin{pmatrix} 0 & -1 & 0 & 0 \\ -1 & 0 & 0 & 0 \\ 0 & 0 & 0 & 0 \\ 0 & 0 & 0 & 0 \end{pmatrix}. \tag{B.22}$$

However we parameterize it, we may denote the transformation matrix coefficients $\partial x'^\mu / \partial x^\nu$ as $\Lambda^\mu{}_\nu$,[8] and write a generic Lorentz transformation as

$$x'^\mu = \Lambda^\mu{}_\nu x^\nu. \tag{B.23}$$

---

[6]Notice the distinction between γ_r for the relative velocity *between* inertial frames, and γ corresponding to the velocity of a particle *within* a given inertial frame.

[7]$v_r/c = \tanh \varepsilon_r$ in conventional units.

[8]The set of coefficients $\Lambda^\mu{}_\nu$ are not components of a tensor. Tensor components transform a certain way *under* a coordinate transformation; the Λ-matrix describes the transformation itself.

Infinitesimal Lorentz transformations may be written as an extension of eq. (B.21), by writing

$$\Lambda^{\mu}{}_{\nu} \approx \delta^{\mu}{}_{\nu} + \varepsilon\omega^{\mu}{}_{\nu} \tag{B.24}$$

for some matrix of constant coefficients $\omega^{\mu}{}_{\nu}$, of which eq. (B.22) forms one example.

The set of all Lorentz transformations forms a group: Each transformation has an inverse, the associative law holds, the set of transformations exhibits closure, and the set contains the identity transformation

$$x'^{\mu} = \delta^{\mu}{}_{\nu}x^{\nu} = x^{\mu}. \tag{B.25}$$

Because there are three components of velocity between reference frames, three axes about which rotations may be made, three directions for translating the spatial origin, and one translation for the origin of the time coordinate, the Lorentz transformation can be enlarged to the ten-parameter Poincaré group, including spacetime translations a^{μ}:

$$x'^{\mu} = \Lambda^{\mu}{}_{\nu}x^{\nu} + a^{\mu}. \tag{B.26}$$

Let us refine the possibilities open to Lorentz transformations. The invariance of the spacetime interval requires

$$g'_{\mu\nu}dx'^{\mu}dx'^{\nu} = g_{\rho\sigma}dx^{\rho}dx^{\sigma}. \tag{B.27}$$

In terms of the $\Lambda^{\mu}{}_{\nu}$, this becomes

$$g'_{\mu\nu}\Lambda^{\mu}{}_{\rho}\Lambda^{\nu}{}_{\sigma}dx^{\rho}dx^{\sigma} = g_{\rho\sigma}dx^{\rho}dx^{\sigma}, \tag{B.28}$$

so that

$$(\Lambda^{t})^{\mu}_{\rho}g'_{\mu\nu}\Lambda^{\nu}_{\sigma} = g_{\rho\sigma} \tag{B.29}$$

or, in terms of the matrices themselves,

$$\Lambda^{t}g'\Lambda = g, \tag{B.30}$$

where Λ^{t} denotes the transpose of Λ. If both coordinate systems employ rectangular Cartesian axes, then their metric tensor coefficients are the same, $\mathrm{diag}(1, -1, -1, -1)$. Evaluating the determinant of eq. (B.30), and noting that $|g'| = |g| = -1$ and that a matrix and its transpose have the same determinant, we find that the Lorentz transformation matrices respect the constraint $|\Lambda|^2 = 1$, or

$$|\Lambda| = \pm 1. \tag{B.31}$$

Let's break $\{\Lambda^{\mu}{}_{\nu}\}$ up into the time-time component $\Lambda^{0}{}_{0}$; the time-space components $\Lambda^{0}{}_{k}(k = 1, 2, 3)$, the latter forming a column vector $\boldsymbol{\lambda}$ (with $\boldsymbol{\lambda}^{t}$

the corresponding row vector); and the space-space components forming a 3×3 rotation matrix R. Therefore we may write

$$\{\Lambda^{\mu}{}_{\nu}\} = \begin{pmatrix} \Lambda^0{}_0 & \boldsymbol{\lambda}^t \\ \boldsymbol{\lambda} & R \end{pmatrix}. \tag{B.32}$$

In terms of R and $\boldsymbol{\lambda}$, the determinant of Λ may be written

$$|\Lambda| = \Lambda^0{}_0|R| - |\boldsymbol{\lambda}|^2 = \pm 1. \tag{B.33}$$

For zero relative velocity between the two frames, $\boldsymbol{\lambda} = \mathbf{0}$. For a rotation of the xy axes clockwise through the angle θ about the z-axis to form the new $x'y'z'$ axes, R becomes

$$R(\theta) = \begin{pmatrix} \cos\theta & \sin\theta & 0 \\ -\sin\theta & \cos\theta & 0 \\ 0 & 0 & 1 \end{pmatrix} \tag{B.34}$$

for which $|R| = +1$. For an inversion of axes, $x' = -x, y' = -y$, and $z' = -z$, so that

$$R_{\text{inversion}} = \begin{pmatrix} -1 & 0 & 0 \\ 0 & -1 & 0 \\ 0 & 0 & -1 \end{pmatrix} \tag{B.35}$$

and in this case $|R| = -1$.

If we return to eq. (B.29) and set $\rho = \sigma = 0$, it gives

$$(\Lambda^0{}_0)^2 = 1 + (\Lambda^1{}_0)^2 + (\Lambda^2{}_0)^2 + (\Lambda^3{}_0)^2 \geq 1, \tag{B.36}$$

so either

$$\Lambda^0{}_0 \geq 1 \tag{B.37}$$

or

$$\Lambda^0{}_0 \leq -1. \tag{B.38}$$

Equation (B.31) and inequalities (B.37) and (B.38) together offer four possibilities: $|\Lambda| = \pm 1$; and $\Lambda^0{}_0 \geq 1$ or $\Lambda^0{}_0 \leq 1$. Consider the four possibilities separately:

a. If $|\Lambda| = +1$ with $\Lambda^0{}_0 = 1$, then from equation (B.33), $|R| = 1$, a rotation.

b. If $|\Lambda| = +1$ with $\Lambda^0{}_0 = -1$, then $|R| = -1$, a combined time reversal and space inversion.

c. If $|\Lambda| = -1$ with $\Lambda^0{}_0 = 1$, then $|R| = -1$, which describes a space inversion.

d. If $|\Lambda| = -1$ with $\Lambda^0{}_0 = -1$, then $|R| = +1$, which describes a time reversal, the exchange of initial and final states.

Space inversions and time reversals are not continuous symmetries. Without a parameter ε for them that can be increased infinitesimally from 0, the conditions of Noether's theorem do not apply. But these discrete transformations do have physical meaning. Processes exist whose spatially inverted counterparts are not identical (e.g., parity violations in beta decay), and processes exist that do not run in reverse (e.g., processes that increase entropy go only one way spontaneously).

Transformations with $|\Lambda| = -1$ are called improper Lorentz transformations. One can also have improper Lorentz transformations even if $|\Lambda| = +1$, when with no boosts $\Lambda^0{}_0 = -1$ and $|R| = -1$ (a time reversal and space inversion, as noted above). But we as Noether theorem enthusiasts concentrate on the proper Lorentz transformations for which $|\Lambda| = +1$ and $\Lambda^0{}_0 \geq 1$. Only proper Lorentz transformations admit continuous transformations derived smoothly from an identity transformation, and fit into the conditions of Noether's theorem.

Appendix C

Equations of Motion in Quantum Mechanics

In quantum theory, making a laboratory measurement on a system corresponds, in the mathematical representation of the event, to an operator operating on the system's quantum state. How does that work in nonrelativistic quantum theory? It implements a calculus of operators beginning with the relation between a free particle's Hamiltonian and momentum,

$$\frac{p^2}{2m} = H. \tag{C.1}$$

Multiply this from the right with the quantum wave function $\Psi(t, x)$, and use the quantum mechanical "first quantization" algorithms

$$H \to -\frac{\hbar}{i} \frac{\partial}{\partial t}, \tag{C.2}$$

$$\mathbf{p} \to \frac{\hbar}{i} \nabla \tag{C.3}$$

to obtain the free-particle, time-dependent Schrödinger equation,

$$-\frac{\hbar^2}{2m} \nabla^2 \Psi = -\frac{\hbar}{i} \frac{\partial \Psi}{\partial t}. \tag{C.4}$$

So far so good, but this cannot be the final story, because special relativity insists that space and time are to be considered equal partners in spacetime; they should be treated "the same" (other than the relative minus sign in the metric). However, the Schrödinger equation deals with space and time asymmetrically, taking the second derivative of Ψ with respect to spatial coordinates, but the first derivative with respect to time. The situation

286

can be redeemed in two ways, depending on how we choose to combine the first quantization algorithms of eqs. (C.2) and (C.3) with the relativistic energy-momentum relation $E^2 - (pc)^2 = (mc^2)^2$.[1] One way, applicable to bosons, takes second derivatives of Ψ with respect to both space and time coordinates. The other way, which describes fermions, takes first derivates of Ψ with respect to both space and time.

So let's return to the relativistic energy-momentum relation of eq. (B.13), $E^2 - p^2 = m^2$ (in units where $c = 1$), that applies to a free particle of mass m. Multiply eq. (B.13) from the right by the wave function, here denoted φ, and use the first quantization algorithms of eqs. (C.2) and (C.3). Thereby we obtain the Klein-Gordon equation,[2]

$$\hbar^2(\nabla^2 - \partial_t^2)\varphi = m^2\varphi, \qquad (C.5)$$

which is applicable to a free particle of integer spin. In terms of the four-dimensional gradient in the spacetime of special relativity, this may be written (upon absorbing the $\hbar$):

$$\partial^\mu \partial_\mu \varphi = m^2 \varphi. \qquad (C.6)$$

Because $\partial^\mu \partial_\mu$ and m^2 are already Lorentz scalars, the transformation properties of the Klein-Gordon equation will be those of φ.

If we include interactions through some potential energy function U, then the inhomogeneous Klein-Gordon equation results:

$$\partial_\mu \partial^\mu \varphi - m^2\varphi = -\frac{\partial U}{\partial \varphi}. \qquad (C.7)$$

Notice that $E^2 - p^2 \neq m^2$ for an interacting particle.[3]

For applications of Noether's theorem, we are interested in the Klein-Gordon equation as an Euler-Lagrange equation. If the quanta are uncharged and φ is a real scalar field, the free-particle Lagrangian density (again, with $\hbar = 1$) is[4]

$$\mathcal{L}(\varphi, \partial_\mu \varphi) = \tfrac{1}{2}(\partial_\mu \varphi)(\partial^\mu \varphi) - \tfrac{1}{2}m^2\varphi^2. \qquad (C.8)$$

---

[1]We assume—because we don't know what else to do—that the first quantization algorithms of eqs. (C.2) and (C.3) hold in both Newtonian relativity and special relativity. Furthermore, in going from the former to the latter, we have no better assumption for a replacement, nor any compelling reason to do so. So we make our assumption, work out the consequences, and let the data speak for itself.

[2]Oskar Klein (1894–1977) and Walter Gordon (1893–1939); the Klein-Gordon equation dates from 1926.

[3]One definition of a "virtual" particle is a particle that interacts, so goes off the "mass-shell," in other words, $E^2 - p^2 \neq m^2$.

[4]In the spacetime of special relativity, where the metric tensor components $g_{\mu\nu}$ are not dynamic, it is not necessary to write both $\partial_\mu \varphi$ and $\partial^\mu \varphi$ in the arguments of the Lagrangian density $\mathcal{L}$, because $(\partial_\mu \varphi)(\partial^\mu \varphi)$ can also be written $g^{\mu\nu}(\partial_\mu \varphi)(\partial_\nu \varphi)$.

But if the particles carry a value of a charge from a set of possibilities, such as $\pm e$, then the Lagrangian density will be the complex-valued quantity (denoting $\partial_\mu \varphi = \varphi_\mu, \partial^\mu \varphi = \varphi^\mu$)

$$\mathcal{L}(\varphi, \varphi^*, \varphi_\mu, \varphi_\mu^*) = \varphi_\mu^* \varphi^\mu - m^2 \varphi^* \varphi. \tag{C.9}$$

This Lagrangian density yields two Klein-Gordon equations, one for φ and another for φ^*. Requiring the functional to be an extremal with respect to φ^* yields the Euler-Lagrange equation

$$\frac{\partial \mathcal{L}}{\partial \varphi^*} = \partial_\mu \left(\frac{\partial \mathcal{L}}{\partial \varphi_\mu^*} \right), \tag{C.10}$$

which produces the homogeneous Klein-Gordon equation for φ as given above. The Euler-Lagrange equation for φ gives the Klein-Gordon equation for φ^*,

$$\partial_\mu \partial^\mu \varphi^* = m^2 \varphi^*, \tag{C.11}$$

the complex conjugate of the Klein-Gordon equation for φ.

The other way to treat time and space equally emerges when the *sum of squares* in the Klein-Gordon equation is factored into the *square of a sum*. This results in the relativistic quantum wave equation for fermions, the Dirac equation.[5] To see how this comes about, introduce a four-vector with components γ^μ to be determined:

$$\{\gamma^\mu\} = (\gamma^0, \gamma^1, \gamma^2, \gamma^3) = (\gamma^0, \boldsymbol{\gamma}), \tag{C.12}$$

and consider what happens when multiplying out the left-hand side of the following postulated expression:

$$(\boldsymbol{\nabla} \cdot \boldsymbol{\gamma} + \gamma^0 \partial_t)^2 \psi = m^2 \psi \tag{C.13}$$

where ψ is the quantum wave function. Suppose we require the quantities γ^μ to be such that, when the square is multiplied out, eq. (C.13) recovers the free-particle Klein-Gordon equation (C.6) for ψ. The γ's must satisfy

$$\gamma^\mu \gamma^\nu + \gamma^\nu \gamma^\mu = 2g^{\mu\nu}. \tag{C.14}$$

Clearly these noncommuting γ^μ cannot be ordinary complex numbers. They offer an example of *hypercomplex numbers*.[6] Because matrix multiplication does not commute in general, these hypercomplex γ-numbers may be represented by matrices with complex entries. Relativistic quantum mechanics textbooks show how the smallest dimensionality of these "Dirac

---

[5] Paul A. M. Dirac (1902–1984) The Dirac equation dates from 1928.
[6] Emmy Noether developed much of the algebra of the hypercomplex numbers.

matrices" is 4×4. A popular representation [Bjorken & Drell (1964)] of the Dirac matrices looks like this:

$$\gamma^0 = \begin{pmatrix} 1 & 0 \\ 0 & -1 \end{pmatrix}, \quad \gamma^k = \begin{pmatrix} 0 & \sigma^k \\ -\sigma^k & 0 \end{pmatrix} \quad (C.15)$$

where $k = 1, 2, 3$, and in the matrix elements 1 denotes the 2×2 unit matrix, 0 the 2×2 zero matrix, and the σ^k are the Pauli matrices (chapter 7). This factorization of the Klein-Gordon equation allows as one possibility (with $\hbar$ absorbed into a redefinition of the mass, $m/\hbar \to m$)

$$i(\boldsymbol{\nabla} \cdot \boldsymbol{\gamma} + \gamma^0 \partial_t)\psi = m\psi, \quad (C.16)$$

the Dirac equation for a noninteracting fermion. In summation notation, the Dirac equation for the free particle reads

$$i\gamma^\mu \partial_\mu \psi = m\psi. \quad (C.17)$$

When one takes the nonrelativistic limit of eq. (C.17), the Schrödinger equation emerges with the Pauli matrices for spin built in. The Dirac equation therefore intrinsically describes particles of spin-$\frac{1}{2}$, fermions such as the electron. Since the Dirac γ-matrices are 4×4, the Dirac wave function ψ is a four-component column vector called a "spinor."[7] The top two components keep track of the spin-up and spin-down quantum numbers for (say) the electron, and the bottom two entries keep track of the spin for its antiparticle (the positron).

The three space component gamma matrices $\boldsymbol{\gamma} = (\gamma^1, \gamma^2, \gamma^3)$ correspond in the nonrelativistic limit (after some manipulation) to the particle's velocity. An electric current density $\mathbf{j}$ may be written in terms of the particle velocity $\mathbf{v}$ as $\rho\mathbf{v}$, where ρ denotes the charge density. In quantum mechanics, the probability density for locating a particle is given in terms of the wave function by $\psi^*\psi$. For a particle carrying some charge q, the charge density, we expect, will be given by $q\psi^*\psi$. This idea is basically correct, but in the case of spin-$\frac{1}{2}$ fermions, other factors besides $\mathbf{v}$ are needed to go between ψ^* and ψ, because of the spinor nature of these wave functions. The probability density and probability current density for a Dirac fermion are $\bar{\psi}\psi$, a scalar, and $\bar{\psi}\gamma^\mu\psi$, a four-vector, where $\bar{\psi} \equiv \psi^\dagger \gamma^0$.

The Lagrangian density for noninteracting Dirac fermions (again with $\hbar$ absorbed into m) is

$$\mathcal{L} = \bar{\psi}(\gamma^\mu i\partial_\mu - m)\psi. \quad (C.18)$$

---

[7]Thanks to the built-in spin, a spinor has Lorentz transformation properties that differ from a vector in spacetime, even though both have four components [Bjorken & Drell (1964)].

The Euler-Lagrange equation for $\bar{\psi}$,

$$\frac{\partial \mathcal{L}}{\partial \bar{\psi}} = \partial_\mu \left(\frac{\partial \mathcal{L}}{\partial (\partial_\mu \bar{\psi})} \right) \tag{C.19}$$

gives the Dirac equation for ψ, eq. (C.17).

When we consider improper Lorentz transformations that allow the coordinate axes to be inverted, we must also consider pseudoscalars and pseudovectors (also called axial vectors). Textbooks describing the weak interaction [Bjorken and Drell (1964) 26] show that the quantity $\gamma_5 \equiv i\gamma^0\gamma^1\gamma^2\gamma^3$ is a pseudoscalar, that is, it maintains invariance under a proper Lorentz transformation, but changes sign under a coordinate inversion. Therefore the four-vector made of fermion wave functions with Dirac matrices γ^μ can be made into an axial vector by replacing γ^μ with $\gamma^\mu\gamma_5$. In the theory of the weak interaction, the current j^μ includes both vector and axial vector terms:

$$j^\mu_{weak} = iG(\alpha\bar{\psi}\gamma^\mu\psi + \beta\bar{\psi}\gamma^\mu\gamma_5\psi) \tag{C.20}$$

where G is a weak interaction coupling constant and α, β are constants. If $\beta = 0$ then the weak interaction conserves parity; if $\beta \neq 0$ then parity conservation is violated. The data shows that $\alpha = \beta$; not only is parity conservation violated, it is maximally violated. Normalized so that $|\alpha|^2 + |\beta|^2 = 1$, the weak current becomes

$$j^\mu_{weak} = i\frac{G}{\sqrt{2}}\bar{\psi}(1 + \gamma_5)\psi. \tag{C.21}$$

Appendix D

Conjugate Variables and Legendre Transformations

The Lagrangian and Hamiltonian are related to each other by a Legendre transformation. Let's review what goes on in a generic Legendre transformation,[1] and why we use them.[2]

Consider a function W of two independent variables x and y, $W = W(x, y)$. Evaluate its differential:

$$dW = \left(\frac{\partial W}{\partial x} \right)_y dx + \left(\frac{\partial W}{\partial y} \right)_x dy \tag{D.1}$$

where the subscripts redundantly remind us which variable is held constant when evaluating a partial derivative. Let

$$P \equiv \left(\frac{\partial W}{\partial x} \right)_y \tag{D.2}$$

and

$$Q \equiv \left(\frac{\partial W}{\partial y} \right)_x . \tag{D.3}$$

Therefore, eq. (D.1) may be written

$$dW = P dx + Q dy. \tag{D.4}$$

---

[1]I gratefully acknowledge the late Ashley Carter of Drew University. His statistical mechanics textbook [Carter (2001)] contains an unusually lucid presentation of Legendre transformations, and was the inspiration for this appendix.

[2]Not only are Legendre transformations important in Hamilton-Lagrangian dynamics, but they are abundantly used in thermodynamics, to relate thermodynamic energy functions such as internal energy, Helmholtz energy, Gibbs energy, and enthalpy.

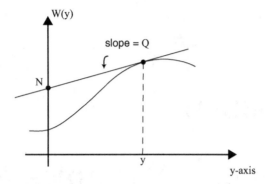

Figure D.1: *The curve $W = W(y)$ and one of its tangent lines, described by the equation $N = W - Qy$.*

P is said to be canonically conjugate to x, and Q conjugate to y. If we picture $W(x, y)$ as a surface over the xy plane, then P and Q measure the slope of this surface's local tangent line above the point (x, y), P with respect to the x-axis, and Q with respect to the y-axis, respectively. The coordinates x and y, and their respective slopes or "momenta" P and Q, come in pairs, like mathematically married couples. Our example has two such couples, (x, P) and (y, Q).

Temporarily suppress the x-dimension (with its P) by holding x fixed (say at $x = 0$), and consider W as a function of y. $W(y)$ describes a curve above the y-axis, as in figure D.1. At some value of y, the slope of the local tangent line has the value Q. Extend this tangent line to cross the W-axis, where the intercept is N. The equation of this tangent line is

$$Q = \frac{W - N}{y - 0},\tag{D.5}$$

which transposed becomes the standard form of a Legendre transformation,

$$N = W - Qy.\tag{D.6}$$

The differential of N yields

$$dN = dW - Qdy - ydQ.\tag{D.7}$$

Using eq. (D.4) for dW (with variable x restored), the Qdy cancels, leaving

$$dN = Pdx - ydQ.\tag{D.8}$$

But if $N = N(x, Q)$, then its differential says

$$dN = \left(\frac{\partial N}{\partial x}\right)_Q dx + \left(\frac{\partial N}{\partial Q}\right)_x dQ.\tag{D.9}$$

Even though N is a function of x and Q, P and y can still be extracted from it, because comparison of eq. (D.8) with (D.9) requires

$$P = \left(\frac{\partial N}{\partial x}\right)_Q \tag{D.10}$$

and

$$y = -\left(\frac{\partial N}{\partial Q}\right)_x . \tag{D.11}$$

We now have two ways to describe the information carried in the W-surface above the xy plane: we can either use $W(x, y)$ itself, or we can trade one of the coordinates for its conjugate slope, as we just did with $N(x, Q)$ where we swapped y for its "momentum" Q. Such trades can be done in a variety of ways that preserve the "x-ness" and "y-ness" in the dependent functions W or N, because we always have a function of one element from each pair of coordinates and its conjugate slope. Our options will therefore be function of $(x, y), (x, Q), (P, y)$, or (P, Q).

The functions of these variables considered so far are $W(x, y)$ and $N(x, Q) = W - Qy$. Functions of the other two parings can be derived by further Legendre transformations, $R \equiv W - Px$ and $S \equiv W - Px - Qy$. Together, these four Legendre pairings grant us the luxury of describing a function's x-ness and y-ness in a variety of ways, by selecting the combination of coordinates and slopes most convenient for the data at hand.

As an illustration, consider the combined first and second laws of thermodynamics. Under a reversible process, the change in a system's internal energy U is related to its change in entropy S and volume V according to

$$dU = TdS - PdV \tag{D.12}$$

where T denotes temperature and P the pressure. Compared to the differential of $U(S, V)$,

$$dU = \left(\frac{\partial U}{\partial S}\right)_V dS + \left(\frac{\partial U}{\partial V}\right)_S dV, \tag{D.13}$$

we see that

$$T = \left(\frac{\partial U}{\partial S}\right)_V \tag{D.14}$$

and

$$P = -\left(\frac{\partial U}{\partial V}\right)_S . \tag{D.15}$$

While eq. (D.12) is fundamental to thermodynamics, perhaps from the available data it would be better to know how the system's energy changes, not with respect to entropy, but with respect to temperature. No problem! Merely perform a Legendre transformation to swap out T for S. Since $U = U(S, V)$, the transformation that will do the trick is

$$U(S, V) \rightarrow U - S\frac{\partial U}{\partial S}$$

$$= U - TS \tag{D.16}$$

$$\equiv F(T, V)$$

where eq. (D.14) has been used. It is easily confirmed that F, the Helmholtz energy,[3] is a function of T and V by evaluating its differential and recalling eq. (D.13):

$$dF = d(U - TS)$$

$$= dU - TdS - SdT$$

$$= TdS - PdV - TdS - SdT \tag{D.17}$$

$$= -SdT - PdV.$$

Even though F is a function of temperature and volume, it still contains information about entropy and pressure because

$$S = -\left(\frac{\partial F}{\partial T}\right)_V \tag{D.18}$$

along with

$$P = -\left(\frac{\partial F}{\partial V}\right)_T. \tag{D.19}$$

The Hamiltonian is defined formally as a Legendre transformation that swaps out coordinate velocities for their canonical momenta:

$$H(t, q^\mu, p_\mu) \equiv \dot{q}^\mu p_\mu - L(t, q^\mu, \dot{q}^\mu) \tag{D.20}$$

where

$$p_\mu \equiv \frac{\partial L}{\partial \dot{q}^\mu}. \tag{D.21}$$

Notice how the partners in a conjugate pair appear in an inverse relation throughout all these Legendre transformations. For example, with the conjugate pair x and P in our generic discussion above, $P \sim \partial/\partial x$. We see this inverse relationship echoed repeatedly in thermodynamics and Lagrangian-Hamiltonian dynamics, and carried over into quantum mechanics.

---

[3]The Helmholtz energy is denoted A by some authors.

Appendix E

The Jacobian

Consider a closed curve C that forms the boundary of a surface S that has area A. Lay out this area on a set of xy axes and consider what must be done when mapping that same area onto another coordinate system x' and y' (Figure E.1.).

How do we calculate the area A by integration using the xy axes? Find the points $x = a$ and $y = b$ where the tangent lines to C are vertical. Let us call the curve above those points of tangency $y_u(x)$ (u for "upper"), and for the curve below let us say $y_l(x)$ (l for "lower"). Therefore, the area A follows from

$$
\begin{aligned}
A &= \int_a^b y_u \, dx - \int_a^b y_l \, dx \\
&= -\int_b^a y_u \, dx - \int_a^b y_l \, dx \\
&= -\oint_C y \, dx.
\end{aligned}
\tag{E.1}
$$

By the same reasoning, using the $x'y'$ axes we obtain

$$
A' = -\oint_C y' \, dx'.
\tag{E.2}
$$

Suppose the $x'y'$ axes are related to the xy axes by the transformations

$$
x' = f(x, y)
\tag{E.3}
$$

$$
y' = g(x, y).
\tag{E.4}
$$

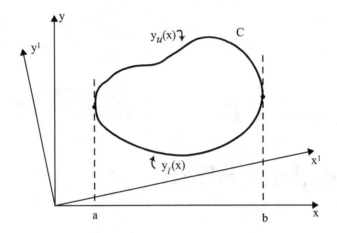

Figure E.1: *The area and coordinates used to illustrate the Jacobian.*

When placed in eq. (E.2) these transformations become

$$A' = - \oint_C g \left(\frac{\partial f}{\partial x} \, dx + \frac{\partial f}{\partial y} \, dy \right)$$

$$\equiv - \oint_C (E_x \, dx + E_y \, dy) \tag{E.5}$$

where

$$E_x \equiv g \left(\frac{\partial f}{\partial x} \right) \tag{E.6}$$

and

$$E_y \equiv g \left(\frac{\partial f}{\partial y} \right). \tag{E.7}$$

Now recall Stokes' theorem,

$$\oint_C \mathbf{E} \cdot d\mathbf{r} = \iint_S (\nabla \times \mathbf{E}) \cdot \hat{\mathbf{n}} \, da, \tag{E.8}$$

where S is the surface bounded by the closed path C and $\hat{\mathbf{n}}$ denotes the unit vector normal to the infinitesimal patch of area da, with the direction of $\hat{\mathbf{n}}$ given by a right-hand rule relative to the sense in which the circuit of C is traversed. For our curve in the xy plane, Stoke's theorem becomes

$$\oint_C (E_x dx + E_y dy) = \iint_{A'} \left(\frac{\partial E_y}{\partial x} - \frac{\partial E_x}{\partial y} \right) \, dx dy. \tag{E.9}$$

Denoting the partial derivatives of f and g with subscripts, we may write eq. (E.5) as

$$A' = -\iint_{A'} \left(\frac{\partial(gf_y)}{\partial x} - \frac{\partial(gf_x)}{\partial y} \right) \, dxdy$$

$$= -\iint_{A'} (g_x f_y - g_y f_x) \, dxdy \qquad \text{(E.10)}$$

$$= \iint_{A'} \left(\frac{\partial y'}{\partial x} \frac{\partial x'}{\partial y} - \frac{\partial y'}{\partial y} \frac{\partial x'}{\partial x} \right) \, dxdy.$$

The Jacobian J of the transformation from (x, y) to (x', y') is by definition the determinant of the transformation coefficients:

$$J \equiv \begin{vmatrix} \dfrac{\partial x'}{\partial x} & \dfrac{\partial x'}{\partial y} \\[2ex] \dfrac{\partial y'}{\partial x} & \dfrac{\partial y'}{\partial y} \end{vmatrix}. \qquad \text{(E.11)}$$

Therefore,

$$A' = \iint_{A'} J \, dxdy. \qquad \text{(E.12)}$$

But we can also write

$$A' = \iint_{A'} dx'dy'. \qquad \text{(E.13)}$$

Comparing the two expressions for A' shows that

$$dx'dy' = J \, dxdy. \qquad \text{(E.14)}$$

Let us relate the Jacobian to the metric tensor with components $g_{\mu\nu}$. As a rank-2 tensor with lower indices, it transforms according to

$$g'_{\mu\nu} = \frac{\partial x^\rho}{\partial x'^\mu} \frac{\partial x^\sigma}{\partial x'^\nu} g_{\rho\sigma}. \qquad \text{(E.15)}$$

Now take the determinant of both sides:

$$|g'| = \frac{|g|}{J^2} \qquad \text{(E.16)}$$

or

$$J = \sqrt{\frac{|g|}{|g'|}}. \qquad \text{(E.17)}$$

Therefore, eq. (E.14) may be written

$$\sqrt{|g'|}dx'dy' = \sqrt{|g|}dxdy \qquad \text{(E.18)}$$

so that $\sqrt{|g|}\,dx\,dy$ is an invariant area.

The argument generalizes to higher-dimensional spaces: the invariant volume element in a space of N dimensions,[1]

$$\sqrt{|g'|}\,dx'^1\,dx'^2 \cdots dx'^N = \sqrt{|g|}\,dx^1\,dx^2 \cdots dx^N. \qquad \text{(E.19)}$$

If the determinant $|g|$ happens to be negative (as in the spacetime of special relativity), then $-|g|$ must appear under the square root of the invariant volume element, or equivalently, the absolute value $||g||$ of the determinant.

---

[1]This applies in Riemannian or pseudo-Riemannian geometries, that is, the metric tensor components may have either sign.

Appendix F

The Covariant Derivative

A vector can be written as a superposition over a set of linearly independent basis vectors $\{\hat{\mathbf{e}}_\mu\}$,

$$\mathbf{V} = V^\mu \hat{\mathbf{e}}_\mu. \tag{F.1}$$

Therefore when we evaluate the derivative of $\mathbf{V}$, we must take into account the rate of change of the basis vectors:

$$\partial_\nu \mathbf{V} = (\partial_\nu V^\mu)\hat{\mathbf{e}}_\mu + V^\mu(\partial_\nu \hat{\mathbf{e}}_\mu). \tag{F.2}$$

For example, in plane polar coordinates (r, θ) with unit basis vectors $\hat{\mathbf{r}}$ and $\hat{\boldsymbol{\theta}}$, a particle's location relative to the origin is $\mathbf{r} = r\hat{\mathbf{r}}$, and its velocity (overdots mean time derivative) is[1]

$$\dot{\mathbf{r}} = \dot{r}\hat{\mathbf{r}} + r\frac{d\hat{\mathbf{r}}}{dt}$$

$$= \dot{r}\hat{\mathbf{r}} + r\dot{\theta}\hat{\boldsymbol{\theta}}.$$

In general, because the rate of change of a basis vector is a superposition of all the basis vectors, we may write

$$\partial_\nu \hat{\mathbf{e}}_\mu \equiv \Gamma^\rho{}_{\nu\mu} \hat{\mathbf{e}}_\rho \tag{F.3}$$

where the coefficient $\Gamma^\rho{}_{\mu\nu}$ is called the "affine connection," or a "Christoffel symbol."[2] In our plane polar coordinate example, the superposition of affine

---

[1] For a geometrical construction of such derivatives see Marion (1970), 30.

[2] After Elwin Bruno Christoffel (1829–1900), whose work in differential geometry laid some groundwork for tensor calculus.

299

connection coefficients becomes

$$\frac{d\hat{\mathbf{r}}}{dt} = \Gamma^{\rho}{}_{tr}\hat{\mathbf{e}}_{\rho}$$

$$= \Gamma^{r}{}_{tr}\hat{\mathbf{r}} + \Gamma^{\theta}{}_{tr}\hat{\boldsymbol{\theta}} \qquad\qquad (F.4)$$

$$= 0\hat{\mathbf{r}} + \dot{\theta}\hat{\boldsymbol{\theta}}.$$

By dotting eq. (F.3) with a dual basis vector $\hat{\mathbf{e}}^{\sigma}$, where by definition of the dual basis $\hat{\mathbf{e}}^{\sigma} \cdot \hat{\mathbf{e}}_{\rho} = \delta^{\sigma}{}_{\rho}$, the affine connection can also be written

$$\Gamma^{\sigma}{}_{\nu\mu} = \hat{\mathbf{e}}^{\sigma} \cdot (\partial_{\nu}\hat{\mathbf{e}}_{\mu}). \qquad\qquad (F.5)$$

In all spaces that concern us in this book, the affine connection coefficients are symmetric in the two lower indices,[3]

$$\Gamma^{\rho}{}_{\mu\nu} = \Gamma^{\rho}{}_{\nu\mu}, \qquad\qquad (F.6)$$

in which case the affine connection coefficients can be calculated from the metric tensor components $g_{\mu\nu} = g_{\nu\mu}$ [Weinberg (1972) 75; Neuenschwander (2015) 108],

$$\Gamma^{\rho}{}_{\mu\nu} = \frac{1}{2}g^{\rho\sigma}[\partial_{\mu}g_{\nu\sigma} + \partial_{\nu}g_{\nu\sigma} - \partial_{\sigma}g_{\mu\nu}]. \qquad\qquad (F.7)$$

Returning to eq. (F.2), in terms of the affine connection coefficients,

$$\partial_{\nu}\mathbf{V} = (\partial_{\nu}V^{\mu})\hat{\mathbf{e}}_{\mu} + V^{\mu}\Gamma^{\rho}{}_{\mu\nu}\hat{\mathbf{e}}_{\rho}. \qquad\qquad (F.8)$$

With some relabeling of the repeated dummy indices, the basis vectors can be factored out:

$$\partial_{\nu}\mathbf{V} = (\partial_{\nu}V^{\mu} + \Gamma^{\mu}{}_{\rho\nu}V^{\rho})\,\hat{\mathbf{e}}_{\mu}. \qquad\qquad (F.9)$$

The quantity in the parentheses is called the "covariant derivative" of the vector component V^{μ} with respect to coordinate x^{ν}, and is denoted $\nabla_{\nu}V^{\mu}$:

$$\nabla_{\nu}V^{\mu} \equiv \partial_{\nu}V^{\mu} + \Gamma^{\mu}{}_{\rho\nu}V^{\rho}. \qquad\qquad (F.10)$$

In terms of the covariant derivatives of the components, the derivative of the full vector may be compactly written

$$\partial_{\nu}\mathbf{V} = (\nabla_{\nu}V^{\mu})\hat{\mathbf{e}}_{\mu}. \qquad\qquad (F.11)$$

The covariant derivative of vector components with lower indices (see appendix A) is

$$\nabla_{\nu}V_{\mu} \equiv \partial_{\nu}V_{\mu} - \Gamma^{\rho}{}_{\mu\nu}V_{\rho}. \qquad\qquad (F.12)$$

---

[3]There exist theories with "torsion" where $\Gamma^{\rho}{}_{\mu\nu} \neq \Gamma^{\rho}{}_{\nu\mu}$, but these are not needed in most physics applications.

The affine connection coefficients are nonzero whenever the set of unit basis vectors change direction from one location to another within the space, as $\hat{\mathbf{r}}$ and $\hat{\boldsymbol{\theta}}$ do in polar coordinates.

A scalar λ depends on no coordinate system or basis vectors, so the covariant derivative of a scalar reduces to the same as the usual partial derivative, $\nabla_\nu \lambda = \partial_\nu \lambda$. Setting $\nu = \mu$ in $\nabla_\nu V^\mu$ produces the covariant divergence:

$$\nabla_\mu V^\mu = \partial_\mu V^\mu + \Gamma^\mu{}_{\rho\mu} V^\rho \tag{F.13}$$

which may also be written

$$\nabla_\mu V^\mu = \frac{1}{\sqrt{||g||}} \partial_\mu (V^\mu \sqrt{||g||}) \tag{F.14}$$

where $|g|$ denotes the determinant of the metric tensor and $||g||$ the determinant's absolute value.[4] A covariant derivative version of Gauss's divergence theorem can be derived [Hobson et al. (2006) 532]:

$$\int_{\mathcal{R}} (\nabla_\nu V^\nu) \sqrt{||g||} d^N x = \oint_{\mathcal{S}} (n_\nu V^\nu) \sqrt{||\gamma||} d^{N-1} x \tag{F.15}$$

where the surface $\mathcal{S}$ forms the boundary of the volume $\mathcal{R}$ in N dimensions, and γ denotes the metric tensor for a coordinate system on the surface $\mathcal{S}$.[5] The proof uses the same argument that demonstrates the ordinary divergence theorem, provided that in expressions such as $[S_z(z + dz) - S_z(z)]dxdy = (\Delta S_z)dxdy$, one writes ΔS_z as $(\nabla_z S_z)dz$. Setting $V^\mu = \nabla^\mu \varphi$ for a scalar function φ produces the covariant Laplacian,

$$\nabla_\mu \nabla^\mu \varphi = \frac{1}{\sqrt{||g||}} \partial_\mu (\sqrt{||g||} \partial^\mu \varphi). \tag{F.16}$$

In the covariant curl, the affine connection coefficients cancel,

$$(\mathbf{curl\ V})_{\mu\nu} \equiv \nabla_\mu V_\nu - \nabla_\nu V_\mu = \partial_\mu V_\nu - \partial_\nu V_\mu. \tag{F.17}$$

When you work out the covariant derivatives of the components of a rank-2 tensor, you find, for example,

$$\nabla_\rho T^{\mu\nu} = \partial_\rho T^{\mu\nu} + \Gamma^\mu{}_{\rho\sigma} T^{\sigma\nu} + \Gamma^\nu{}_{\rho\sigma} T^{\mu\sigma}. \tag{F.18}$$

Significantly, the covariant derivative of the metric tensor vanishes identically, $\nabla_\rho g_{\mu\nu} = 0$.

---

[4]Sometimes, $|g| < 0$, as in flat spacetime with metric $diag(1, -1, -1, -1)$, so that $|g| = -1$.

[5]The invariant volume element $\sqrt{||g||} d^N x$ has been used (see appendix E).

The covariant derivative of components of a third-rank tensor with two upstairs and one downstairs indices works out to be

$$\nabla_\rho S^{\mu\nu}{}_\sigma = \partial_\rho S^{\mu\nu}{}_\sigma + \Gamma^\mu{}_{\alpha\rho} S^{\alpha\nu}{}_\sigma + \Gamma^\nu{}_{\alpha\rho} S^{\mu\alpha}{}_\sigma - \Gamma^\alpha{}_{\sigma\rho} S^{\mu\nu}{}_\alpha, \qquad \text{(F.19)}$$

and so on, with a $+\Gamma$ term for each upstairs index, and a $-\Gamma$ term for each downstairs index.

One can show (see ex. 8.3) that the covariant derivative respects the usual rule for differentiating a product:

$$\nabla_\rho(T^\mu R^\nu) = (\nabla_\rho T^\mu)R^\nu + T^\mu(\nabla_\rho R^\nu). \qquad \text{(F.20)}$$

The affine connections coefficients are not rank-3 tensors, because they transform from one coordinate system to another according to

$$\Gamma'^\mu{}_{\alpha\beta} = \frac{\partial x'^\mu}{\partial x^\nu} \frac{\partial x^\sigma}{\partial x'^\alpha} \frac{\partial x^\gamma}{\partial x'^\beta} \Gamma^\nu{}_{\sigma\gamma} + \frac{\partial x'^\mu}{\partial x^\nu} \frac{\partial^2 x^\nu}{\partial x'^\alpha \partial x'^\beta}. \qquad \text{(F.21)}$$

The first term on the right-hand side is the transformation rule of a rank-3 tensor, but the second term spoils the tensor transformation rule. However, this is useful, because it turns out that, in curved spaces, the derivative of a tensor is not necessarily another tensor. For instance, the derivative of a vector component with respect to a coordinate transforms according to

$$\frac{\partial V'^\mu}{\partial x'^\nu} = \frac{\partial x'^\mu}{\partial x^\alpha} \frac{\partial x^\beta}{\partial x'^\nu} \frac{\partial V^\alpha}{\partial x^\beta} + \frac{\partial x^\beta}{\partial x'^\nu} \frac{\partial^2 x'^\mu}{\partial x^\beta \partial x^\alpha} V^\alpha. \qquad \text{(F.22)}$$

By adding to the usual derivative a Γ-term to make a covariant derivative, the covariant derivative transforms as a proper tensor:

$$\nabla_{\nu'} V'^\mu = \frac{\partial x'^\mu}{\partial x^\alpha} \frac{\partial x^\beta}{\partial x'^\nu} (\nabla_\beta V^\alpha). \qquad \text{(F.23)}$$

By the principle of general covariance, the laws of mathematics and physics must transcend the choice of reference frame. If written in terms of tensors, the equations can be moved from one coordinate system to another without loss of content or meaning. This is the strategy behind the general theory of relativity. One writes the laws of physics in the local free-fall Minkowskian spacetime, where there is no gravity. We transform them to an arbitrary coordinate system, accelerated relative to the first, where by the principle of the equivalence of gravitational and inertial mass, the acceleration is locally indistinguishable from a gravitational field. Because most physics equations are differential equations, when describing the interaction of matter and radiation with gravitation, one replaces usual derivatives with covariant derivatives. Then derivatives are tensors and transform properly.

A convenient notation for usual partial derivatives employs a comma,

$$\frac{\partial V^\mu}{\partial x^\nu} \equiv \partial_\nu V^\mu \equiv V^\mu{}_{,\nu}. \tag{F.24}$$

When replacing partial derivatives with covariant derivatives, $\partial_\nu \to \nabla_\nu$, an alternative notation changes the comma to a semicolon:

$$V^\mu{}_{,\nu} \to V^\mu{}_{;\nu} \equiv \nabla_\nu V^\mu. \tag{F.25}$$

Bibliography

Abers, E. S. & Lee, B. W. (1973), "Gauge Theories." *Phys. Rep.* 96: 1–141.

Aitchison, I. J. R. & Hey, A. J. G. (1982), *Gauge Theories in Particle Physics*. Bristol: Adam Hilger.

Berry, M. V. (1984), "Quantal Phase Factors Accompanying Adiabatic Changes." Proceedings of the Royal Society A 39, 45–57.

Bjorken, J. D. & Drell, S. D. (1964), *Relativistic Quantum Mechanics*. New York: McGraw-Hill.

Bohm, D. (1951), *Quantum Theory*. New York: Prentice Hall.

Brewer, J. W. & Smith, Martha K., eds. (1982), *Emmy Noether: A Tribute to Her Life and Work*. New York: Marcel Dekker.

Bronowski, J. (1956), *Science and Human Values*. New York: Harper & Row.

Burton, D. M. (2011), *The History of Mathematics*, 7th ed. New York: McGraw Hill.

Byers, N. (1999), "E. Noether's Discovery of the Deep Connection between Symmetries and Conservation Laws." Israel Mathematical Conference Proceedings 12. Bar Ilan University, Tel Aviv, Israel Dec. 2–3, 1996.

Calaprice, A., ed. (1996), *The Quotable Einstein*. Princeton, NJ: Princeton University Press.

Campbell, J. W. (1947), *An Introduction to Mechanics*. London: Pitman.

Carter, A. H. (2001), *Classical and Statistical Thermodynamics*. Upper Saddle River, NJ: Prentice Hall.

Dallen, L. & Neuenschwander, D. E. (2011), "Noether's Theorem in a Rotating Reference Frame." *Am. J. Phys.* 79(3): 326–332.

Dick, A. (1981), *Emmy Noether: 1882-1935*. Translated by H. I. Blocher. Basel, Switzerland: Birkhäuser Verlag.

Dirac, P. A. M. (1947), *The Principles of Quantum Mechanics*, 3rd ed. Oxford, UK: Oxford University Press.

Einstein, A. (1916), "Die grundlage der allgemeinen relativitätstheorie." *Annalen der Physik* 354: 769–822.

Einstein, A. (1974), *The Meaning of Relativity*, 5th ed. Princeton, NJ: Princeton University Press.

Einstein, A., Lorentz, H., Weyl, H. & Minkowski, H. (1952), *The Principle of Relativity*, New York: Dover.

Feynman, R. P. & Hibbs, A. R. (1965), *Quantum Mechanics and Path Integrals*. New York: McGraw-Hill.

Finkelstein, R. J. (1973), *Nonrelativistic Mechanics*. Reading, MA: W.A. Benjamin.

Gelfand, I. M. & Fomin, S. V. (1963), *Calculus of Variations*. Edited and translated by R. A. Silverman. Mineola, NY: Dover.

Georgi, H. (1982), *Lie Algebras in Particle Physics: From Isospin to Unified Theories*. Reading, MA: Benjamin/Cummings.

Georgi, H. & Glashow, S. (1974), "Unity of All Elementary-Particle Forces." *Phys. Rev. Lett.* 32(8): 438.

Gleick, J. (1987), *Chaos: Making a New Science*. New York: Penguin Books.

Godfrey-Smith, P. (2003), *Theory and Reality: An Introduction to the Philosophy of Science*. Chicago: University of Chicago Press.

Goldstein, H. (1965), *Classical Mechanics*. Reading, MA: Addison-Wesley.

Gray, C. G. & Taylor, E. F. (2007), "When Action Is not Least." *Am. J. Phys.* 75(5): 434–438.

Griffiths, D. J. (2005), *Introduction to Quantum Mechanics*, 2nd ed. Upper Saddle River, NJ: Pearson Prentice Hall.

Hecht, E. (2002), *Optics*, 4th ed. San Francisco: Pearson Addison Wesley.

Hobson, M. P., Efstathiou, G., & Lasenby, A. N. (2006), *General Relativity: An Introduction for Physicists*. Cambridge, UK: Cambridge University Press.

Holmes, W. (1994), "Perturbative Solution for Nonlinear Waves on a Stretched String." *J. Undergrad. Res. Phys.* 12(2): 1–5.

Huang, K. (1982), *Quarks, Leptons, and Gauge Fields.* Singapore: World Scientific.

Jackson, J. D. (1975), *Classical Electrodynamics Second Edition.* New York: Wiley.

James, I. (2002), *Remarkable Mathematicians from Euler to von Neumann.* Cambridge, UK: Cambridge University Press.

Jenkins, F. A. & White, H. E. (1950), *Fundamentals of Optics Second Edition.* New York: McGraw-Hill.

Kennedy, R. E. (2012), *A Student's Guide to Einstein's Major Papers.* Oxford, UK: Oxford University Press.

Kobe, D. (1993), "Canonical Transformation to Energy and 'Tempus' in Classical Mechanics." *Am. J. Phys.* 61(11): 1031–1037.

Kosmann-Schwarzbach, Y. (2010), *The Noether Theorems: Invariance and Conservation Laws in the Twentieth Century.* Translated by B. E. Schwarzback. New York: Springer.

Landau, L. D. & Lifshitz, E. M. (1962), *The Classical Theory of Fields.* Reading, MA: Addison-Wesley.

Laugwitz, D. (1965), *Differential and Riemannian Geometry.* New York: Academic Press.

Lindsay, R. B. (1950), *Physical Mechanics.* Princeton, NJ: Van Nostrand.

Logan, J. D. (1974), "Conformal Invariance of Multiple Integrals in the Calculus of Variations." *Journal of Mathematical Analysis and Applications* 48(2): 618–631.

Logan, J. D. (1977), *Invariant Variational Principles.* New York Academic Press.

Marciano, W. & Pagels, H. (1978), "Quantum Chromodynamics." *Phys. Rep.* 36(3): 137–276.

Marion, J. B. (1970), *Classical Dynamics of Particles and Systems.* New York: Academic Press.

Marion, J. B. & Thornton, S. T. (2004), *Classical Dynamics of Particles and Systems Fifth Edition.* Belmont, CA: Thomson Brooks Cole.

Millikan, R. A., Roller, D. & Watson, E. C. (1937), *Mechanics, Molecular Physics, Heat, and Sound.* Boston: Ginn.

Misner, C. W., Thorne, K. S. & Wheeler, J. A. (1971), *Gravitation.* San Francisco: Freeman.

Moore, J. T. (1967), *Elements of Abstract Algebra Second Edition.* Toronto: Collier-Macmillan.

Nambu, Y. & Jona-Lasinio, G. (1961a), "Dynamical Model of Elementary Particles Based on an Analogy with Superconductivity. I." *Phys. Rev.* 122: 345.

Nambu, Y. & Jona-Lasinio, G. (1961b), "Dynamical Model of Elementary Particles Based on an Analogy with Superconductivity. II." *Phys. Rev.* 124: 246.

Neuenschwander, D. E. (2014a), "Light, the nexus in physics." *Radiations* 20(2); 20–25.

Neuenschwander, D. E. (2014b), "Resource Letter NTUC-1: Noether's Theorem in the Undergraduate Curriculum." *Am. J. Phys* 82(3): 183–188.

Neuenschwander, D. E. (2015), *Tensor Calculus for Physics.* Baltimore, MD: Johns Hopkins University Press.

Neuenschwander, D. E., Taylor, E. F. & Tuleja, S. (2006), "Action: Forcing Energy to Predict Motion." *Phys. Teach.* 44(3): 146–152.

Noether, E. (1918), "Invariante Variationsprobleme." *Nachr. Akad. Wiss. Göttingen, Math.-Phys. Kl. II*: 235–257.

Noether, E. (1971), "Invariant Variation Problems." Translated by M. A. Tavel. *Transp. Theory Stat. Phys.* 1: 186–207.

Ohanian, H. C. (1976), *Gravitation and Spacetime.* New York: W. W. Norton.

Pais, A. (1982), *Subtle Is the Lord: The Science and the Life of Albert Einstein.* Oxford, UK: Oxford University Press.

Panofsky, W. K. H. & Phillips, M. (1962), *Classical Electricity and Magnetism.* Reading, MA: Addison-Wesley.

Perl, M. L. (1974), *High Energy Hadron Physics.* New York: Wiley.

Poincaré, H. (2004), *Science and Method.* New York: Barnes and Noble.

Rainville, E. D. & Bedient, P. E. (1974), *Elementary Differential Equations*, 5th ed. New York: Macmillan.

Ramond, P. (1981), *Field Theory: A Modern Primer*. Reading, MA: Benjamin/Cummings.

Rashed, R. (1990), "A Pioneer in Anaclastics: Ibn Sahl on Burning Mirrors and Lenses." *Isis* 81: 464–491.

Reid, C. (1972), *Hilbert*. New York: Springer-Verlag.

Ritger, P. D. & Rose, N. J. (1968), *Differential Equations with Applications*. New York: McGraw-Hill.

Roman, P. (1968), *Introduction to Quantum Field Theory*. New York: John Wiley.

Rund, H. (1972), "A Direct Approach to Noether's Theorem in the Calculus of Variations." *Util. Math.* 2: 205–214.

Sakurai, J. J. (1967), *Advanced Quantum Mechanics*. Reading, MA: Addison-Wesley.

Saxon, D. S. (1968), *Elementary Quantum Mechanics*. San Francisco: Holden-Day.

Schutz, B. F. (2005), *A First Course in General Relativity*. Cambridge, UK: Cambridge University Press.

Srednicki, M. (2007), *Quantum Field Theory*. Cambridge, UK: Cambridge University Press.

Starkey, S. & Neuenschwander, D. E. (1993), "Adiabatic Invariance Derived from the Rund-Trautman Identity and Noether's Theorem." *Am. J. Phys.* 61(11): 1008–1013.

Symon, K. R. (1953), *Mechanics*. Cambridge, MA: Addison-Wesley.

Taylor, E. F. & Wheeler, J. A. (2000), *Exploring Black Holes: Introduction to General Relativity*. San Francisco: Addison Wesley Longman.

Taylor, E. F. & Wheeler, J. A. (1966), *Spacetime Physics*. San Francisco: Freeman.

Taylor, G. & Neuenschwander, D. E. (1996), "The Adiabatic Invariants of Plasma Physics Derived from the Rund-Trautman Identity and Noether's Theorem." *Am. J. Phys.* 64(11): 1428–1430.

Taylor, G., Neuenschwander, D. E. & Ferrario, C. (1998), "Comments on 'the Adiabatic Invariants of Plasma Physics Derived from the Rund-Trautman Identity and Noether's Theorem.'" *Am. J. Phys.* 66(11): 1016–1017.

Taylor, J. R. (2005), *Classical Mechanics.* Sausalito, CA: University Science Books.

Tent, M. B. W. (2008), *Emmy Noether: The Mother of Modern Algebra.* Wellesley, MA: A. K. Peters.

Trautman, A. (1967), "Noether's Equations and Conservation Laws." *Commun. Math. Phys.* 6: 248–261.

Turner, B. (1991), "A Hamiltonian for geometrical optics." *J. Undergrad. Res. Phys.* 10(1): 23–28.

Wechsler, J., ed. (1981), *On Aesthetics in Science.* Cambridge, MA: MIT Press.

Weinberg, S. (1967), "A Model of Leptons." *Phys. Rev. Lett.* 19(21): 1264–1266.

Weinberg, S. (1972), *Gravitation and Cosmology: Principles and Applications of the General Theory of Relativity.* New York: Wiley.

Weinberg, S. (1995), *The Quantum Theory of Fields*, vol. 1. Cambridge, UK: Cambridge University Press.

Whitfield, S. (1999), *Life Along the Silk Road.* Berkeley, CA: University of California Press.

Yourgrau, W. & Mandelstam, S. (1968), *Variational Principles in Dynamics and Quantum Theory.* Mineola, NY: Dover.

Index